UCF
$4.95

RF TECHNOLOGIES FOR LOW POWER WIRELESS COMMUNICATIONS

RF TECHNOLOGIES FOR LOW POWER WIRELESS COMMUNICATIONS

Edited by

TATSUO ITOH
University of California—Los Angeles, California

GEORGE HADDAD
University of Michigan, Ann Arbor, Michigan

JAMES HARVEY
U.S. Army Research Office, Research Triangle Park, North Carolina

The Institute of Electrical and Electronics Engineers, Inc., New York

A JOHN WILEY & SONS, INC., PUBLICATION
New York • Chichester • Weinheim • Brisbane • Singapore • Toronto

This book is printed on acid-free paper. ⊛

Copyright © 2001 by John Wiley & Sons, Inc. All rights reserved.

Published simultaneously in Canada.

No part of this publication may be reproduced, stored in a retrieval system or transmitted in any form or by any means, electronic, mechanical, photocopying, recording, scanning or otherwise, except as permitted under Sections 107 or 108 of the 1976 United States Copyright Act, without either the prior written permission of the Publisher, or authorization through payment of the appropriate per-copy fee to the Copyright Clearance Center, 222 Rosewood Drive, Danvers, MA 01923, (978) 750-8400, fax (978) 750-4744. Requests to the Publisher for permission should be addressed to the Permissions Department, John Wiley & Sons, Inc., 605 Third Avenue, New York, NY 10158-0012, (212) 850-6011, fax (212) 850–6008, E-Mail:PERMREQ @ WILEY.COM.

For ordering and customer service, call 1-800-CALL-WILEY.

Library of Congress Cataloging-in-Publication Data:

ISBN 0-471-38267-1

Printed in the United States of America.

10 9 8 7 6 5 4 3 2 1

To my father Brigadier General Clarence C. Harvey, Jr.
Formerly of the field artillery

The caissons go rolling along ...

JAMES HARVEY

CONTRIBUTORS

Peter M. Asbeck, Department of Electrical and Computer Engineering, University of California—San Diego, 9500 Gilman Drive, La Jolla, CA 92093-0407

Alexander Balandin, Department of Electrical Engineering, University of California—Riverside, 3401 Watkins Drive, Riverside, CA 92521-0403

Andrew R. Brown, Department of Electrical Engineering and Computer Science, The University of Michigan, 2105 Lurie Engineering Center, 1221 Beal Avenue, Ann Arbor, MI 48109-2122

M. Frank Chang, Device Research Laboratory, Department of Electrical Engineering, University of California—Los Angeles, 405 Hilgard Avenue, Los Angeles, CA 90095-1594

William R. Deal, Malibu Networks, Inc., 26637 Agoura Road, Calabasas, CA 91302

Jack East, Department of Electrical Engineering and Computer Science, University of Michigan, 2105 Lurie Engineering Center, 1221 Beal Avenue, Ann Arbor, MI 48109-2122

George I. Haddad, Department of Electrical Engineering and Computer Science, University of Michigan, 2105 Lurie Engineering Center, 1221 Beal Avenue, Ann Arbor, MI 48109-2122

James F. Harvey, U.S. Army Research Office, P.O. Box 12211, Research Triangle Park, NC 27709-2211

Tatsuo Itoh, Device Research Laboratory, Department of Electrical Engineering, University of California—Los Angeles, 405 Hilgard Avenue, Los Angeles, CA 90095-1594

Linda P. B. Katehi, Department of Electrical Engineering and Computer Science, University of Michigan, 2105 Lurie Engineering Center, 1221 Beal Avenue, Ann Arbor, MI 48109-2122

Larry Larson, Department of Electrical and Computer Engineering, University of California—San Diego, 9500 Gilman Drive, La Jolla, CA 92093-0407

Larry Milstein, Department of Electrical and Computer Engineering, University of California—San Diego, 9500 Gilman Drive, La Jolla, CA 92093-0407

Clark T.-C. Nguyen, Center for Integrated Microsystems, Department of Electrical Engineering and Computer Science, University of Michigan, Ann Arbor, MI 48109-2122

Sergio P. Pacheco, Radiation Laboratory, Department of Electrical Engineering and Computer Science, University of Michigan, Ann Arbor, MI 48109-2122

Dimitris Pavlidis, Department of Electrical Engineering and Computer Science, University of Michigan, 2105 Lurie Engineering Center, 1221 Beal Avenue, Ann Arbor, MI 48109-2122

Zoya Popovic, Department of Electrical Engineering, University of Colorado, Campus Box 425, Boulder, CO 80309-0425

Yongxi Qian, Device Research Laboratory, Department of Electrical Engineering, University of California—Los Angeles, 405 Hilgard Avenue, Los Angeles, CA 90095-1594

Vesna Radisic, HRL Laboratories, 3011 Malibu Canyon Road, Malibu, CA 90265-4799

Gabriel M. Rebeiz, Department of Electrical Engineering and Computer Science, University of Michigan, 2105 Lurie Engineering Center, 1221 Beal Avenue, Ann Arbor, MI 48109-2122

Donald Sawdai, Department of Electrical Engineering and Computer Science, University of Michigan, 2105 Lurie Engineering Center, 1221 Beal Avenue, Ann Arbor, MI 48109-2122

Wayne Stark, Department of Electrical Engineering and Computer Science, University of Michigan, 2105 Lurie Engineering Center, 1221 Beal Avenue, Ann Arbor, MI 48109-2122

Robert J. Trew, U.S. Department of Defense, 4015 Wilson, Suite 209, Arlington, VA 22203

Kang L. Wang, Device Research Laboratory, Department of Electrical Engineering, University of California—Los Angeles, 405 Hilgard Avenue, Los Angeles, CA 90095-1594

Dwight L. Woolard, U.S. Army Research Office, P.O. Box 12211, Research Triangle Part, NC 27709-2211

CONTENTS

Introduction   1
James F. Harvey, Robert J. Trew, and Dwight L. Woolard

1   **Wireless Communications System Architecture and Performance**   9
     Wayne Stark and Larry Milstein

2   **Advanced GaAs-Based HBT Designs for Wireless Communications Systems**   39
     M. Frank Chang and Peter M. Asbeck

3   **InP-Based Devices and Circuits**   79
     Dimitris Pavlidis, Donald Sawdai, and George I. Haddad

4   **Si/SiGe HBT Technology for Low-Power Mobile Communications System Applications**   125
     Larry Larson and M. Frank Chang

5   **Flicker Noise Reduction in GaN Field-Effect Transistors**   159
     Kang L. Wang and Alexander Balandin

6   **Power Amplifier Approaches for High Efficiency and Linearity**   189
     Peter M. Asbeck, Zoya Popovic, Tatsuo Itoh, and Larry Larson

7   **Characterization of Amplifier Nonlinearities and Their Effects in Communications Systems**   229
     Jack East, Wayne Stark, and George I. Haddad

8	**Planar-Oriented Passive Components** *Yongxi Qian and Tatsuo Itoh*	**265**
9	**Active and High-Performance Antennas** *William R. Deal, Vesna Radisic, Yongxi Qian, and Tatsuo Itoh*	**305**
10	**Microelectromechanical Switches for RF Applications** *Sergio P. Pacheco and Linda P. B. Katehi*	**349**
11	**Micromachined K-Band High-Q Resonators, Filters, and Low Phase Noise Oscillators** *Andrew R. Brown and Gabriel M. Rebeiz*	**383**
12	**Transceiver Front-End Architectures Using Vibrating Micromechanical Signal Processors** *Clark T.-C. Nguyen*	**411**

Index   **463**

RF TECHNOLOGIES FOR LOW POWER WIRELESS COMMUNICATIONS

INTRODUCTION

JAMES F. HARVEY AND DWIGHT L. WOOLARD
U. S. Army Research Office, Research Triangle Park, NC

ROBERT J. TREW
U. S. Department of Defense, Arlington, VA

The driving purpose of recent advances in communications technology has been to untether users, allowing them complete mobility and freedom of movement while maintaining their connection to electronic services. The wireless revolution has led to an expectation that voice, fax, and data services, and even internet access, can be available anywhere, without recourse to specific locations in a fixed infrastructure and even while moving or traveling. The last requirement tying the user to a fixed infrastructure is the requirement for power, either wall plug or battery recharging. For a commercial system, this requirement is manifested in the time between battery recharging, which is the only time the user is truly free of the fixed infrastructure. There have been very impressive advances in battery technology, resulting in longer times between battery recharges. However, battery technology is beginning to approach practical limits, still short of the real physical limits dictated by physical chemistry. Many technologists doubt that further advances in storage battery technology will produce more than a factor of 2 improvement in battery lifetime. The other end of this issue is the electronics systems that consume the power. If electronics systems can be designed to consume less power to accomplish the same functionality, then batteries will last longer without recharge. For military systems the situation is more complicated. Military operators of most manpacked and man portable electronic systems are accustomed to the use of disposable batteries in order to avoid the requirement for recharging during a combat operation. However,

---

RF Technologies for Low Power Wireless Communications
Edited by Tatsuo Itoh, George Haddad, and James Harvey
ISBN 0-471-38267-1   Copyright © 2001 by John Wiley & Sons, Inc.

transport of the batteries required for missions of more than a day become a significant load on the soldier, particularly as new concepts such as the Land Warrior and future soldier systems add significant electronic functionality to the individual soldier. A major concern is the weight in both the electronic equipment and batteries that the soldier must carry in combat. In addition, there are huge logistics requirements generated throughout the supply chain by the need to supply batteries in large quantities to front line troops. This issue is a major concern affecting plans for strategic airlift, strategic mobility, and the ability to project military force throughout the world. It affects transportation requirements, adds administrative effort just to keep track of the batteries through the system, and is a large procurement expense. There are also battery issues for unpiloted aerial vehicles (UAVs) and loitering missiles with mission times exceeding a few minutes. Battery requirements must trade off against the aerial vehicle payload or against its range and maximum mission time. Even in helicopters, with a large capacity power source from the engines, there are concerns. The greater the power usage in the electronics equipment, the heavier the equipment becomes. Also the power conditioning equipment for the electronics systems adds weight in proportion to the power required. For helicopters, weight trades off against lift, which can be critical in combat, or against the other payload.

Several years ago, a program in low power electronics was initiated by the Army Research Office. This program was focused on addressing the issue of RF and microwave systems with a major concern for the prime power required for wireless transmitters. At about the same time DARPA (Defense Advanced Research Programs Agency) initiated a program to address the reduction of power in digital and computing systems. The DARPA program was directed toward techniques to reduce processing power in CMOS-based electronics. One thrust was to reduce the bias voltage of CMOS transitors. Adiabatic switching techniques were also explored. As a complement to these programs, five years ago the Office of the Secretary of Defense (OSD) initiated a multidisciplinary university research initiative (MURI) program to augment the Army program in RF and microwave systems. This program ran for five years and involved researchers in four universities: the University of California Los Angeles (UCLA), the University of California San Diego (UCSD), the University of Michigan, and the University of Colorado at Boulder. The office of Deputy Undersecretary of Defense for Science and Technology provided the funding and program oversight for the MURI, while the day-to-day technical management was exercised by the Army Research Laboratory's Army Research Office (ARO). The principal investigators from that MURI are the authors of this book, which presents the results of the sponsored research. The presentation is coherent, placing the advances made during the program in perspective for a reader with a general electrical engineering background. The material in the book is presented to the design community in order to take advantage of the research in reducing power consumption in RF systems. The MURI effort focused primarily on communications systems. However, most of the research concepts can be applied to other RF systems, such as radar or target seekers in missiles. Currently most of the wireless market is in the high megahertz to low gigahertz frequency range, although satellite, wireless local area network (WLAN), and local multipoint distribution service (LMDS) systems utilize higher frequencies, up to 60 GHz. New concepts have been proposed, such as high altitude, long operation (HALO) platform communications in the 48 GHz range. Hunger for bandwidth and spectrum availability will drive both commercial and military communications systems to higher wireless frequencies. For this reason the research was not limited to the traditional cell phone/PCS frequencies. It was not possible in this book to go into as much technical detail in each topical area as is contained in the many technical publications resulting from the research. The book attempts to present the concepts and conclusions in an understandable manner and to allow the reader to reference the detailed publications for more in-depth information as required.

The goal of this research program was to develop techniques to accomplish the RF functions at the lowest expenditure of energy. Certain RF functions require a disproportionate fraction of the system power. One such example is the power amplifier stage of a radio transmitter. Here the focus of the research was directed toward reduction of the power losses, rather than the power itself. A primary goal was determination of an optimal solution within system constraints. However, the intent of the research was not to address circuit optimization in isolation, but to consider an RF system as an interacting network of subsystems that could be optimized both on the subsystem level and on a global basis. The resulting comprehensive approach requires a highly interdisciplinary effort involving device and semiconductor materials science, circuit engineering, electromagnetics, antenna engineering, and communications systems engineering. As this introduction is being written, even a good cell phone is limited by very low efficiency in transmit mode. We believe that the concepts described in this book can open the door to efficiencies approaching 20%. Although the power consumed by a cell phone peaks in the transmit mode, the receive or standby mode is also very important because it is typically used for long periods of time, resulting in significant power drain. Receive mode issues are also addressed in this book.

Chapter 1 addresses low power RF issues from a system architecture point of view. It examines the power and energy usage implications of modulation (including spread spectrum) and coding techniques, including such trade-offs as bandwidth versus efficiency and bandwidth versus energy. It discusses frequency hopping, direct sequence, and multicarrier direct sequence spread spectrum techniques and examines the effects of amplifier nonlinearities on the power requirements for multicarrier transmitters and on receiver architectures. In order to achieve the linearity needed for low error rate modulation and low noise receiver operation, it is necessary to operate amplifiers with a narrower range of voltage or current swings. This results in lower efficiencies. The trade-offs between efficiency, linearity, bit error rate, and the modulation and coding schemes are complex, and these issues are introduced in this chapter.

Chapters 2–5 focus on issues of device physics, materials science, fabrication processes, and circuit issues for the active device building blocks for RF components. Although CMOS technology has made impressive advances in RF capability, this area was not included in the MURI research program because there was already significant effort being made in commercial industry. These four chapters deal with GaAs, InP, SiGe, and GaN technologies, respectively. GaAs HBTs are currently in widespread use in commercial wireless systems because of their attractive performance, circuit integration, and fabrication characteristics. GaAs devices also represent a relatively mature technology. Chapter 2 examines the issues of emitter design and collector design on GaAs HBT performance and reliability. A unique on-ledge Schottky diode potentiometer is presented that is capable of direct, quantitative, in-place monitoring of the emitter ledge passivation. An analytic model is discussed to explain the physics of the potentiometer and to relate its measurements to the HBT performance. The effect of the ledge passivation on performance, noise characteristics, and failure mechanisms is explored. The effect of collector design on performance and reliability is also examined in Chapter 2. The DHBT (double heterojunction bipolar transistor) structure with a GaInP collector is shown to have significant potential advantages over single heterojunction designs, including better breakdown voltages, lower offset voltages, and lower knee voltage. Innovative designs are proposed to mitigate some of the disadvantages of the DHBT design. In Chapter 3, InP devices and circuits are discussed. InP devices will operate at higher frequencies than GaAs-based devices. HEMTs made in this technology generally have better noise performance, while HBTs demonstrate higher gain and better scaling features due to lower surface recombination, better process control due to etching selectivity, and better heat dissipation for power devices due to higher thermal conductivity. Moreover, the offset voltage and lower contact and sheet resistances of the emitter cap and collector layers of InP-based HBTs lead to smaller knee voltage. The smaller knee and turn-on voltages allow the use of low voltage batteries and increase the amplifier efficiency. However, the InP technology is newer and the available substrates are smaller (4 in. vs. 6 in.) and more expensive. Most InP HBT research has focused on NPN devices, that is, device structures doped N-type in the emitter and collector layers and P-type in the base, because of their speed. The MURI research focused on developing a complementary PNP InP HBT technology, in order to facilitate efficient, linear Class B power amplifier or output buffer circuits. To place the technology issues in perspective, the physics of NPN and PNP InP HBTs is also discussed and comparisons are made to GaAs technology. Finally, push–pull operation of complementary NPN and PNP InP HBT circuits is demonstrated. Chapter 4 is a discussion of Si/SiGe HBT technology. In general, SiGe technology has greater limitations in frequency range and breakdown voltage (restricting its power applications) than GaAs or InP technology, but it is compatible with silicon planar technology. It has the desirable characteristics of providing greater frequency and gain performance, and higher power efficiency than silicon BJT devices. Si/SiGe HBTs perform quite well in the low gigahertz frequency region, which is the high market volume personal communications application region. This technology offers the potential for low cost systems integrating analog and digital functions on a single die for lower frequency wireless applications. The specific contribution of the MURI research is in analyzing the device physics and in formulating the design rules for power amplifier circuits, although this chapter contains substantial additional perspective of the SiGe technology. Research into GaN devices has been conducted under a number of governmental programs because they promise the generation of significantly higher power levels at high microwave or millimeter wave frequencies than single GaAs or InP devices. At higher frequencies, solid state sources of moderate power must use some kind of spatial or corporate combining structure, which inevitably introduces losses. By reducing the degree of combining required for a given power level, GaN RF power sources can be much more efficient than comparable sources based on other semiconductor technologies. One of the main barriers to the use of GaN in communications systems is its relative noisiness. The MURI research focused on this noise issue, and the results are reported in Chapter 5.

Chapters 6 and 7 focus on the power amplifier stage, where signal power is raised to a highest RF level in the transmitter. The efficiency of this stage is the upper bound for the efficiency of the overall system, and considerable attention has been paid to improving efficiency in the power amplifier. Amplifiers are generally much more efficient when operated in their power saturation region. This results in a trade-off of efficiency and linearity, with the high linearity requirements of modern communications systems pushing conventional amplifier circuits into an inefficient mode of operation. Chapter 6 presents several unconventional approaches to efficient power amplifier concepts. The use of a dc–dc converter to provide a continuously optimized supply voltage is discussed. The use of Class E and F switching amplifiers in microwave systems is presented, and the trade-off with linearity is examined. Techniques to preserve efficiency and linearity simultaneously, the LINC amplifier (linear amplification with nonlinear components) and Class S amplifiers, are also considered. And a novel approach to the self-consistent design of the amplifier and the antenna structure is applied to eliminate the conventional matching network, and its losses, between these transmitter stages. The possibility of using antennas for harmonic filters, in addition to radiation, is presented for increasing the amplifier efficiency. Chapter 7 presents an analysis of the nonlinearities in a power amplifier and new analytical tools to quantitatively address the complex nonlinear effects on the wide band of frequencies inherent in digital signals.

Passive components can be major sources of loss and inefficiency in planar RF circuits. Particularly at higher frequencies, interconnects can be very lossy, with losses to the substrate and to radiation, as well as ohmic losses in the metal. Planar antennas can have major losses to substrate modes, which can also seriously degrade the antenna patterns of arrays, effectively further reducing the efficiency of the antenna as well as complicating interfering antenna problems by radiating in unwanted directions and reception through sidelobes. The control of unwanted frequencies and spectral regrowth presents a special problem for truly planar fabricated or wafer scale integrated circuits. On-wafer approaches to the reduction in interference and frequency problems result in more complicated circuitry, with associated additional power consumption.

Chapter 8 presents two concepts that have the potential for a significant effect on these components. The SIMPOL technique provides very low loss interconnects for the integration of high performance microwave RF components with CMOS digital circuits. This technique opens the door to system-on-a-chip concepts, which have many system advantages in addition to reduced connection losses. The second concept is based on the so-called photonic bandgap structures (or electromagnetic bandgap structures). Periodic passive structures can provide planar approaches to harmonic tuning of high efficiency microwave amplifiers, reduced transmission line leakage, low loss slow wave structures, improved planar filters, the elimination of antenna substrate modes, and a perfect magnetic impedance surface, which affords flexibility in the design of high efficiency antennas.

Chapter 9 reviews planar antenna approaches, including some innovative applications of the older concept of the quasi-Yagi antenna, and discusses the design of active integrated antennas. Active integrated antennas are active semiconductor devices or circuits integrated directly within the planar antenna structure. This type of integrated antenna circuit presents the opportunity to reduce losses between the power amplifier and the antenna due to impedance matching circuits. It also enables the design of an antenna array consisting of essentially nonlinear, nonreciprocal antenna elements, for application, for example, in phase conjugating, retroreflective arrays.

Micromachining fabrication methods harness the manufacturing processes responsible for the VLSI planar IC industry for RF circuits and circuit components. These techniques can have orders-of-magnitude impact on the size, weight, and cost of RF systems and can enable a corresponding significant reduction in power dissipation. Micromachining techniques form an overarching circuit integration technology based on extremely low loss transmission lines and metallic component structures, an inexpensive self-packaging process that eliminates spurious electromagnetic packaging effects, monolithically integrated high Q filters and resonators, the wafer-scale integration of circuits based on different substrate materials, and a natural three-dimensional layered integration capability. These techniques can essentially eliminate radiation and substrate losses from transmission lines and other passive structures, reducing losses to solely ohmic losses. Thus a planar circuit structure can approach waveguide performance, although the planar structures cannot equal the waveguide performance because the waveguide structure has more metal and therefore smaller ohmic losses. Micromachining fabrication techniques also offer the opportunity for entirely new device structures, such as the combination of RF electrical and mechanical functions in a single device, the RF MEMS (micro electro mechanical systems) devices. The micromachined and MEMS devices, such as high Q filters and switches, can replace one-for-one components in existing radio or radar architectures, resulting in simple, low loss, on-chip planar circuits. Of more interest is the ability to engineer entirely new planar monolithic architectures with reduced power requirements. These micromachining and MEMS techniques are an enabling technology for such architectures as fully duplex communications, radar simultaneous transmit/receive, common aperture and common electronics, cognitive radio, and reconfigurable aperture systems. And planar high Q components can be used to increase RF circuit selectivity, thereby reducing power consumption. The high Q components also reduce the specifications for dynamic range and phase noise in the active circuit components, allowing lower-power-consuming designs of the active components. The research under the MURI program focused on some of the critical issues of this new technology, and these results are discussed in Chapters 10, 11, and 12.

Chapter 10 deals with MEMS switches for RF applications. Mechanical and electrical design considerations for fixed beam, compliant beam, and cantilever beam switches are discussed and concepts for high isolation switching are introduced. RF MEMS switches have relatively low RF insertion loss (on the order of 0.2 dB or less), virtually zero dc power consumption, small size, and are constructed using a batch planar fabrication process. MEMS devices have switched several watts of RF power in laboratory experiments, providing the hope that research into the basic physical mechanical, thermal, and electrical mechanisms of operation will lead to reliable switching of moderate power levels by single MEMS switches. Conventional RF MEMS switches require between 40 and 80 volts to activate reliably, which is useful for some applications. Compliant switches can activate with as little as 5 volts, but other performance features must be traded off to achieve these low activation voltages. These MEMS switches can be used in place of many semiconductor switches in RF circuits that can tolerate the slower MEMS switching times (on the order of milliseconds), for example, in phased array beam steering and reconfigurable antenna structures. Chapter 11 describes innovative concepts in micromachined circuits to integrate high Q filters directly with an active semiconductor device to produce a planar circuit low phase noise oscillator. The MEMS devices described in Chapter 10 are basically switches, while the micromachined resonators and filters in Chapter 11 are nonmechanical filters based on purely electrical resonators. In contrast, Chapter 12 describes RF MEMS devices based on very high Q mechanical resonators, which couple to the electrical signal. The result is a very small (on the order of 100 microns in size), very high Q (greater than 10,000 in vacuum) MEMS filter. These filters have been demonstrated at VHF and have the potential for application at UHF. The filters are ultra low loss and require ultra small dc activation energies. Because of their extremely small size, they provide the potential for their massive use in entirely new RF architectures that utilize frequency selectivity to achieve low power consumption. On the other hand, the small size and the inherently mechanical nature of operation place a significantly increased emphasis on packaging issues.

Individual research areas started under this MURI program continue under various other government programs. The research on power amplifiers inspired a workshop on this subject, which evolved into an annual IEEE Topical Workshop on Power Amplifiers for Wireless Communications. These individual topical areas of research continue to be of strong interest to the military and commercial RF sectors. However, the editors strongly feel that the success of many of these areas was due to their being conducted and managed in a university environment with a strong multidisciplinary and interdisciplinary structure.

This book is written for graduate students and engineering professionals with general background of electrical engineering. Although it is assumed that they are familiar with the background subjects such as electromagnetic fields, antennas, microwave devices, and communications systems, no detailed knowledge is expected. Although the contents are coherently organized, individual chapters can also be read independently. Although reasonably extensive reference lists are included in each chapter, the wealth in information in related subjects is enormous. Readers with interest in specific subjects may refer to the latest publications such as *IEEE Transactions*.

1

WIRELESS COMMUNICATIONS SYSTEM ARCHITECTURE AND PERFORMANCE

WAYNE STARK
Department of Electrical Engineering and Computer Science
The University of Michigan, Ann Arbor

LARRY MILSTEIN
Department of Electrical and Computer Engineering
University of California—San Diego

1.1 INTRODUCTION

Low power consumption has recently become an important consideration in the design of commercial and military communications systems. In a commercial cellular system, low power consumption means long talk time or standby time. In a military communications system, low power is necessary to maximize a mission time or equivalently reduce the weight due to batteries that a soldier must carry. This book focuses attention on critical devices and system design for low power communications systems. Most of the remaining parts of this book consider particular devices for achieving low power design of a wireless communications system. This includes mixers, oscillators, filters, and other circuitry. In this chapter, however, we focus on some of the higher level system architecture issues for low power design of a wireless communications system. To begin we discuss some of the goals in a wireless communications system along with some of the challenges posed by a wireless medium used for communications.

RF Technologies for Low Power Wireless Communications
Edited by Tatsuo Itoh, George Haddad, and James Harvey
ISBN 0-471-38267-1   Copyright © 2001 by John Wiley & Sons, Inc.

1.2 PERFORMANCE GOALS AND WIRELESS MEDIUM CHALLENGES

A system level (functional) block diagram of a wireless communications system is shown in Figure 1.1. In this figure the source of information could be a voice signal, a video signal, situation awareness information (e.g., position information of a soldier), an image, a data file, or command and control data. The source encoder processes the information and formats the information into a sequence of information bits $\in \{\pm 1\}$. The goal of the source encoder is to remove the unstructured redundancy from the source so that the rate of information bits at the output of the source encoder is as small as possible within a constraint on complexity. The channel encoder adds structured redundancy to the information bits for the purpose of protecting the data from distortion and noise in the channel. The modulator maps the sequence of coded bits into waveforms that are suitable for transmission over the channel. In some systems the modulated waveform is also spread over a bandwidth much larger than the data rate. These systems, called spread-spectrum systems, achieve a certain robustness to fading and interference not possible with narrowband systems. The channel distorts the signal in several ways. First, the signal amplitude decreases due to the distance between the transmitter and receiver. This is generally referred to as propagation loss. Second, due to obstacles the signal amplitude is attenuated. This is called shadowing. Finally, because of multiple propagation paths between the transmitter antenna and the receiver antenna, the signal waveform is distorted. Multipath fading can be either constructive, if the phases of different paths are the same, or destructive, if the phases of the different paths cause cancellation. The destructive or constructive nature of the fading depends on the carrier frequency of the signal and is thus called frequency selective fading. For a narrowband signal (signal bandwidth small relative to the inverse delay spread of the channel), multipath fading acts like a random attenuation of the signal. When the fading is constructive the bit error probability can be very small. When the fading is destructive the bit error probability becomes quite large. The average overall received amplitude value causes a significant loss in performance (on the order of 30–40 dB loss). However, with proper error control coding or diversity this loss in performance can essentially be eliminated.

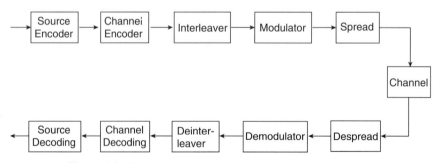

Figure 1.1. Block diagram of a digital communications system.

In addition to propagation effects, typically there is noise at the receiver that is uncorrelated with the transmitted signal. Thermal (shot) noise due to motion of the electrons in the receiver is one form of this noise. Other users occupying the same frequency band or in adjacent bands with interfering sidelobes is another source of this noise. In commercial as well as military communications systems interference from other users using the same frequency band (perhaps geographically separated) can be a dominant source of noise. In a military communications system hostile jamming is also a possibility that must be considered. Hostile jamming can easily thwart conventional communications system design and must be considered in a military communications scenario.

The receiver's goal is to reproduce at the output of the source decoder the information-bearing signal, be it a voice signal or a data file, as accurately as possible with minimal delay and minimal power consumed by the transmitter and receiver. The structure of the receiver is that of a demodulator, channel decoder, and source decoder. The demodulator maps a received waveform into a sequence of decision variables for the coded data. The channel decoder attempts to determine the information bits using the knowledge of the codebook (set of possible encoded sequences) of the encoder. The source decoder then attempts to reproduce the information.

In this chapter we limit discussion to an information source that is random data with equal probability of being 0 or 1 with no memory; that is, the bit sequence is a sequence of independent, identically distributed binary random variables. For this source there is no redundancy in the source, so no redundancy can be removed by a source encoder.

There are important parameters when designing a communications system. These include data rate R_b (bits/s, or bps), at the input to the channel encoder, the bandwidth W (Hz), received signal power P (watts), noise power density $N_0/2$ (W/Hz), and bit error rate $P_{e,b}$. There are fundamental trade-offs between the amount of power or equivalently the signal-to-noise ratio used and the data rate possible for a given bit error probability, $P_{e,b}$. For ideal additive white Gaussian noise channels with no multipath fading and infinite delay and complexity, the relation between data rate, received power, noise power, and bandwidth for $P_{e,b}$ approaching zero was determined by Shannon as [1]

$$R_b < W \log_2 \left(1 + \frac{P}{N_0 W}\right). \tag{1.1}$$

If we let $E_b = P/R_b$ represent the energy used per data bit (joules per bit), then an equivalent condition for reliable communication is

$$\frac{E_b}{N_0} > \frac{2^{R_b/W} - 1}{R_b/W}.$$

This relation determines the minimum received signal energy for reliable communications as a function of the spectral efficiency R_b/W (bps/Hz). The

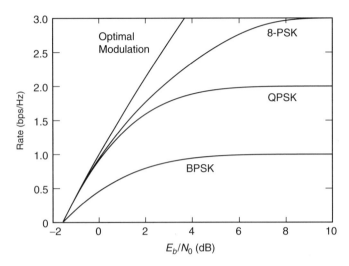

Figure 1.2. Possible data rate for a given energy efficiency.

interpretation of this condition is that for lower spectral efficiency, lower signal energy is required for reliable communications. The trade-off between bandwidth efficiency and energy efficiency is illustrated in Figure 1.2. Besides the trade-off for an optimal modulation scheme, the trade-off is also shown for three modulation techniques: binary phase shift keying (BPSK), quaternary phase shift keying (QPSK), and 8-ary phase shift keying (8PSK).

In this figure the only channel impairment is additive white Gaussian noise. Other factors in a realistic environment are multipath fading, interference from other users, and adjacent channel interference. In addition, the energy is the received signal energy and does not take into account the energy consumed by the processing circuitry. For example, power consumption of signal processing algorithms (demodulation, decoding) are not included. Inefficiencies of power amplifiers and low noise amplifiers are not included. These will be discussed in subsequent sections and chapters. These fundamental trade-offs between energy consumed for transmission and data rate were discovered more than 50 years ago by Shannon (see Cover and Thomas) [1]. It has been the goal of communications engineers to come close to achieving the upper bound on data rate (called the channel capacity) or equivalently the lower bound on the signal-to-noise ratio.

To come close to achieving the goals of minimum energy consumption, channel coding and modulation techniques as well as demodulation and decoding techniques must be carefully designed. These techniques are discussed in the next two sections.

1.3 MODULATION TECHNIQUES

In this section we describe several different modulation schemes. We begin with narrowband techniques whereby the signal bandwidth and the data rate are roughly

MODULATION TECHNIQUES   13

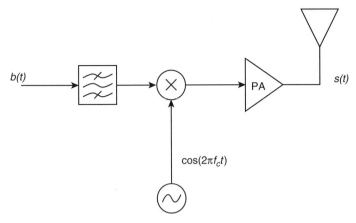

Figure 1.3. Transmitter.

equal. In wideband techniques, or spread-spectrum techniques, the signal bandwidth is much larger than the data rate. These techniques are able to exploit the frequency-selective fading of the channel. For more details see Proakis [2].

1.3.1 Narrowband Techniques

A very simple narrowband modulation scheme is binary phase shift keying (BPSK). The transmitter and receiver for BPSK are shown in Figures 1.3 and 1.4, respectively. A sequence of data bits $b_l \in \pm 1$ is mapped into a data stream and filtered. The filtered data stream is modulated onto a carrier and is amplified before being radiated by the antenna. The purpose of the filter is to confine the spectrum of the signal to the bandwidth mask for the allocated frequency. The signal is converted from baseband by the mixer to the desired center or carrier frequency (upconversion). The signal is then amplified before transmission. With ideal devices (mixers, filters, amplifiers) this is all that is needed for transmission. However, the

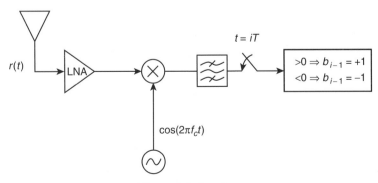

Figure 1.4. Receiver.

mixers and amplifiers typically introduce some additional problems. The amplifier, for example, may not be completely linear. The nonlinearity can cause the bandwidth of the signal to increase (spectral regrowth), as will be discussed later.

For now, assume that the filter, mixer, and amplifier are ideal devices. In this case the transmitted (radiated) signal can be written as

$$s(t) = \sqrt{2P} \sum_{l=-\infty}^{\infty} b_l h(t - lT) \cos(2\pi f_c t), \quad (1.2)$$

where P is the transmitted power, T is the duration of a data bit or the inverse of the data rate R_b, f_c is the carrier frequency, and $h(t)$ is the impulse response of the pulse-shaping filter. There are various choices for the pulse-shaping filter. A filter with impulse response being a rectangular pulse of duration T seconds results in a constant envelope signal (peak-to-mean envelope ratio of 1) but has large spectral splatter, whereas a Nyquist-type pulse has high envelope variation and no spectral splatter. The disadvantage of high envelope variation is that it will be distorted by an amplifier operating in a power efficient mode because of the amplifier's nonlinear characteristics. Thus there is a trade-off between power efficiency and bandwidth efficiency in the design of the modulation.

The simplest channel model is called the additive white Gaussian noise (AWGN) channel. In this model the received signal is the transmitted signal (appropriately attenuated) plus additive white Gaussian noise:

$$r(t) = \alpha s(t) + n(t). \quad (1.3)$$

The noise is assumed to be white with two-sided power spectral density $N_0/2$ W/Hz.

The receiver for BPSK is shown in Figure 1.4. The front end low noise amplifier sets the internal noise figure for the receiver. The mixer converts the radio frequency (RF) signal to baseband. The filter rejects out-of-band noise while passing the desired signal. The optimal filter in the presence of additive white Gaussian noise alone is the matched filter (a filter matched to the transmitter filter). This very simplified diagram ignores many problems associated with nonideal devices. For the case of ideal amplifiers and a transmit filter and receiver filter satisfying the Nyquist criteria for no intersymbol interference [2], the receiver filter output can be expressed as

$$X_l = \sqrt{\bar{E}} b_{l-1} + \eta_l,$$

where $\bar{E}$ is the received energy ($\bar{E} = \alpha^2 PT$) and η_l is a Gaussian distributed random variable with mean zero and variance $N_0/2$. The decision rule is to decide $b_{l-1} = +1$ if $X_l > 0$ and to decide $b_{l-1} = -1$ otherwise. For the simple case of an additive white Gaussian noise channel, the error probability is

$$P_{e,b} = Q\left(\sqrt{\frac{2\bar{E}}{N_0}}\right),$$

where $Q(x) = \int_x^\infty (1/\sqrt{2\pi}) \exp(-u^2/2) \, du$. This is shown in Figure 1.5.

MODULATION TECHNIQUES

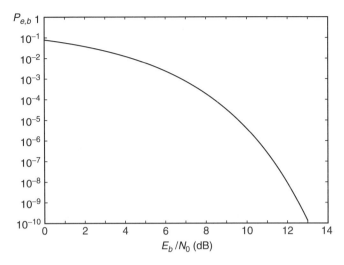

Figure 1.5. Bit error probability for BPSK in AWGN.

From Figure 1.5 it can be seen that in order to provide error probabilities around 10^{-5} it is necessary for the received signal-to-noise ratio to $\bar{E}/N_0 = 9.6$ dB. The capacity curve for BPSK in Figure 1.2, however, indicates that if we are willing to lower the rate of transmission we can significantly save on energy. For example, it is possible to have a nearly 0 dB signal-to-noise ratio if we are willing to reduce the rate of transmission by 50%. Thus about 9.6 dB decrease in signal power is possible with a 50% reduction in transmission rate.

The above analysis is for the case of additive white Gaussian noise channels. Unfortunately, wireless channels are not accurately modeled by just additive white Gaussian noise. A reasonable model for a wireless channel with relatively small bandwidth is that of a flat Rayleigh fading channel. While there are more complex models, the Rayleigh fading channel model is a model that provides the essential effect. In the Rayleigh fading model the received signal is still given by Eq. (1.3). However, α is a Rayleigh distributed random variable that is sometimes large (constructive addition of multiple propagation paths) and sometimes small (destructive addition of multiple propagation paths). However, the small values of α cause the signal-to-noise ratio to drop and thus the error probability to increase significantly. The large values of α corresponding to constructive addition of the multiple propagation paths result in the error probability being very small. However, when the average error probability is determined there is significant loss in performance. The average error probability with Rayleigh fading and BPSK is

$$\bar{P}_{e,b} = \int_{r=0}^{\infty} f(\alpha) Q\left(\sqrt{\frac{2E\alpha}{N_0}}\right) d\alpha = \frac{1}{2} - \frac{1}{2}\sqrt{\frac{\bar{E}/N_0}{1+\bar{E}/N_0}}, \qquad (1.4)$$

16 WIRELESS COMMUNICATIONS SYSTEM ARCHITECTURE AND PERFORMANCE

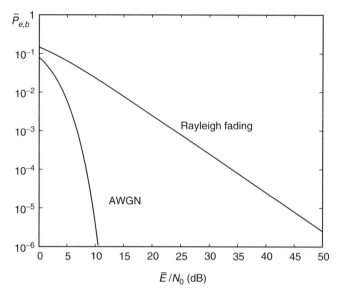

Figure 1.6. Bit error probability for BPSK with Rayleigh fading.

where $f(r)$ is the Rayleigh density and $\bar{E}$ is the average received energy. The average error probability as a function of the average received energy is shown in Figure 1.6. Included in this figure is the performance with just white Gaussian noise. As can be seen from the figure, there is a significant loss in performance with Rayleigh fading. At a bit error rate of 10^{-5} the loss in performance is about 35 dB. In early communications systems the transmitted power was boosted to compensate for fading. However, there are more energy efficient ways to compensate for fading.

For a Rayleigh fading channel the fundamental limits on performance are known as well. The equations determining the limits are significantly more complicated [3]. Nevertheless, they can be evaluated and are shown in Figure 1.7. By examining the curve for BPSK we can see that it is possible to reduce the loss in performance to about 2 dB (rather than 35 dB) if proper signal (coding) design is used. Thus by reducing the data rate by 50% with proper coding a 33 dB savings in energy is possible. One method of signaling that reduces this performance loss is by the use of wideband signals, as discussed in the next section.

1.3.2 Wideband Techniques

Wideband signals have the potential of overcoming the problem of fading [4]. This is because the fading characteristics are frequency dependent. Different frequencies fade differently because the phase relationships of different paths change as the frequency changes. In addition, wideband techniques are able to handle interference from jammers or from other users. There are several techniques that are employed for wideband communications systems. Two popular techniques are direct-sequence

MODULATION TECHNIQUES

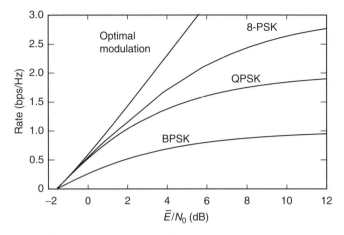

Figure 1.7. Fundamental limits for Rayleigh fading.

(DS) spread spectrum and frequency-hopped spread spectrum. We first discuss direct-sequence spread spectrum.

1.3.2.1 DS Spread Spectrum Conceptually, a direct-sequence spread spectrum works as shown in Figure 1.8. A data sequence is first encoded for error protection. The encoded waveform is then modulated using a standard narrowband modulation technique (e.g., BPSK). The narrowband waveform is then spread over a wider bandwidth with a spreading code as shown in the figure. At the receiver the received signal is despread by mixing with an identical spreading code followed by a narrowband demodulator. The result is then used for decoding or error correction. If there are multiple users using the same bandwidth at the same time but with

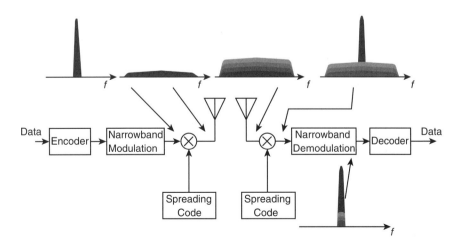

Figure 1.8. DS code division multiple access (CDMA) system.

different spreading codes then the received waveform consists of the sum of the signals of the different users (as shown in the figure). However, after mixing with the spreading code of the desired user, only the signal of the desired user becomes narrowband. The other users' signals, since they have a different spreading code than that used by the receiver, remain wideband signals. The narrowband demodulator then removes much of the interference. The amount by which the interference signals are reduced in power is roughly equal to the ratio of the bandwidth of the wideband signal to the narrowband signal. This is often referred to as the processing gain of the system.

The situation for a jamming signal is similar. Consider a jammer that transmits energy in a narrow bandwidth directly in the band of the desired user. The received signal will consist of a sum of the desired signal and the jamming signal, as shown in Figure 1.9. The receiver processes the signal by first mixing with a replica of the user's spreading code (assumed not available to the jammer). The desired signal gets despread while the jamming signal becomes spread. After demodulation the jamming signal power gets reduced by a factor equal to the processing gain.

The above are conceptual descriptions of the way in which a direct-sequence spread-spectrum system rejects interference. In both cases the receiver depicted is not the optimal receiver. The receivers depicted are optimal only if the interference is white Gaussian noise. Better receivers exist for multiuser interference and for jamming interference. In addition, for jamming signals, we have not considered the worst possible jamming signal of a given power. Power consumption occurs mainly in two places in this conceptual diagram. The first is at the transmitter in amplifying the signal. If the signal is not constant envelope, then nonlinearities of the amplifier cause distortion. When the amplifier is operating in the linear region the power efficiency of the amplifiers is small. If the envelope is constant, then, while the nonlinearities do not affect the signal, the signal has higher sidelobes than a nonconstant envelope signal. Significant power is also needed for the despreading operation. For example, if the despreading is done digitally then an analog-to-digital converter is needed with dynamic range equal to that of the jamming signal and

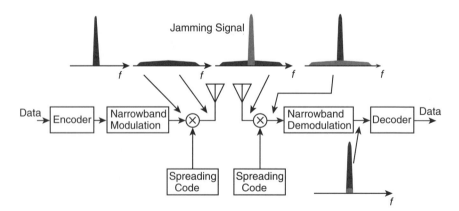

Figure 1.9. DS spread spectrum with jamming.

MODULATION TECHNIQUES

bandwidth of the desired signal. This combination of high dynamic range and bandwidth leads to high power consumption for the digital circuitry needed to despread. Thus despreading a spread-spectrum signal and thereby rejecting unwanted interference requires circuitry with significant power consumption.

Below we specify a simple direct-sequence spread-spectrum system. We analyze the performance in the presence of jamming, in a fading channel, and in multiple-access interference. The transmitter for a direct-sequence system is shown in Figure 1.10. The data sequence $b(t)$ consists of a sequence of data bits of duration T. The data sequence is multiplied with a binary spreading sequence $a(t)$, which has N components called chips per data bit.

In Figure 1.10, $b(t)$ represents the data and can be expressed as

$$b(t) = \sum_{l=-\infty}^{\infty} b_l \, p_T(t - lT), \qquad b_l \in \{+1, -1\},$$

where $p_T(t)$ is a rectangular pulse of unit amplitude and duration T beginning at $t = 0$. Similarly, the spreading code $a(t)$ is written as

$$a(t) = \sum_{l=-\infty}^{\infty} a_l \, p_{T_c}(t - lT_c), \qquad a_l \in \{+1, -1\},$$

where a_l is a binary symbol $\in \pm 1$ and $T_c = T/N$. In this case it is useful to model a_i as a sequence of independent, identically distributed binary random variables equally likely to be ± 1. The number of chips per bit is often referred to as the "processing gain." It is the factor by which the signal is spread. The transmitted signal is then

$$s(t) = \sqrt{2P} a(t) \, b(t) \cos(2\pi f_c t).$$

The transmitted signal has power P. In Figure 1.11 we show a data signal and the result of multiplying by a spreading signal with 7 chips per bit.

The receiver consists of a mixer followed by a filter matched to the spreading code of the transmitter as shown in Figure 1.12. A typical filter output is shown in

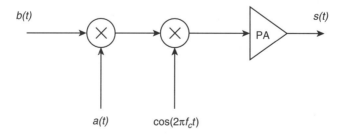

Figure 1.10. Block diagram of direct-sequence spread-spectrum transmitter.

20 WIRELESS COMMUNICATIONS SYSTEM ARCHITECTURE AND PERFORMANCE

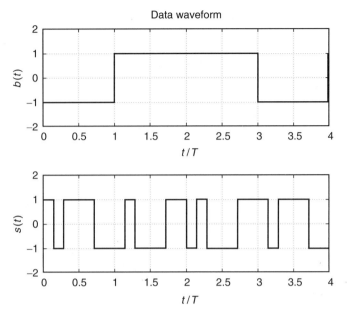

Figure 1.11. Waveforms $b(t)$ and $a(t)b(t)$.

Figure 1.13. In this figure the filter output is sampled every T seconds. At these times the filter impulse response is completely correlated with the incoming desired signal and produces a large amplitude (positive or negative depending on the sign of the data bit) signal. The sample output corresponds to a data sequence consisting of two positive polarity data bits followed by two negative polarity data bits followed by a positive polarity bit. For the case of an additive channel (no fading) with interference and a lowpass filter matched to the transmitted pulse shape (rectangular assumed here), the output of the filter at time iT can be expressed as

$$Z(iT) = \sqrt{E}b_{i-1} + I + \eta_i,$$

where I represents the contribution due to the interference and η_i is the output at time iT due to background noise (e.g., thermal noise).

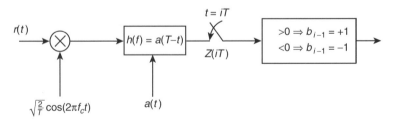

Figure 1.12. Block diagram of direct-sequence spread-spectrum receiver.

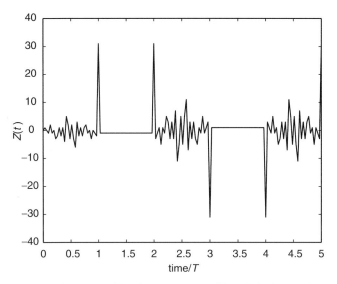

Figure 1.13. Output of matched filter for a sequence of length 31 in the absence of noise.

The component of the output of the receiver due to interference will depend on the nature of the interference (jamming, multiuser, etc.) as well as the spreading code $a(t)$ used. For simplicity in what follows we will assume that the spreading code consists of a sequence of independent and identically distributed random variables with equal probability of being $+1$ and -1.

One simple measure of performance of a communications system is the signal-to-noise ratio at the output of the receiver. The signal-to-noise ratio (SNR) is defined as

$$\text{SNR} = \frac{(E[Z(iT)])^2}{\text{Var}[Z(iT)]}.$$

Consider a direct-sequence system operating in the presence of a jamming signal $j(t)$, which is a pure tone operating at the exact carrier frequency as the desired signal (and with the same phase):

$$j(t) = \sqrt{2J}\cos(2\pi f_c t).$$

Then the signal-to-noise ratio at the output of the receiver is given by

$$\text{SNR} = \frac{P}{J/N}.$$

From this we can see the effect of spreading is to reduce the effective jamming power by a factor of N, the processing gain. If the jamming signal has a random phase offset from the desired signal then the SNR would be increased by a factor of 2. The error probability can be approximated from the SNR by assuming the output of the receiver due to the interfering signal has a Gaussian density. With this assumption,

$$P_{e,b} \approx Q(\sqrt{\text{SNR}}).$$

Now consider the case of multiuser interference. Assume that there are K users with unique spreading codes (modeled as a random sequence). In addition, assume there are random delays and phases between the users. With these assumptions the signal-to-noise ratio for user 1 is given by

$$\text{SNR} = \frac{E_1}{\sum_{k=2}^{K} E_i/(3N) + N_0/2},$$

where K is the number of users, E_i is the received energy per bit for user i, and $N_0/2$ is the two-sided power spectral density of the background noise. Again it is clear how the spreading reduces the effect of the interference. The factor of 3 arises due to the random phase and delays between users.

Finally, consider the case of multipath interference. Because multipath is not additive but rather multiplicative the model needs updating. Consider a simple model whereby the received signal is a sum of delayed versions of the transmitted signal plus additive white Gaussian noise. That is,

$$r(t) = \sum_{l=1}^{L} \alpha_l s(t - \tau_l) + n(t).$$

For simplicity, assume that the delays are separated by at least T_c seconds. In this case the paths are said to be resolvable. This implies that the output of the matched filter consists of peaks that are nonoverlapping and thus can be resolved. Consider the ideal case of a spread-spectrum signal with zero sidelobes. In this case the output of the filter consists of a peak for each multipath present. If we assume that the correlations are ideal (so that the output consists of just impulses) and that the amplitude of each path is independent and Rayleigh distributed, then the performance improves dramatically compared to a single-path system as shown in Figure 1.14.

1.3.2.2 Frequency-Hopped Spread-Spectrum Frequency-hopped spread spectrum works by pseudorandomly changing the center frequency of the carrier over a set of frequencies. The sequence of frequencies used is called the frequency-hopping pattern. In a jamming environment this can force the jammer to spread its power over a very wide bandwidth in order to guarantee that the transmitted signal is disrupted (to some extent). When the jammer spreads its signal over the whole bandwidth the amount of power in each frequency slot is a small fraction of the total

MODULATION TECHNIQUES

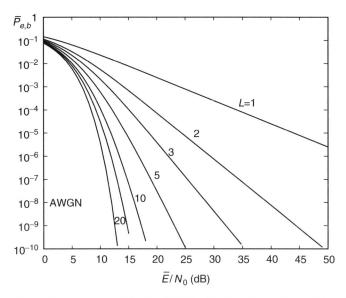

Figure 1.14. Bit error probability for BPSK with diversity and Rayleigh fading.

power. Thus the effectiveness of a wideband jammer is reduced in proportion to the bandwidth over which the signal is spread. If the spreading is large enough the jammer is not effective at disrupting communications. If the jammer only jams a fraction of the band but with high power in certain slots then the performance can be severely degraded (as opposed to a wideband jammer with the same total power). However, in this case a proper error control code and decoding algorithm that corrects errors in slots jammed can change the optimal jamming strategy from narrowband to broadband and thus regain the advantage of spreading the spectrum.

For multiple access, different users have different frequency-hopping patterns and different users will collide occasionally. These collisions can be handled with the use of appropriate error control coding. A code with a very low rate is needed for a large number of interferers while for a small number of interferers a large rate code should be employed. To maximize the information throughput per unit bandwidth an optimal code rate and number of users can be found.

For fading channels, provided the frequency separation between slots is larger than the coherence bandwidth, different frequencies will fade independently. If the bandwidth within a slot is small compared to the coherence bandwidth then the fading will be nonselective within a hop. An error control code will be able to correct errors from a badly faded hop. It is interesting to note that the (uncoded) performance in a fading environment is actually worse than the performance in a jamming environment. In both cases the performance degradation (without coding) is on the order of 30–40 dB compared to additive white Gaussian noise. This can be reduced somewhat in the partial-band jamming case by spreading over a very large bandwidth. However, with the proper combination of coding and spreading a jammer's optimum strategy is to jam the whole bandwidth. With this optimum jamming strategy the performance of a spread-spectrum system with worst-case jamming becomes the same as an additive noise channel with the same average power. In this sense then, a jammer can be defeated with the right combination of spreading and coding. In fact, the required signal-to-noise ratio with proper spreading and coding with a partial-band jammer is lower than the required signal-to-noise ratio for an uncoded spread-spectrum system with just additive noise of the same average power.

1.3.2.3 Multicarrier Techniques Multicarrier modulation techniques have recently gained significant popularity in the United States and Europe. In the United States multicarrier is used for digital subscriber loop (DSL) applications, while in Europe it is used in digital audio broadcasting (DAB). Multicarrier works by employing more than one carrier simultaneously. Consider an encoded data stream $b(t)$ consisting of data bits with rate $R_b = 1/T_b$ bits per second. The data stream is converted from a single stream into M separate streams via a serial-to-parallel converter as shown in Figure 1.15. After the serial-to-parallel converter the data streams are mixed onto M different carriers before being combined and amplified. At the receiver the inverse process in used. The received signal is mixed down to baseband using M different carriers. After mixing down to baseband the signal is filtered before a decision is made regarding each bit.

There are several motivations for considering multicarrier modulation techniques and several disadvantages of multicarrier techniques. For high data rate (relative to the inverse delay spread of the channel) applications, single-carrier systems suffer from severe intersymbol interference. This interference can be mitigated by sufficient equalization but requires significant complexity at the receiver to do this.

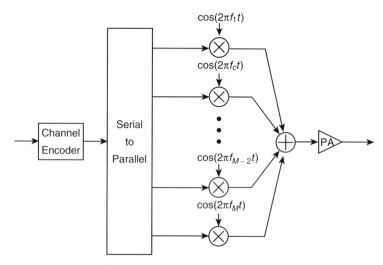

Figure 1.15. Multicarrier modulation.

By using multiple carriers the data rate on each carrier is reduced by a factor M. This means that the intersymbol interference is reduced by a factor of M and the receiver complexity is also reduced. However, the peak-to-mean power ratio at the input to the power amplifier now becomes large. For amplifiers operating with high backoff (not near saturation) this is not a problem. However, operating an amplifier with high backoff is typically not very efficient. If the amplifier is operating with low backoff then the large envelope variations cause distortion because of intermodulation products that fall in band or in adjacent channels. Another advantage of multicarrier modulation is that the frequency occupancy can be rather flexible. That is, we can build a system that occupies noncontiguous frequency bands. If we consider a linear amplifier and additive white Gaussian noise alone then the performance on multicarrier modulation is identical to a single-carrier system. Multicarrier techniques can also be applied to direct-sequence spread-spectrum systems. In such a system the data stream on each carrier is spread with a spreading code. This gives a multicarrier system the advantage with respect to interference that a direct-sequence system has. Later we will show the performance of multicarrier direct-sequence systems with nonideal amplifier characteristics.

1.4 CODING TECHNIQUES

Coding techniques are crucial to reducing the power consumption of digital communications systems. The basic idea of coding is to add redundancy to the transmitted data. For example, for every four information bits we might want to transmit four information bits and three parity bits or redundant bits. In this way the four information bits are encoded into seven coded bits. If we represent the information bits by b_0, b_1, b_2, b_3, which are 0 or 1, then the coded bits are determined by $p_4 = b_0 + b_1 + b_2, p_5 = b_0 + b_1 + b_3$, and $p_6 = b_1 + b_2 + b_3$, where the equations are interpreted to mean mod 2 addition. The transmitted codeword is $(b_0, b_1, b_2, b_3, p_4, p_5, p_6)$ and is said to have block length 7. At the receiver the information bits can be determined even if a bit is received in error by recomputing the parity equations. For example, if the third bit (b_2) is received in error then the first and last parity equations are not satisfied. Because b_2 is the only bit that participates in only those two parity equations, it is found to be the bit in error and the decoder can correct the error. The code described above is called the Hamming code and can correct any pattern of a single error. If the energy used to transmit a single bit of information is denoted by E_b and the energy used to transmit an encoded bit is E, then for this example $4E_b = 7E$ or $E = 4E_b/7$. So each coded bit actually has less energy than what is allocated for an information bit. Because of this, the signal-to-noise ratio for each coded bit is reduced by a factor of 4/7 from what could be used in an uncoded system. In spite of this, the error correction capability of the code makes up for this loss when the signal-to-noise ratio is reasonably large.

In practice, convolutional codes are used in many communications systems because of their excellent performance. In Figure 1.16 we show the bit error

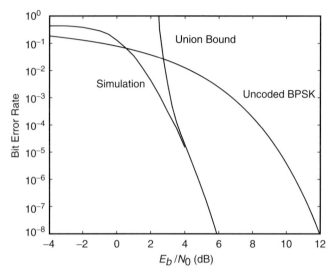

Figure 1.16. Error probability of BPSK with convolutional coding in additive white Gaussian noise.

probability for a typical convolutional code compared to an uncoded system. In this example the modulation is BPSK and the channel has just additive white Gaussian noise. This code has rate 1/2 meaning that for each information bit two coded bits are transmitted. From this figure we can see a 5 dB power performance improvement compared to an uncoded system. The disadvantage is that the bandwidth efficiency has been reduced by a factor of 2.

In Figure 1.17 the performance in Rayleigh fading (a more realistic environment) is shown. In this case the improvement is about 37 dB compared to an uncoded system. This illustrates the benefits in energy efficiency of coding relative to an uncoded system. The penalty paid for this improved energy efficiency is a decrease in bandwidth efficiency plus additional complexity at the receiver. Recently, another coding technique called turbo codes was invented. These codes achieve even better performance provided their block length is sufficiently long.

1.5 EFFECT OF NONLINEAR AMPLIFICATION ON DIRECT-SEQUENCE MULTICARRIER WAVEFORMS

1.5.1 Introduction

Multicarrier (MC) direct-sequence (DS) signaling, as described in Section 1.3, has certain desirable properties relative to single-carrier DS, such as flexibility in deploying the waveform over a noncontiguous bandwidth. That is, if certain segments of a given frequency band are occupied with narrowband signals, the

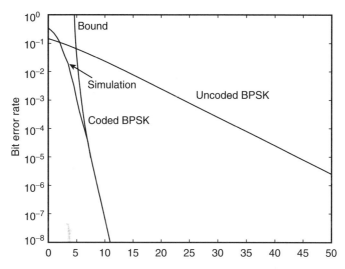

Figure 1.17. Error probability of BPSK with convolutional coding in additive white Gaussian noise and Rayleigh fading.

subcarriers that constitute the MC waveform can be interspersed between the narrowband signals so as not to cause any interference. However, MC signaling is not without its drawbacks, and arguably first among them is the fact that the composite envelope is not constant. Indeed, one can have a very high ratio of peak-to-average power and thus be susceptible to degradation due to nonlinear amplification.

In this section, we attempt to quantify the extent of this degradation; we also present a technique that can regain some of the lost performance by suppressing, at the receiver, some of the intermodulation products that were generated at the transmitter. In particular, we consider a binary communications system operating over a frequency-selective Rayleigh fading channel. The communications system employs convolutional coding with soft decision decoding.

1.5.2 System and Channel Description

A block diagram of the transmitter is shown in Figure 1.18 (see also Xu and Milstein [5]); it consists of a rate-$1/M$ convolutional encoder, followed by an interleaver, a serial-to-parallel converter, an MC modulator, and a power amplifier. The input–output characteristic for the power amplifier when the input is a single sine wave is shown in Figure 1.19. Note that this particular amplifier is assumed to exhibit only AM/AM conversion (i.e., it has no AM/PM conversion); this type of characteristic is typical of a solid state amplifier, as opposed to a traveling-wave tube.

The receiver is shown in Figure 1.20; the incoming waveform is demodulated, deinterleaved, and finally decoded.

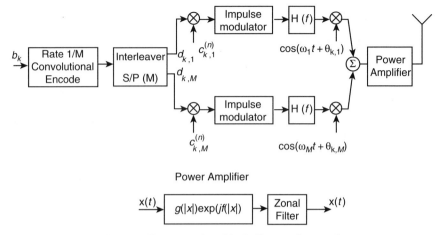

Figure 1.18. Transmitter block diagram for user k.

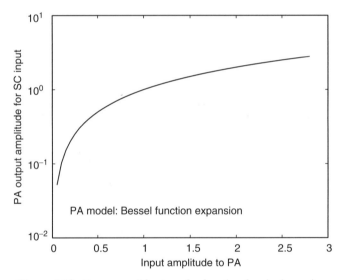

Figure 1.19. Power amplifier transfer function for single carrier.

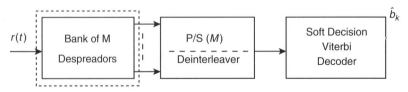

Figure 1.20. Receiver block diagram for user k.

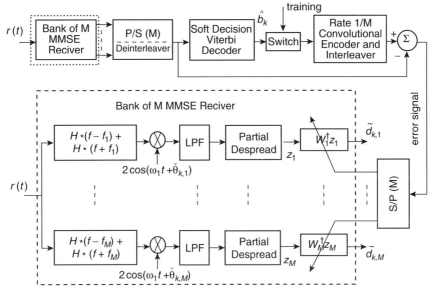

Figure 1.21. Receiver block diagram for user k [5].

As indicated above, an enhanced receiver will also be considered, whereby the enhancement will come from appropriate signal processing at the receiver so that the intermodulation distortion caused by the nonlinear amplifier at the transmitter is reduced. A block diagram of this enhanced receiver is shown in Figure 1.21, and it is seen that the only change from the original receiver is the insertion, after each of the M despreaders, of an interference suppression filter designed via a minimum mean-squared error criterion.

Finally, the channel introduces additive white Gaussian noise (AWGN), as well as flat Rayleigh fading on each subcarrier. The fading is assumed to be sufficiently slow so that it remains essentially constant over several symbols.

1.5.3 Performance Results

In the results that follow, the number of subcarriers, denoted by M, is taken to be 4. Figure 1.22 shows the levels of both third order and fifth order intermodulation (IM) products as a function of the drive level into the amplifier. Also shown is the desired signal response at each input level. It is seen that when the amplifier is driven heavily into saturation, the output levels of both third and fifth order intermodulation products can approach the level of the desired output.

When the above amplifier output is transmitted across the channel, and the receiver of Figure 1.20 is used, the resulting average probability of error is shown in Figure 1.23. It is seen that as the level of the signal into the amplifier is continually increased, a point is reached whereby the performance experiences rapid

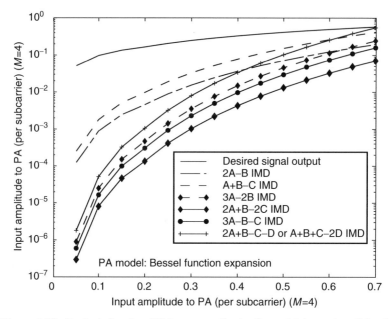

Figure 1.22. Desired signal and IM term amplitudes for multiple carriers $(M=4)$.

degradation. In order to reduce that degradation, the receiver of Figure 1.21 can be used, and its performance is also shown in Figure 1.23. Note that when the initial degradation is small, the enhanced receiver provides no noticeable benefit. However, when the initial degradation is large, the enhanced receiver reduces the average probability of error by about an order of magnitude.

As another illustration of the effect of the nonlinear amplifier, consider the use of the MC DS waveform in a code division multiple access (CDMA) environment. Assume that each user in the system transmits asynchronously using identical power amplifiers. The signal at any receiver then is the sum of all the intermodulation products coming out of each of the transmitters. In Figure 1.24, average probability of error curves are shown plotted against the number of simultaneously active users in the system. There are curves corresponding to two distinct drive levels (per subcarrier) into the amplifier: 0.5 and 0.6. Also, the performances for the two receivers of Figures 1.20 and 1.21 are shown. System performance corresponding to a perfectly linear amplifier is also presented.

It can be seen that, as the number of users in the system increases, the curves corresponding to the presence of intermodulation distortion diverge from the linear amplifier curves. To see what is happening, it is necessary to understand the operation of the suppression filters. Each filter is implemented as a tapped delay line, whereby the number of taps has been set equal to the processing gain per subcarrier; in this particular example, the processing gain is 32. When the total number of interfering waveforms is less than the number of taps, the suppression filter has a sufficient number of degrees of freedom to provide some measure of attenuation to

NONLINEAR AMPLIFICATION ON DIRECT-SEQUENCE MC WAVEFORMS

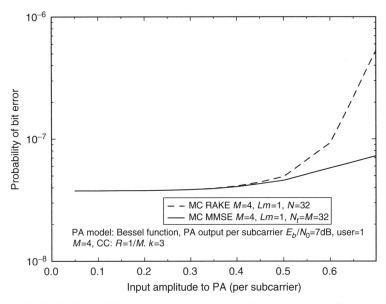

Figure 1.23. Probability of bit error in the presence of intermodulation distortion (IMD) (different PN (DPN) code for each subcarrier).

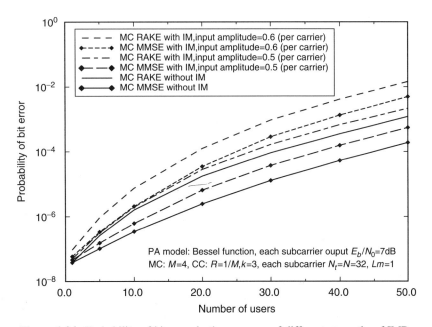

Figure 1.24. Probability of bit error in the presence of different strengths of IMD.

each of the interferers. However, when the number of such interferers exceeds the number of taps, the suppression filter is typically overwhelmed.

Now consider the curves shown in Figure 1.24. As can be seen, when the number of users is small, the receiver of Figure 1.21 can suppress a reasonable percentage of the intermodulation products from all the users and thus results in better performance than does the receiver of Figure 1.20. However, the large number of intermodulation products present as the number of transmitters grows presents too many terms for the suppression filter to handle. That is, the number of terms that the suppression filter must attenuate is not only the total number of first order intermodulation products (i.e., the desired signal term out of the amplifier of each interfering user), but all the significant higher order terms as well (e.g., those corresponding to third and fifth order products). Thus, while the use of the suppression filter still results in enhanced performance over the simple correlation receiver, its use does not result in performance comparable to that when intermodulation products do not exist.

As a final example we consider a scenario where 30 users in a cellular system are communicating with a base station. The different users might be at different distances from the base. Power control is assumed so that the received power is the same for all users. We will assume that the user's received power decreases with distance as d^4. Users close to the base station will transmit with lower power. Because of this the amplifier will be operating in the linear range. However, users far away will need to transmit substantially more power in order for all the signals to arrive with the same power level. Thus distant users will be operating closer to the nonlinear range of the amplifier. These users then will create more intermodulation products that interfere with the signals of all the other users.

The model for the channel we employ is flat Rayleigh fading on each of the subcarriers. The amplifier output is assumed to saturate for some input power level. The ouput power level at which we operate the amplifier is reduced from the saturation output power level by some amount (in dB) called the output backoff (OBO). In addition, a rate 1/2 constraint length 7 convolutional code is considered with spreading by a factor of 80. In the numerical results that follow there are 30 users in a cell and the users are at distance r_1 or r_2 from the base station where $r_2/r_1 = 20$. The performance is evaluated for a particular user called user A.

In Figure 1.25 the bit error probability for 30 users each using 10 carriers is shown. In this figure all the users are at location r_2. The receiver in this case is the conventional matched filter for each carrier without any interference suppression filter. The degradation due to the nonlinearity is seen to be small for large output backoff, while for small output backoff the amplifier is operating closer to the nonlinearity and thus the degradation in performance is larger. The performance for the case where users are at different locations is shown in Figure 1.26. In this figure we plot the additional E_b/N_0 required to achieve the same error probability (e.g., $P_b = 10^{-4}$) due to the nonlinearity. The degradation when the user-of-interest (user A) is located at the closer location (r_1) while the 29 other users are located at the further location (r_2) is quite minimal. In this case the intermodulation products from users at distance r_2 cause only a very small degradation (less than 1 dB). Also

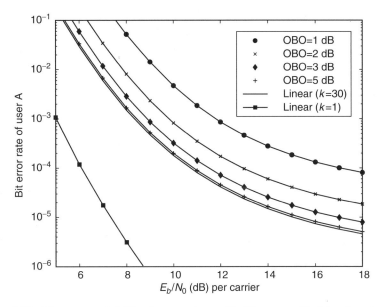

Figure 1.25. Bit error probability for a system with 30 users with 10 subcarriers and spreading 80 for different amplifier operating points.

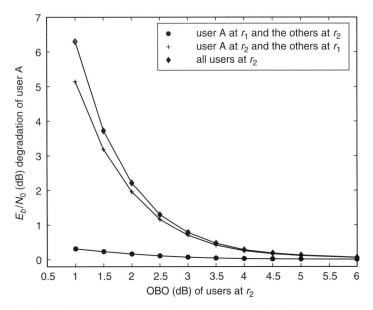

Figure 1.26. Degradation in performance due to nonlinearities for 30 users and spreading 80 for different amplifier operating points.

the desired user, because of power control, has a very significant backoff and does not generate any substantial intermodulation products. However, when the desired user is at distance r_2 the intermodulation products cause a degradation in performance by more than 5 dB at very low backoff. For additional details see Jong and Stark [6].

In the preceding discussion the performance curves were plotted as a function of the received energy per bit. However, in amplifying a signal the power amplifier, besides generating intermodulation products due to the nonlinearity, also has a varying power added efficiency. Power added efficiency is a measure of the amount of power converted from the dc power supply to output power. When the output backoff is large the power added efficiency is very small, while when the backoff is small the power added efficiency is much larger. Thus for large backoff the amplifier does not generate significant intermodulation products but is inefficient in converting dc power to output power. When the backoff is small the amplifier generates intermodulation products that distort the signal but, on the other hand, is able to convert the dc power to output power more efficiently. As a result of this, there is an optimal operating point in order to minimize the energy needed for a given performance (bit error probability). To quantify the overall performance we define the total degradation to be the sum of the output backoff (dB) plus the E_b/N_0 (dB) degradation due to the nonlinearity as shown in Figure 1.26. The output backoff is an accurate measure of the inefficiency of the amplifier (large output backoff implies large inefficiency), while E_b/N_0 degradation is a measure of the distortion due to the nonlinearity. This is shown in Figure 1.27 for the scenario of 30 users all operating at

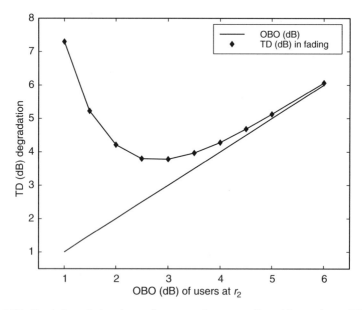

Figure 1.27. Total degradation in performance due to nonlinearities and amplifier inefficiency.

a distance of r_2 using multicarrier modulation and convolutional coding. As can be seen there is an optimal backoff that minimizes the total degradation in performance. This optimal backoff is about 3 dB and results in a performance that is degraded from the optimal by a little less than 4 dB. This incorporates both the degradation due to the intermodulation products and the inefficiency of the amplifier. This topic will be discussed in further detail in Chapter 7.

1.6 RECEIVER ARCHITECTURE

Besides containing the desired signal, the received signal contains unwanted interfering signals in adjacent frequency bands. The adjacent channel interfering signals may be much higher in power (by as much as 150 dB in certain systems) than the desired signal as seen in Figure 1.28. So one goal of the receiver is to select the desired signal and to reject the adjacent channel signals. Because the bandwidth is typically much smaller than the center frequency, rejecting the interference usually occurs in stages. First a wideband bandpass filter removes signals that are significantly distant (in frequency) from the desired signal. Then the signal is downconverted to a lower intermediate frequency (IF), where another filter with narrower bandwidth removes signals adjacent to the desired signal. Since the IF is typically much lower than the carrier frequency it much easier to design a filter of a given bandwidth centered at the IF than a filter of the same bandwidth at the carrier frequency.

Another goal in designing a receiver is to minimize the thermal noise of the receiver. For this reason a low noise amplifier is usually incorporated into the receiver very close to the antenna to set the noise figure of the receiver. Another goal is to minimize the receiver size. Filters are generally hard to incorporate into

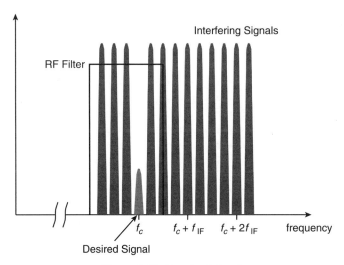

Figure 1.28. Received signal.

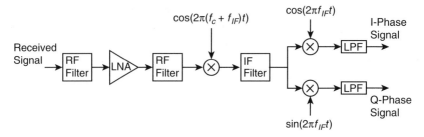

Figure 1.29. Superheterodyne receiver.

integrated circuits and thus are usually off-chip. A key question then is how to reject the interfering signals and have a low noise figure with minimal energy used by the receiver and minimal size. In the remainder of this chapter we explore various options and discuss the advantages and disadvantages of each.

The conventional design of a receiver is known as a superheterodyne receiver and is shown in Figure 1.29. The desired signal at frequency f_c is filtered, amplified, and filtered. The goal of the first filter is to reject the signal at frequency $f_c + 2f_{IF}$. This is known as the image frequency. A signal at this frequency when mixed with a local oscillator frequency of $f_c + f_{IF}$ will be translated to frequency f_{IF}. The desired signal will also be translated to this frequency so it is necessary to first remove the image frequency before mixing down to the intermediate frequency. The second RF filter is used to reject any signals generated by a nonlinearity of the low noise amplifier. The output of the second RF filter is then mixed down to a much lower frequency (f_{IF}) known as the intermediate frequency. It is easier at this (lower) frequency to design a narrow bandwidth filter that rejects the adjacent frequencies than it is at the higher RF frequency. Finally, the signal is mixed either to baseband by a pair of mixers (as shown) or to a very low frequency before being sampled. After sampling, digital processing is used to recover the desired information as described earlier. With conventional filters the superheterodyne architecture has the disadvantage that the filters are not "on-chip" filters and thus the size of the receiver is limited by these filters (which for certain frequency bands can be a limiting factor in the size of the receiver). The low noise amplifier consumes energy in order to amplify the signal. As with the transmitter amplifier, the low noise amplifier needs to operate well within the linear region in order not to generate intermodulation products of the input signal. However, the power efficiency of the amplifier is very small when the backoff is very large. So there is a trade-off here of performance versus energy. Better performance can be achieved if the amplifier operates with larger backoff but at the expense of increased energy consumption. Better performance and lower energy consumption are possible if the RF filters could do a better job at removing all the interference before amplifying. Such a filter may be possible with the advent of microelectromechanical system (MEMS) filters. In addition many MEMS filters can be incorporated into an integrated circuit with the same technology. The design of such filters is the topic of Chapter 12.

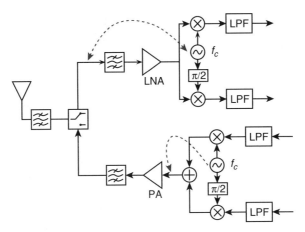

Figure 1.30. Direct conversion transmitter/receiver.

Another potential architecture shown in Figure 1.30 is a direct conversion architecture. In this receiver there is no intermediate frequency. Because of this, all of the components associated with the IF section are eliminated and thus the power consumption should be smaller than that of a superheterodyne architecture. However, the drawback to this architecture is that the signal out of the local oscillator (LO) may leak into the input to the low noise amplifier (LNA). Because the LO signal has a large amplitude relative to the received signal, the part of the LO signal that is coupled into the amplifier input could dominate the desired signal. Because the LO signal and the desired signal are at the same frequency the leakage can cause significant interference to the desired signal. In addition, a strong interfering signal at a frequency different from the carrier might leak into the LO signal input to the mixer. If this happens the interfering signal will get mixed to the same frequency as the desired signal by the part of the interfering signal that has leaked into the LO input to the mixer.

Another architecture is shown in Figure 1.31 and is called direct sampling. In direct sampling the output of the RF section is sampled at a sufficiently high rate and then digital filtering is used to filter out the undesired signals.

A direct sampling architecture requires a large dynamic range for analog-to-digital conversion. Large dynamic range is necessary as an interferer might be 50–150 dB stronger than the desired signal. Either very high Q filters are needed at the front end so that the interference is rejected or very high dynamic range conversion is necessary so that the jammer does not saturate the analog-to-digital converter (ADC) and the digital processing can filter out the interference. In addition, a sampling rate proportional to the carrier frequency is needed. This combination of high number of bits of quantization and high sampling rate is very power hungry and thus not a suitable choice for a low power design. However, for a cellular system this might be a good choice for the front end of the base station receiver since only one front end is needed for all users and power consumption is not as critical as in a mobile.

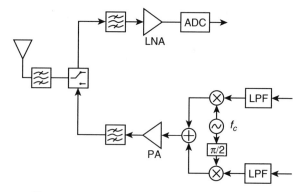

Figure 1.31. Direct sampling transmitter/receiver.

ACKNOWLEDGMENT

The authors gratefully acknowledge the assistance of Je-hong Jong and Weiping Xu in generating some of the numerical results presented here. In addition, the support of the Department of Defense under the MURI programs on Low-Power Electronics and Low Energy Electronics Design of Mobile Platforms DAAH04-96-0001, DAAH04-96-0005, and DAAH04-96-0377 is gratefully acknowledged.

REFERENCES

1. T. M. Cover and J. A. Thomas, *Information Theory*, Wiley-Interscience, New York, 1991.
2. J. G. Proakis, *Digital Communications*, 3rd ed., McGraw-Hill, New York, 1995.
3. S. G. Wilson, *Digital Modulation and Coding*, Prentice-Hall, Englewood Cliffs, NJ, 1996.
4. R. L. Peterson, R. E. Ziemer, and D. E. Borth, *Introduction to Spread Spectrum Communications*, Prentice-Hall, Englewood Cliffs, NJ, 1995.
5. W. Xu and L. Milstein, "On the use of interference suppression to reduce intermodulation distortion in multicarrier CDMA systems" *IEEE Transactions on Communications*, Vol. 49, No. 1, pp. 130–141 (2001).
6. Je-hong Jong and W. E. Stark, "Power-controlled MC-CDMA in the presence of nonlinearities—Part II: Coded system performance," in *Proceedings of the 1999 IEEE Military Communications Conference*, 1999.

2

ADVANCED GaAs-BASED HBT DESIGNS FOR WIRELESS COMMUNICATIONS SYSTEMS

M. FRANK CHANG
Department of Electrical Engineering, University of California—Los Angeles

PETER M. ASBECK
Department of Electrical and Computer Engineering, University of California—San Diego

2.1 INTRODUCTION

Recent advances in wireless communications systems demand very high performance amplifiers for high efficiency and high linearity microwave power transmission. GaAs-based heterojunction bipolar transistors (HBTs) combine many of the needed characteristics for such power amplifiers. They offer excellent microwave gain, efficiency, linearity, and ruggedness with single-supply operation in a compact structure and potentially manufacturable at very low cost. In comparison with silicon (or SiGe) devices, GaAs HBTs benefit from high electron mobility and the availability of semi-insulating substrates. These characteristics lead to low forward voltages at high currents and enable several stages of amplification to be cointegrated with appropriate reactive impedance matching elements. Accordingly, GaAs-based HBTs are presently in widespread use in commercial wireless systems, with annual manufacturing volume above 20 million units.

The GaAs-based HBT technology has advanced very rapidly in the past several years. This chapter addresses critical issues in GaAs-based HBT device design and

---

RF Technologies for Low Power Wireless Communications
Edited by Tatsuo Itoh, George Haddad, and James Harvey
ISBN 0-471-38267-1 Copyright © 2001 by John Wiley & Sons, Inc.

offers effective and practical solutions to enhance the linearity and efficiency in microwave power amplification and the reliability in device operation. The individual device layer designs (primarily, the emitter and collector layer structures) will be discussed first. The optimization of the overall device design will be discussed according to the scaling rules of HBTs and the wireless communications system requirements. Novel HBT device structures based on advanced designs, processes, and/or materials will be discussed as well.

2.2 EMITTER STRUCTURE DESIGN

The emitter structure design is extremely crucial to the HBT's performance and reliability. An optimized emitter design should provide HBTs with high current gain, low junction capacitance, and sufficient device reliability.

Since the surface recombination velocity of GaAs is rather high, it degrades the HBT's dc current gain as the emitter dimensions decrease [1]. The same cause is also believed to be an important factor that affects the device reliability [2]. A popular method for improving the device current gain and reliability in small devices is to leave a heteroguard ring or "ledge" of thin depleted emitter layer to passivate the emitter–base junction, as shown in Figure 2.1. While there has been a great deal of research work which studied the effects of emitter ledge on HBT current gain [3,4], little work has been explored in design algorithms and monitoring techniques to ensure the quality of emitter ledge passivation. In this section, we present a new method in emitter structure design based on information obtained from an "on-ledge" Schottky diode potentiometer. Since this potentiometer contacts the emitter ledge directly, it provides in-place and quantitative information in determining the effectiveness of the emitter ledge passivation.

2.2.1 On-Ledge Schottky Potentiometer

The effectiveness of the emitter ledge passivation determines the HBT's current gain and reliability. If the emitter ledge is not completely pinched-off (or depleted) during

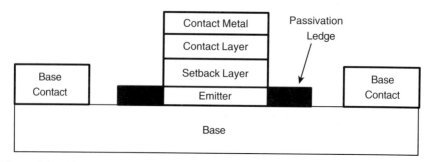

Figure 2.1. Thin and depleted heteroguard-ring ledge used to passivate the emitter–base junction periphery.

the device operation, HBTs may suffer serious current gain degradation and poor long-term reliability [5,6]. There were techniques used in the past to examine the extent of the ledge depletion by comparing the gain of HBTs with different ledge widths and P/A (periphery-to-area) ratios [7]. Those techniques were indirect and only qualitative. Here we discuss the feasibility of using an on-ledge Schottky potentiometer to determine the extent of the ledge depletion under any base–emitter V_{BE} bias conditions. With extensive measurement, analysis, and modeling, we find the ledge potentiometer to be extremely sensitive and quantitative in determining the effectiveness of the emitter ledge passivation.

We have developed an analytical model to explain the physics and operation of the ledge potentiometer. Figure 2.2(a) shows the cross section of an HBT with one-side emitter ledge (500–600 Å thick and 1.1 μm wide) and a base contact. The potentiometer is made of a Ti/Pt/Au Schottky diode (0.8 μm wide), which contacts the emitter close to the end of the passivation ledge. With a given V_{BE}, the region under the ledge potentiometer (d_2–d_3) can be biased into partially or fully depleted states. The equivalent circuit configuration of the emitter ledge is illustrated in Figure 2.2(b). For simplicity, the region under the Schottky metal contact (d_2–d_m) is assumed to be equipotential. The same assumption applies to the heavily doped p^+ base. The electrical property of the "free ledge," the region (d_1–d_2), is modeled by distributed resistors R_L and diodes D_{EX1}. Under the potentiometer, the electrical property is modeled by the Schottky diode D_{SH}, diode D_{EX2}, and the space-charge resistor R_d.

Our analytic model is to relate V_{Ledge} to the externally applied voltage V_{BE}. The first step is to calculate the voltage distribution in the free ledge region (d_1–d_2). As shown in Figure 2.2(a), the lateral current density $J_L(x)$ is governed by Eq. (2.1) [8]:

$$\frac{d^2 J_L(x)}{dx^2} + \frac{\rho}{V_t} \cdot \frac{dJ_L(x)}{dx} \cdot J_L(x) = 0, \quad (2.1)$$

where V_t is thermal voltage and ρ is the ledge resistivity. Using nonzero boundary conditions as follows and assuming Λ is the width of the undepleted ledge in region d_1–d_2, we find

$$J_L(x)|_{x=d_2} = J(d_2) \quad \text{and} \quad \frac{dJ_L(x)}{dx}\bigg|_{x=d_1} = -\frac{J_E(d_1)}{\Lambda}.$$

$J_L(x)$ can be derived as follows:

$$J_L(x) = \frac{2cV_t}{\rho(d_2 - d_1)} \cdot \tan\left[c\left(1 - \frac{x - d_1}{d_2 - d_1}\right) + \xi\right] \quad (2.2)$$

with

$$\xi = \arctan\left(\frac{\rho(d_2 - d_1)}{2cV_t} \cdot J(d_2)\right), \quad (2.3)$$

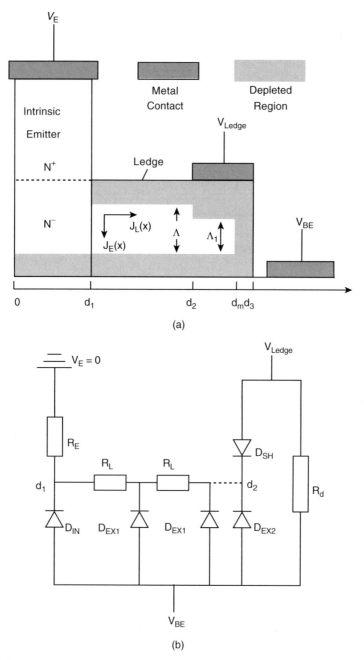

Figure 2.2. (a) A schematic of HBT cross section showing the geometry of the emitter passivation ledge and the Schottky diode potentiometer positioned near the end of the ledge. (b) Equivalent circuit schematic configuration used to explain the electrical properties of the emitter ledge and the ledge potentiometer. R_E is the emitter resistance. D_{IN} is the intrinsic diode. D_{EX1} and D_{EX2} are extrinsic diodes in free and Schottky metal covered ledge regions, respectively. D_{SH} is the Schottky diode of the ledge potentiometer. R_d is the space-charge resistor.

EMITTER STRUCTURE DESIGN

where ξ and c are related to each other by

$$c^2 \sec^2(c+\xi) = \frac{\rho(d_2-d_1)^2}{2V_t \Lambda} \cdot J_E(d_1). \qquad (2.4)$$

The voltage $V(d_2)$ can be written as

$$V(d_2) = V(d_1) + 2V_t \ln\left(\frac{|\cos(\xi)|}{|\cos(c+\xi)|}\right) \qquad (2.5)$$

At the node d_2, the current continuity requires

$$I(d_2) = I_{EX2} + I_{SH}, \qquad (2.6)$$

where I_{EX2} is the current of the heterojunction diode D_{EX2}, and I_{SH} is the thermionic emission current of the Schottky diode D_{SH} [9] and also can be expressed as

$$I_{SH} = (V_{BE} - V_{Ledge})/R_d \qquad (2.7)$$

with the space-charge resistance R_d defined in Sze and Shockley [10]. Giving $A(\Lambda)$ as the area where the current flows into the region (d_1–d_2), the voltage $V(d_1)$ becomes

$$V(d_1) = R_E[I_{IN} + A(\Lambda) \cdot J_L(d_1)], \qquad (2.8)$$

where I_{IN} is the current flowing through the intrinsic diode D_{IN}. By designating η as the ratio

$$\eta = \frac{d_3 - d_2}{d_3 - d_m} \qquad (2.9)$$

we can deduce the ledge potential V_{Ledge} as a function of V_{BE} by using a self-consistent calculation process based on Eqs. (2.2)–(2.9).

In order to measure V_{Ledge} versus V_{BE}, we have set up AlGaAs/GaAs HBTs in a common-emitter configuration and biased V_{CE} at 1.5 V. The measurement results are shown in Figure 2.3, with three distinguished types of device characteristics. Among them, type-A and type-B HBTs belong to the same wafer with 500 Å thick ledge and type-C HBT belongs to a different wafer with 600 Å thick ledge. In type-A HBTs, V_{Ledge} always equals V_{BE} as V_{BE} varies from 0 to 1.6 V. In type-B HBTs, V_{Ledge} follows V_{BE} initially but soon reaches a plateau as V_{BE} approaches about 0.8 V. However, it curves up again when V_{BE} reaches about 1.25 V. In type-C HBTs, V_{Ledge} does not track V_{BE} initially but increases very gradually and finally curves up at about $V_{BE} = 1.35$ V.

The behavior of the three types of HBTs can be explained with an analytical model by assuming various undepleted widths in emitter ledges. In case the free

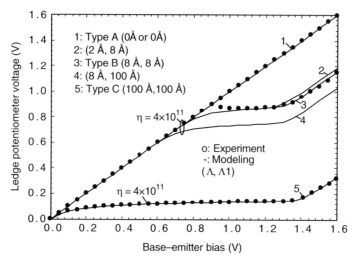

Figure 2.3. Comparing measured and modeled V_{Ledge} versus V_{BE}, with various undepletion widths of Λ and Λ_1 for different types of HBTs, where the $\eta = (d_3-d_2)/(d_3-d_m)$.

ledge (region d_1–d_2) or the ledge under the potentiometer (region d_2–d_3) is fully depleted (Λ or Λ_1 equals zero) under any bias condition of V_{BE}, there is no current flowing through the ledge, which results in an equipotential of V_{Ledge} and V_{BE}.

In case the ledge is slightly undepleted (Λ and Λ_1 are nonzero in regions d_1–d_m), as V_{BE} increases, the Schottky diode D_{SH} is turned on at about 0.7–0.8 V, which allows the leakage current flowing through the conductive ledge into the grounded emitter. When this happens, the ledge potential V_{Ledge} reaches a plateau for being clamped by the turn-on voltage of the Schottky diode. As V_{BE} continues to increase beyond 1.3 V, the voltage at node d_2 rises suddenly due to the gradual turn-on of the heterojunction diodes of D_{EX1} and D_{IN}. The injected heterojunction currents add voltage drop across distributed resistors R_{L} and R_{E}, which again enforces a synchronized increase between V_{Ledge} and V_{BE}. This interesting relationship between V_{Ledge} and V_{BE} is successfully simulated in Figure 2.3 by assuming the ledge is slightly undepleted with both Λ and Λ_1 equal to 8 Å. We also let $\eta = 10$ by assuming the ledge doping concentration to be $N_D = 3 \times 10^{17}$ /cm^3 and the Fermi level pinned to the midgap (0.77 eV) at the surface of the ledge.

The characteristics of V_{Ledge} versus V_{BE} can be analyzed further by varying Λ and Λ_1 values in simulation. By changing (Λ, Λ_1) from (8 Å, 8 Å) to (2 Å, 8 Å), V_{Ledge} reaches the same plateau as previously but curves up later at a slightly lower V_{BE}. On the contrary, changing from (8 Å, 8 Å) to (8 Å, 100 Å), V_{Ledge} reaches a slightly lower plateau at about 0.7 V but curves up at the same voltage at 1.3 V. It is evident that the magnitude of Λ_1 affects the V_{Ledge} clamping (plateau) voltage and the magnitude of Λ affects the V_{Ledge} curving up voltage.

The proposed potentiometer is extremely sensitive in differentiating different types of HBTs. For example, type-A and type-B HBTs are indistinguishable in their

EMITTER STRUCTURE DESIGN

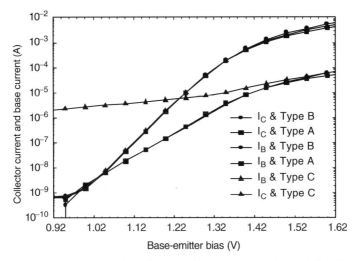

Figure 2.4. Measured Gummel plots of HBTs with V_{Ledge} versus V_{BE} relationship shown in Figure 2.3.

"Gummel" characteristics (Fig. 2.4) but very differentiable with the proposed new method. The potentiometer also reveals more insight about the possible failure mechanism of the type-C HBT. In its Gummel plot, the type-C HBT simply shows no current gain at low V_{BE}. Through our simulations, we find the HBT's current gain degradation caused primarily by an undepleted emitter ledge ($\Lambda = \Lambda_1 = 100$ Å) and/ or an unpinned ledge surface ($\eta = 4 \times 10^{11}$).

By using an on-ledge Schottky potentiometer, one can determine the extent of the ledge depletion down to a few angstroms (<10 Å) in precision. The outstanding sensitivity of the ledge potentiometer makes itself a unique tool in guiding HBT emitter structure design to achieve better emitter passivation for enhanced device gain and reliability.

2.2.2 Dependence of HBT Current Gain on Ledge Potential

In Section 2.2.1, we described the use of an on-ledge Schottky potentiometer to monitor the effectiveness of an HBT's emitter–base junction passivation by relating the ledge potential (V_{Ledge}) to the emitter–base bias voltage (V_{BE}). In this section, we report that the current gain of HBTs can be modulated effectively by biasing the on-ledge Schottky potentiometer externally, and the characteristics of the current gain modulation depend on the degree of the emitter ledge depletion.

To explain the observed current gain modulation effect, we use type-A and type-B HBTs as examples. Based on our previous discussion, the emitter ledge is fully depleted ($\Lambda = 0$) in a type-A HBT and partially depleted ($\Lambda > 0$) in a type-B HBT.

Setting up InGaP/GaAs HBTs in a common-emitter configuration with collector–base bias voltage (V_{CB}) biased at 0 V, we investigate effects of the external ledge bias on the HBTs' current gain and Gummel I–V characteristics. First we note that the gain modulation behaviors are quite different in type-A and type-B HBTs. As shown in Figure 2.5, the current gain of a type-B HBT can only be modulated in the low V_{BE} (<1.35 V) region (Fig. 2.5(a)). On the contrary, the current gain of a type-A HBT can be modulated in both low and high V_{BE} up to 1.6 V (Fig. 2.5(b)).

When we externally bias the ledge at $V_L < V_{BE}$ (e.g., let V_L be 0.3–0.5 V), the base currents increase dramatically from their original current level (corresponding to an unbiased or floating ledge) in the low V_{BE} range (<1.35 V). On the contrary, the base current remains unchanged under the same V_L bias conditions in the high V_{BE} range (>1.35 V). Since the collector current remains unchanged, type-B HBT's current gain decreases at $V_{BE} < 1.35$ V but remains unchanged at $V_{BE} > 1.35$ V.

The situation is slightly different when we bias the ledge of a type-B HBT at $V_L > V_{BE}$ (e.g., $V_L = 2.9$ V). In this case, the base current decreases from its original value and the HBT current gain increases as $V_{BE} < 1.15$ V. When $V_{BE} > 1.15$ V, the HBT's current gain again remains unchanged.

In type-A HBTs, as shown in Figure 2.5(b), the base current increases in the whole operating range of V_{BE} (at least up to 1.6 V) as V_L is biased at voltages lower than V_{BE} (e.g., $V_L = 0.9$–1.0 V). When the ledge is biased at a higher value, $V_L = 1.2$ V, the base current decreases in the region of $V_{BE} < V_L$ but remains unchanged in the region of $V_{BE} > V_L$.

The observed behaviors in the HBT's current gain modulation may be explained through an equivalent circuit model as shown in Figure 2.6, with different types of parasitic transistors shunting between the HBT's emitter and base according to the HBT's emitter ledge pinch-off conditions.

The ledge of type-B HBTs is partially depleted according to our previous experiment [11]. This provides a physical basis for a multiemitter parasitic transistor to be connected as shown in Figure 2.6(a), where the HBT's base becomes the parasitic transistor's emitter (D_{EX}); the HBT's emitter ledge becomes the parasitic transistor's base; and the HBT's Schottky on-ledge potentiometer becomes the parasitic transistor's collector (D_{SH}). The proposed parasitic transistor structure with resistors (R_L's) situated between distributive emitters can be justified through the nature of a partially depleted and conductive ledge.

The ledge of type-A HBTs is fully depleted. This leads to a different parasitic transistor structure as shown in Figure 2.6(b), of which the base is not only completely punched-through but also floating or isolated from any external bias because of the fully depleted and nonconductive ledge.

Based on this proposed equivalent circuit model, the HBT's effective current gain can be expressed as

$$\beta = \frac{I_C}{I_B} = \frac{\beta_0}{1 + [(I_L + I_L^*)/I_{Bin}]}, \qquad (2.10)$$

where $\beta_0 = I_C/I_{Bin}$ is the HBT's intrinsic current gain; I_{Bin} represents the HBT's intrinsic base current with an unbiased emitter ledge; and I_L and I_L^* represent leakage currents that flow through the collector and base of the parasitic transistor,

EMITTER STRUCTURE DESIGN

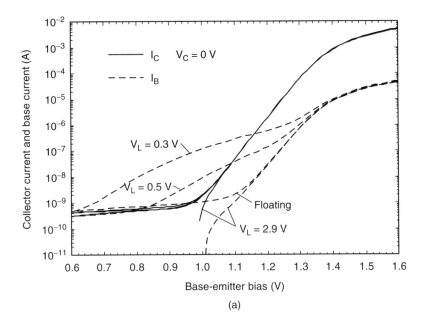

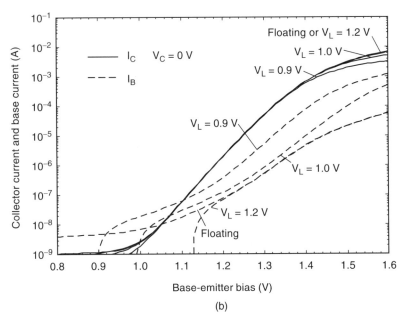

Figure 2.5. Representative Gummel plots measured for (a) Type-A HBTs and (b) type-B HBTs with various V_L bias voltages.

respectively. According to Eq. (2.10), typically as $I_L \gg I_L^*$, the leakage current I_L (or strictly, the ratio of I_L/I_{Bin}) determines the degree of the current gain modulation.

To further analyze the current gain modulation mechanism, we measure $|I_L|$ and the ratio of $|I_L/I_{Bin}|$ as a function of $V_{BE} - V_L$ as shown in Figures 2.7(a) and 2.7(b),

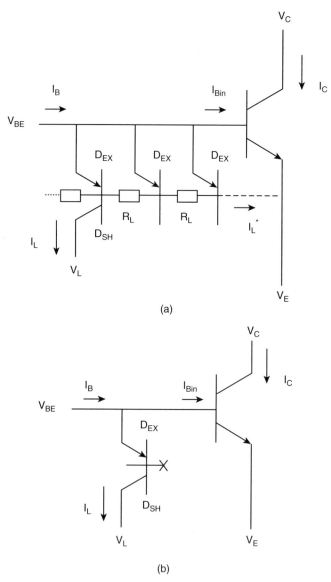

Figure 2.6. Equivalent circuit models for (a) type-A HBTs and (b) type-B HBTs, respectively, where D_{EX} represents a pn diode between the extrinsic base and the emitter-ledge as Nakajima et al. [1] show. D_{SH} represents the on-ledge Schottky diode and R_L is the distributive channel resistance along the undepleted emitter ledge.

EMITTER STRUCTURE DESIGN

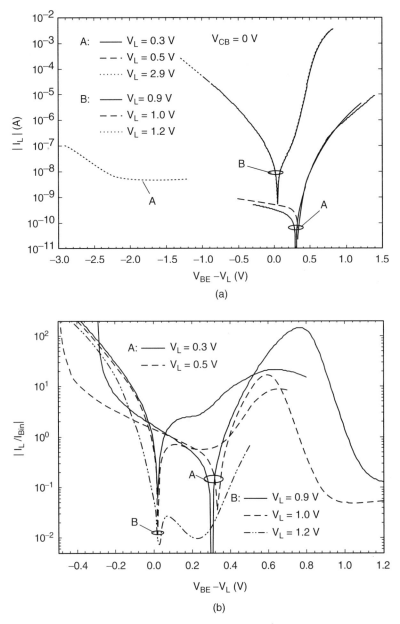

Figure 2.7. (a) Leakage current $|I_L|$ and (b) the ratio of $|I_L/I_{Bin}|$ measured through the on-ledge Schottky potentiometer versus the offset voltage of $V_{BE}-V_L$.

respectively. When $V_{BE} - V_L$ is positive, the parasitic transistor is under a forward mode operation; otherwise it is operated under the reverse mode.

In type-A HBTs, since the parasitic transistor's base is completely punched-through, the leakage current (I_L) turns on at $V_{BE} - V_L = 0$ V under both forward and reverse mode operations (Fig. 2.7(a)). Further measurement of $|I_L/I_{Bin}|$ in Figure 2.7(b) shows that the leakage current I_L of type-A HBTs quickly rises to current levels that overwhelm I_{Bin} as V_L is biased at 0.9 to 1.0 V. This explains why the current gain of type-A HBTs can easily be modulated in both low and high V_{BE} bias ranges.

In type-B HBTs, since the parasitic transistor's base is not totally depleted, the leakage current turns on at an offset voltage of $V_{BE} - V_L = 0.3$ V (Fig. 2.7(a)), which is caused by the different turn-on voltages of the heterojunction diode (D_{EX}) and the on-ledge Schottky diode (D_{SH}). When $V_{BE} < 1.35$ V and we operate the parasitic transistor of type-B HBT in the forward mode, we find the leakage current I_L exceeding I_{Bin} and degrading the HBT's current gain as shown in Figure 2.7(b). When biased at $V_{BE} > 1.35$ V, the lateral ledge leakage current I_L^* increases and causes significant voltage drop across the distributive resistance (R_L's) along the ledge. This effect reduces the forward bias voltage of D_{EX}, decreases the ratio $|I_L/I_{Bin}|$, and ultimately prohibits the current gain modulation in the type-B HBT when $V_{BE} > 1.35$ V.

In summary, we have observed the effect of current gain modulation in GaAs-based HBTs by biasing on-ledge Schottky diodes. Two types of current gain modulation are discovered according to the HBT's Gummel I–V characteristics. In a type-A HBT, the current gain can be modulated in both low and high voltage regions at least up to 1.6 V. In a type-B HBT, the device current gain can be modulated only in the region of low base–emitter bias voltages ($V_{BE} < 1.35$ V). Using the same on-ledge Schottky diode as a potentiometer, we find the behaviors of HBT gain modulation being determined by the extent of the emitter ledge depletion. This discovery also provides new insights to the mechanism of HBT gain degradation that results in similarly degraded Gummel characteristics after high current stress [11–14]. More discussions about the failure mechanism of HBTs will be given in Section 2.2.4.

2.2.3 Dependence of $1/f$ Noise on Ledge Potential

In wireless communications systems, the complex frequency, bandwidth, and signal-to-noise (S/N) ratio requirements make the traditional heterodyne radio frequency (RF) transceivers extremely complicated and costly. A better approach is to adopt a direct conversion (or homodyne) architecture that can eliminate most of the external filters and reduce the total number of local oscillators for frequency conversion [15]. However, the direct conversion transceiver demands semiconductor integrated circuits (ICs) with extremely low $1/f$ noise. GaAs-based HBTs are known for their relatively low $1/f$ noise compared to other competing high speed IC technologies such as Si CMOS, GaAs MESFETs, and HEMTs. However, the GaAs-based HBT's $1/f$ noise level is found to be extremely sensitive to the emitter ledge depletion

EMITTER STRUCTURE DESIGN   51

conditions [16]. By varying the ledge potential to change the degree of emitter ledge depletion, we can greatly vary the magnitude of the HBT's $1/f$ noise level.

As shown in Figure 2.8, when the ledge potential of a type-B HBT is varied from its floating (or unbiased) status to a biased status of $V_L = 0.5$ to 1.2 V, the HBT's base currents increase dramatically as $V_L < V_{BE}$ (Fig. 2.8(a)). Under the same ledge-bias range of V_L, the corresponding $1/f$ noise levels are also increased dramatically (approximately 20–30 dB) from the original unbiased noise level (Fig. 2.8(b)). On the contrary, when the ledge is biased at a higher voltage of $V_L > V_{BE}$, such as $V_L = 1.55$ V, both base current and $1/f$ noise decrease.

Figure 2.9 shows the $1/f$ base noise spectra density S_{ib} measured as a function of I_B. With a fully depleted and unbiased (floating) ledge, the noise spectra density S_{ib} of a type-B HBT is measured to be proportional to $I_B^{1.42}$. When the ledge is biased at $V_L < V_{BE}$ (i.e., $V_L = 0.5$–1.2 V), the base currents and noise spectra densities increase and show stronger dependence on the base current (proportional to $I_B^{1.52}$ to $I_B^{1.84}$, respectively). However, when the ledge is biased at $V_L > V_{BE}$ (i.e., $V_L = 1.55$ V), both base current and $1/f$ noise spectra density decrease and show very little dependence on each other.

These measurement results indicate that the $1/f$ noise levels of HBTs, just like their base currents, are strongly influenced by the degree of the ledge depletion. A fully depleted or well passivated ledge can reduce HBT's $1/f$ noise significantly. By biasing the ledge at $V_L > V_{BE}$, we can even further reduce the $1/f$ noise of HBTs and minimize their dependence on the base current. The lower $1/f$ noise and weaker dependence of S_{ib} on I_B are both beneficial to the high performance and low cost direct conversion transceiver circuit design.

2.2.4 HBT Reliability Enhancement

GaAs-based HBTs often suffer from continuous degradation at high current density and high temperature. This was a severe barrier for the production of high performance GaAs-based HBT power amplifiers and integrated circuits. Due to its recombination-enhanced nature, the median lifetime of HBTs, t_{50}, can be expressed as follows [14]:

$$t_{50} = j_c^{-m} \exp(E_a/kT_j), \qquad (2.11)$$

where E_a is the activation energy, m is a constant, j_c is the density of collector current, k is the Boltzmann constant, and T_j is the junction temperature. The m value typically ranges from 2 to 3 and the activation energy E_a varies from 0.35 to 1.3 eV, depending on the quality of the heterojunction and the emitter/base surface passivation [18]. When $E_a = 1.3$ eV, the best mean time to failure (MTTF) extrapolated for carbon-doped AlGaAs/GaAs HBTs is about $>10^7$ hours at $J_c < 2.5 \times 10^4$ A/cm^2. The MTTF, however, degrades to $<10^7$ hours at $J_c > 5 \times 10^4$ A/cm^2, which may not be sufficient for some wireless communications systems.

Two different HBT failure types are observed after the high current and high temperature stress. Figure 2.10 shows the Gummel plot of a type-1 HBT before and

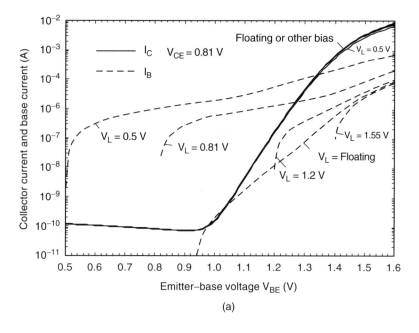

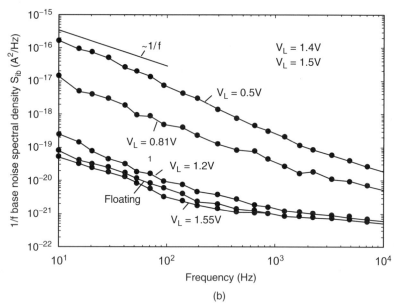

Figure 2.8. (a) Representative Gummel plot and (b) $1/f$ base noise characteristic measured with various V_L bias voltages.

EMITTER STRUCTURE DESIGN 53

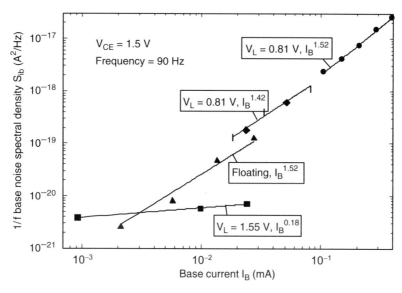

Figure 2.9. Typical $1/f$ base noise density S_{ib} measured with different V_L bias using the same HBT as in Figure 2.8.

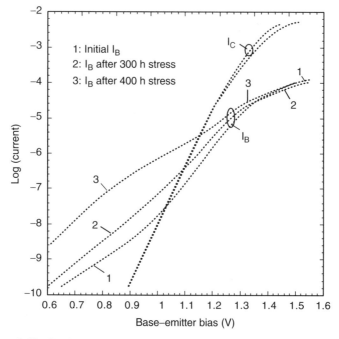

Figure 2.10. Gradual current gain degradation starts from the low bias of V_{BE}.

after stress at $T_j = 185\,^\circ\text{C}$ and $J_c = 5 \times 10^4\,\text{A/cm}^2$ for 0, 300, and 400 hours. An anomalous excess base current appears at the low V_{BE} region initially. As time passes the excess current gradually moves up to the higher bias current regions. This increase in the base current is eventually followed by a drastic reduction in the device current gain. It is noted that the turn-on voltage of the collector current shows no change after the stress, which indicates there is no carbon redistribution from the base to the emitter. Figure 2.11 shows the Gummel plot of a type-2 HBT before and after stress at the same junction temperature and current density. Unlike the type-1 HBT, the current gain of the type-2 HBT degrades incidentally at both low and high V_{BE} regions after the stress. It is interesting to note that the Gummel characteristics of the type-1 HBT after stress resemble those of the type-A HBT with the ledge potential externally biased at $V_L < V_{BE}$, as shown in Figure 2.5(a). On the other hand, the Gummel characteristics of the type-2 HBT after stress resemble those of the type-B HBT with the emitter ledge externally biased again at $V_L < V_{BE}$, as shown in Figure 2.5(b). Based on these resemblances, we believe both types of failure modes appearing in post-stress HBTs are strongly affected by the degree of the emitter ledge depletion. If the ledge is partially depleted like that in the type-1 HBT, the device is relatively short-lived; if the ledge is fully depleted like the type-2 HBT, the device is relatively long-lived [14].

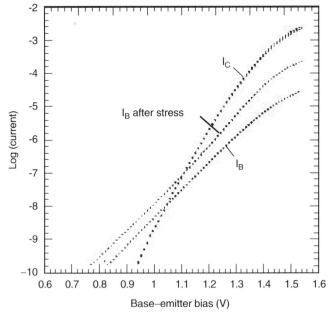

Figure 2.11. Gradual current gain degradation starts from both low and high bias of V_{BE}.

2.3 COLLECTOR DESIGN

The choice of device to be used in microwave power amplifiers is dictated by many requirements, including high microwave gain, high breakdown voltage and current handling capability, low turn-on voltage, good linearity, ruggedness, and low cost of manufacturing. Collector design is a major determinant of all these factors. In this section, design considerations for the collector structure are first reviewed. Subsequently, it is shown that double heterojunction bipolar transistors (DHBTs) on GaAs substrates utilizing GaInP collectors (and their derivatives) offer potentially significant solutions to the problems encountered.

Representative output current versus voltage characteristics for a device used in a power amplifier is shown in Figure 2.12. For a representative power amplifier, the power-added efficiency η depends on the ratio of the maximum voltage across the transistor, V_{max}, to the minimum voltage, V_{min}, through the relation

$$\eta = \gamma \left(1 - \frac{1}{G}\right) \frac{V_{max} - V_{min}}{V_{max} + V_{min}} \tag{2.12}$$

where G is the microwave power gain and γ is a factor of order unity that depends on the operation class of the amplifier (with $\gamma = 0.5$ for Class A, $\gamma = 0.78$ for Class B, etc.). Thus to obtain high efficiency, the V_{max}/V_{min} ratio must be maximized.

V_{max} is dictated either by breakdown voltage or by the power supply voltage that is employed. With a Class AB amplifier (often used in wireless transmitters), the maximum voltage attained is of the order of $V_{min} + \kappa V_{batt}$, where κ has a value of order 1.8 to 2. Under adverse circumstances (associated with abnormal impedance mismatch of the transistor, such as when the antenna is in contact with a large conductor, or even broken off), the output voltage may rise to $V_{min} + 3V_{batt}$. It is

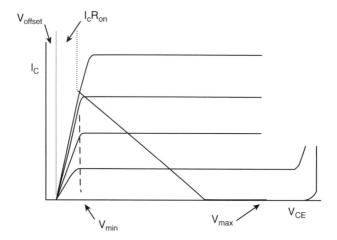

Figure 2.12. Schematic I–V curve of HBT for power amplifier application.

critical that the transistor breakdown voltage is well above this maximum voltage. This requirement translates into a collector–emitter breakdown voltage of order 14 V for operation with a nominal 3.6 V battery. Such a breakdown voltage is difficult to achieve within many high speed device technologies. It represents a significant challenge for SiGe HBT technology and InP high electron mobility transistor (HEMT) technology, for example.

It is important to note that the collector–emitter breakdown of an HBT is a strong function of the impedance value attached to the base terminal. With a short circuit at the base, the breakdown voltage attains the highest value, BV_{CEs}. By contrast, with the base open-circuited, the breakdown voltage has the lower value BV_{CEo}. This value is reduced from BV_{CEs} because the holes generated by avalanching at the base–collector junction flow into the base and constitute an effective input base current, which gets further amplified by the current gain of the transistor.

V_{min} is limited by the collector-up voltage of the transistor in saturation, labeled V_{CEsat} (although to avoid reduction in gain from saturation charge storage, often operation at V_{min} is avoided, and the practical value for V_{min} is somewhat higher than V_{CEsat}). For HBTs, it is conventional to consider two contributions to V_{CEsat}, the "offset voltage" V_{offset} (which applies for $I_C = 0$), and a current-dependent portion, $I_C R_{on}$, as shown in Figure 2.12. The "knee voltage" V_{knee} is the sum of the two for a given transistor. V_{offset} is appreciable in HBTs (since their emitter–base junction structure is significantly different from their collector–base structure), and it can be an important factor in the overall efficiency. For example, $V_{offset} = 0.25$ V for representative AlGaAs/GaAs HBTs used in power amplifiers. It would be desirable to reduce this voltage, particularly if the power supply (battery) voltage is reduced to 3 V or below. The offset voltage is determined, in rough terms, by the difference between the base–emitter voltage needed to produce a given output current I_C, and the base–collector voltage to produce an equal magnitude of current to compensate it (as shown in Fig. 2.13(a)). The currents associated with the two junctions balance under the condition $I_C = 0$, which has associated with it a voltage $V_{CE} = V_{BE} - V_{BC}$. As is evident from Figure 2.13(b), there can be many factors that lead to different relationships between I_C and junction voltage for the base–emitter junction and base–collector junction. For HBTs, a key difference is the different energy band diagrams for the two cases. As discussed below, there is hole current in addition to electron current for the BC junction, but not for the BE junction. The doping ratios and area and perimeter-to-area ratios are different for the BE and BC junction currents. In some cases, $I_B R_B$ drops associated with the base current I_B also alter the characteristics between the "extrinsic" BC junction (under the base contacts) and the "intrinsic" device.

For next-generation power amplifiers, challenges must be met in maximizing efficiency without sacrificing linearity even when the battery supply voltage is decreased, improving thermal management in small die sizes, and increasing ruggedness. The on-voltage must be reduced compared to present devices. GaInP DHBTs promise to provide a solution to a number of these issues and are compatible with present manufacturing practices. In the following, the structure and technology for GaInP DHBTs are first described, followed by a discussion of their

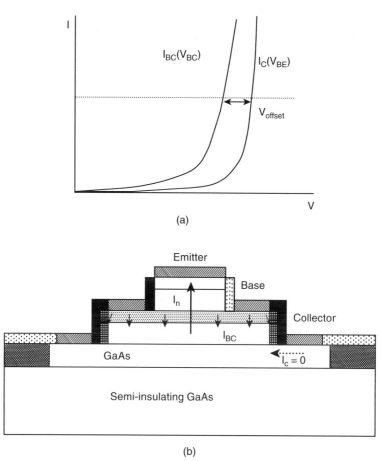

Figure 2.13. (a) Representative collector current versus V_{BE} (at $V_{BC} = 0$) and base–collector diode current I_{BC} versus voltage V_{BC}, showing origin of offset voltage. (b) Current flow components in an HBT under saturation condition ($I_C = 0$).

characteristics in a power amplifier context. New concepts for further improvement of the device characteristics are then presented.

2.3.1 GaInP DHBT Structure and Technology

Figure 2.14 contrasts the band diagrams of GaInP/GaAs/GaAs single heterojunction bipolar transistors (SHBTs) and GaInP/GaAs/GaInP double heterojunction bipolar transistors (DHBTs). GaInP lattice-matched to GaAs has a bandgap in the range 1.85–1.89 eV (with a value dependent on the degree to which the group III elements In and Ga form an ordered sublattice, which in turns depends on epitaxial growth parameters). The bandgap is in a convenient range for blocking minority carrier flow from p-type bases of GaAs (whose bandgap is 1.42 eV). The use of GaInP in the

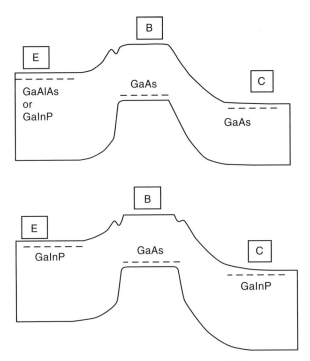

Figure 2.14. Schematic band diagrams of GaInP/GaAs SHBT and GaInP/GaAs/GaInP DHBT. Epitaxial layer structure of DHBT is used in this work.

collector can provide a variety of potential benefits, including lower offset voltage, higher breakdown voltage and optimized trade-off of breakdown voltage with transit time or collector resistance, and reduced saturation charge storage.

GaInP can also be used as a replacement for AlGaAs in the emitter of HBTs fabricated on GaAs substrates. For the emitter material, GaInP has been shown to have advantages due to wider valence band offset, lower deep level densities, and better selective etching than corresponding AlGaAs layers [19–22]. The conduction band offset with respect to GaAs, although believed to be technology dependent, is relatively small $\Delta E_c = 0.18$ eV for representative samples [23]). With carbon doping of p^+ GaAs bases, it is straightforward to achieve p-n junctions that coincide with the position of the heterojunction. The application of GaInP/GaAs to the emitter of HBTs has already led to demonstrations of high reliability [22,24]. Devices fabricated with metallorganic chemical vapor deposition (MOCVD)-grown materials including GaInP emitter and GaAs base (C-doped at 4×10^{19} cm^{-3}) were shown to remain unchanged after 10000 hours operation at 264 °C junction temperature and 2.5×10^4 A/cm^2 [24].

An additional significant advantage of GaInP over AlGaAs as emitter derives from the improved ability to form passivation ledges in an HBT structure. As discussed in the previous section, the emitter-size effect, brought about by recombination at emitter edges, is detrimental for current gain and reliability. Frequently, the problem is addressed with the use of thin layers ("ledges") of emitter material fabricated at the edges of the emitter. These layers are fully depleted and non-conducting and block minority carriers from reaching the surface of the base. GaInP, with larger valence band offset than AlGaAs, permits using thinner emitters that can readily be depleted to form effective ledges.

A variety of experimental results are reported in the discussion below, based on DHBTs fabricated with epitaxial layer structures shown in Figure 2.15. The overall thickness of the collector was 0.4 μm. The conduction band offset was reduced by means of a setback layer of GaAs doped at 3×10^{16} cm^{-3} and thickness 300 Å, followed by a doping pulse 100 Å of GaInP with 5×10^{17} cm^{-3} doping (as discussed below). The collector doping concentration was selected to be relatively high, to increase the current density attainable without significant saturation effects, and to decrease series resistance when the collector is undepleted. Materials were grown by MOCVD. Devices were fabricated into transistors using a self-aligned base contact process. Emitter dimensions were typically 2 μm by 10 μm for each finger. The dc current gain of the transistors ranged between 50 and 100. Key characteristics of importance for power amplifiers are discussed in the following sections.

2.3.2 Breakdown Voltage

The impact ionization coefficients in GaInP have been the subject of several studies [26,27]. In Figure 2.16 the ionization rates are compared with those of GaAs and InP, as a function of electric field. Results indicate that the breakdown electric field E is higher than that of GaAs by a factor of about 1.6 to 1.8 at high ionization rates (which tend to directly apply to the BV$_{CB0}$ value of HBTs, and to p-n junctions). Figure 2.17 shows computed p-n junction breakdown voltages for GaInP one-sided junctions based on the measured ionization coefficients. For comparison are also shown experimental values of breakdown voltage for fabricated p-n junctions, which agree well with predictions. At low electric fields, the ratio of ionization coefficients between materials is even larger, and GaInP provides a correspondingly greater

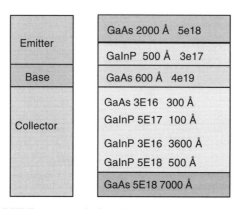

Figure 2.15. DHBT structure designed to reduce the conduction band offset.

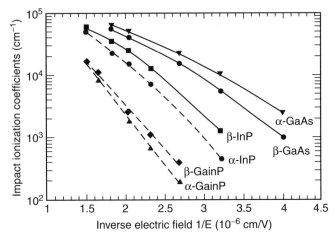

Figure 2.16. Measured impact ionization coefficients for electrons and holes in GaInP as a function of electric field.

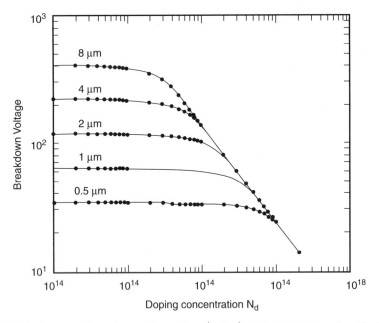

Figure 2.17. Computed breakdown voltages for $p^+n^-n^+$ diodes (or BC junctions) in GaInP. Measured I_C versus V_{CE} for experimental GaInP/GaAs/GaInP DHBT shows low offset voltage.

benefit. The low electric field regime is of central importance for estimation of BV_{CE0}. With an open base, the condition corresponding to breakdown is given by

$$M = 1 + 1/\beta, \tag{2.13}$$

where β is the current gain of the transistor, and M is the multiplication factor for carriers crossing the BC junction, resulting from the impact ionization process. M is thus not much larger than unity at BV_{CE0} (while it is substantially larger than unity at BV_{CB0}). It is typical to represent M as a function of the voltage applied across the junction in the form

$$M = 1/[1 - (V/V_{crit})^n]. \tag{2.14}$$

The rapid falloff of the ionization coefficients with increasing electric field found for GaInP implies that the exponent n relating multiplication factor to V_{crit} is higher than for GaAs (it is of order 8 for GaInP, versus 4–5 for GaAs and 2–3 for Si). Thus BV_{CE0}/BV_{CB0} is higher with GaInP than with GaAs collectors. For comparable doping and thickness, BV_{CE0} in GaInP is expected to be more than two times higher than in GaAs. For the epitaxial structures described above, values of BV_{CE0} were found to be about 15 V, and BV_{CB0} values were about 30 V.

2.3.3 Offset Voltage

The offset voltage of the DHBT is lower than that of the HBT because of the enhanced symmetry between BE and BC junctions. As noted above, the offset voltage can be estimated as the difference between V_{BE} and V_{BC} to produce a given magnitude of current. If both junctions are GaInP/GaAs heterojunctions, their associated I–V characteristics will be closely matched. In particular, the current flow component associated with the flow of holes from the heavily doped base into the collector, which is dominant in the transistor with GaAs collector when the device goes into saturation, is suppressed in the DHBT. There are a few remaining factors that can lead to a slight offset voltage, including the difference in area between BC and BE junctions, and differences in their defect densities (particularly in the area underneath the base contacts). Figure 2.18 illustrates representative I–V curves for a fabricated GaInP DHBT (emitter area 112 μm^2), showing small "offset" voltage V_{CE} of order 0.08 V. This improvement over conventional GaAs collectors (having offset voltage of typically 0.25–0.30 V) leads to improved amplifier efficiency, particularly if the power supply (battery) voltage is low. For 3 V battery voltage, the improvement in efficiency, η, is expected to be about 8%, based on the relationship $\eta \approx (V_{max} - V_{min})/(V_{max} + V_{min})$, given above.

2.3.4 Knee Voltage

In addition to the offset voltage, the knee voltage results from the effective on-resistance of the device in saturation, R_{on}, which in turn depends on a number of

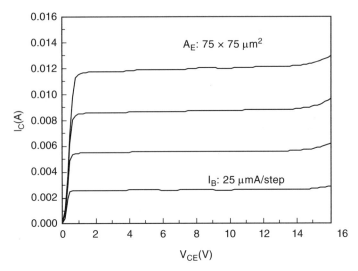

Figure 2.18. Representative *I–V* curves for a fabricated GaInP DHBT showing small "offset" voltage V_{CE}, of order 0.08 V.

factors. These include the emitter resistance, subcollector resistance, collector contact resistance, undepleted collector resistance, and BC barrier resistance (as discussed below). For high voltage or power transistors, the resistive drop across the (undepleted) collector, R_c, is often a major portion of the overall R_{on} since the collector must be designed to be wide (to drop large voltages without exceeding the breakdown field) and lightly doped (so that it can be fully depleted). For a given collector area A_c, the resistance is given by $R_{on} = w/qN_D\mu_n A_c$, where N_D is the collector doping, and μ_n is the electron mobility in this region (assuming there is no significant amount of conductivity modulation in the collector). The high breakdown electric field strength, E_b, of GaInP allows shorter collectors (smaller w) to be used to achieve a given breakdown voltage (BV), and this can lead to lower values of R_{on}. The low field mobility of electrons in GaInP is, however, lower than in GaAs. It is interesting to consider the relative importance of μ and E_b in determining the collector resistance. It has been shown that the maximum attainable value of $(BV)^2/R_c$ tends to be a material constant, given by $\alpha E_b^3/\mu_n$, where α is a constant of order unity dependent on device design, E_b is the breakdown electric field, and E_b^3/μ_e is the Baliga figure of merit [25]. GaInP provides a Baliga figure-of-merit higher than that of GaAs by a factor of approximately 1.6, as shown in Table 2.1. This suggests that for high voltage switching applications, GaInP DHBTs will be of considerable value.

2.3.5 Base–Collector Barrier Effects

The effective value of R_{on} in many GaInP/GaAs DHBTs is significantly increased by BC barrier effects. Such effects are associated with the potential barrier to electron

COLLECTOR DESIGN

TABLE 2.1. Baliga Figure of Merit for HBT Collectors of Different Materials

	Baliga Figure of Merit	$\dfrac{V_b^2}{R_{on}} = \dfrac{\varepsilon \mu E_b^3}{4}$	
Values	Si	GaAs	GaInP
E_b (kV/cm)	350	400	660
μ (cm^2/V·s)	1400	5000	2000
ε	11.8	13	12
FOM	1.58E7	9.26E7	1.54E8

flow between base and collector, arising from the difference in conduction band energy ΔE_c between GaAs and GaInP, as indicated in the band diagram of Figure 2.19. As noted above, ΔE_c is of the order of 0.16 to 0.18 eV. However, its magnitude is dependent on growth technique, on the degree of ordering of the GaInP group III sublattice, and even the growth direction (i.e., whether the structure is produced as a GaInP layer on GaAs, or a GaAs layer of GaInP). It is believed that ΔE_c is significantly influenced by the degree of exchange of P and As during the growth of the interface, which results from the larger binding of As to the growing solid.

Since ΔE_c is large compared with KT/q, if no precautions are taken, the flow of electrons from base to collector will be blocked, resulting in very small output current, small current gain, transconductance, and f_t. In practical devices, however, the barrier height is considerably smaller than ΔE_c. Various strategies have been

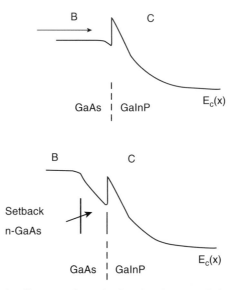

Figure 2.19. Schematic diagram of conduction band near BC junction, indicating the potential barrier formed as a result of ΔE_c (with and without a setback layer of GaAs).

researched extensively for reduction of the barrier effects [30–32], including the following:

1. Use of a "setback layer" of n-type or undoped GaAs at the start of the collector region, as illustrated in Figure 2.19.
2. Use of a doping dipole, generated primarily by an n^+ doping pulse a short distance into the GaInP collector.
3. Grading of the collector composition near the base–collector junction. This is often accomplished with a chirped short-period superlattice, to emulate a graded composition material without the complexity of maintaining lattice-match with quaternary alloys.
4. Use of $Ga_{0.89}Al_{0.11}As$ in the base, such that the conduction band offset energy is eliminated [32]. This strategy has allowed device measurements (such as breakdown voltage and Kirk effect) to be made in a situation where the offset is entirely absent.

The penalty for incomplete reduction of the potential barrier effects is typically a reduction in the maximum J_c value before the onset of current saturation. Figure 2.20 illustrates, for example, measured results of the abrupt falloff of current gain (at about $J_c = 2 \times 10^4$ A/cm^2) and a corresponding fall off of f_t and f_{max} as current is increased, for a device where the BC potential is only incompletely removed by the setback layer. At lower current density, the potential barrier can also cause a significant perturbation to device characteristics. As shown in Figure 2.21, the HBT I_C–V_{CE} curves near the origin have an apparent excess resistance. The excess resistance is in good agreement with a model in which the added voltage required to support a given current is equal to the amount needed to keep the conduction band barrier suppressed.

2.3.6 Maximum Current Density

The maximum collector current density that can be maintained in an HBT is limited by the Kirk effect and related problems. At high current density the space charge associated with the electrons traversing the BC depletion region can become large enough to significantly alter the electric field distribution, leading to a reduction and eventual elimination of the electric field at the edge of the base (as shown in Fig. 2.22). For a model in which the electron velocity v_{sat} is constant across the space-charge layer, and there is uniform collector doping N_d, then the current density J_{Kirk} for the onset of this condition is given by

$$J_{Kirk} = qv_{sat}[2\varepsilon(V + V_{bi})/qw_c^2 + N_d], \qquad (2.15)$$

where w_c is the thickness of the collector and V_{bi} is the built-in potential of the junction. In a standard bipolar transistor with a BC homojunction, base pushout occurs when the current density reaches J_{Kirk}; that is, holes from the base stream into

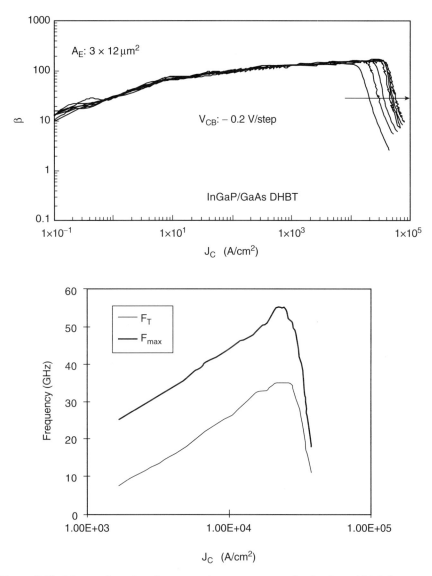

Figure 2.20. Measured results of current gain versus current density J_c, and high frequency response versus J_c, showing rolloff above critical current density.

the metallurgical collector region and produce a quasineutral region that effectively extends the base over a portion of the metallurgical collector. There is an accompanying decrease in f_t (as a result of the increased base transit time for electrons) and often a decrease in current gain (as a result of the increased recombination). Nonetheless, the falloff of gain and f_t is gradual, and current density can reach values significantly higher than J_{Kirk}. For a DHBT, the situation differs at the high current

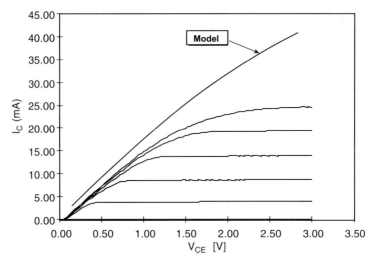

Figure 2.21. HBT I_C–V_{CE} for a DHBT with incompletely eliminated BC potential barrier, showing high effective resistance.

density. The barrier in the valence band prevents holes from flowing into the collector, and although the electric field at the edge of the base decreases to zero, base pushout cannot occur. Instead, the collector current simply shuts off (without extra charge storage). If there is a conduction band barrier (ΔE_c) in addition to ΔE_v, the maximum current of the device will be limited to a smaller value than that determined by J_{Kirk}.

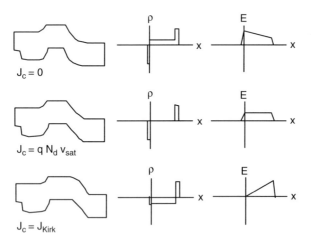

Figure 2.22. Schematic variation of band diagram and electric field for various values of collector current density, showing the falloff of the electric field at the edge of the base. Waveforms in BC diode reverse recovery measurements for SHBTs and DHBTs, showing effects of charge storage.

2.3.7 High Frequency Characteristics and Microwave Gain

GaInP collectors are characterized by lower electron velocity than corresponding GaAs layers. Reduced velocity can lead to increased collector transit time, degrading f_t. It can also lead to reduced current density for the onset of the Kirk effect (which leads to reduced current gain, reduced f_t, and reduced power density). Various estimates have been made of the saturation velocity of electrons in GaInP, based on measurements of f_t and of the Kirk effect onset for GaInP DHBTs [28]. There is uncertainty in such measurements, however, due to the possible influence of electrons trapped in the potential barrier between base and collector (as discussed below). In this work we have found that high f_t and f_{max} in GaInP DHBTs can be achieved, as illustrated in the data of Figure 2.23 (showing f_t as high as 70 GHz, measured at a current density of 5×10^4 A/cm^2, for a device with an emitter area of 40 μm^2). Extracted values of saturation velocity for GaInP are in the range of $(5.5–6) \times 10^6$ cm/s (which is lower than corresponding values of $(8–10) \times 10^6$ cm/s for GaAs).

2.3.8 Saturation Charge Storage

When the base-collector junction is forward biased in transistor saturation, charge is stored in the BC depletion region. In SHBTs, the majority of the stored charge is associated with holes injected into the collector, as a result of the very large ratio of doping density in the two sides of the junction. In a DHBT, the larger bandgap of the collector can suppress the injection of holes when the junction is forward biased, in

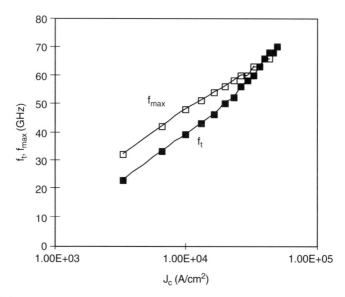

Figure 2.23. Measured f_t versus current density for a GaInP/GaAs/GaInP DHBT that has a BC barrier successfully eliminated.

the same fashion as for the base–emitter junction. Accordingly, dramatic reduction in saturation charge storage is expected. DHBTs are thus expected to enter and go out of the saturation regime without the delays associated with supplying and removing base charge, as is necessary in other bipolar transistors. Experimental results with GaInP DHBTs support this picture [29]. Figure 2.24 shows, for example, data of reverse recovery for base–collector junctions of conventional and double HBTs with sinusoidal excitation. Measurements were based on the experimental setup of Figure 2.25(a), which relies on application of sinusoidal signals (which are experimentally simpler to produce than voltage step functions used traditionally in charge storage evaluation). The presence of large negative transients for the SHBTs and their absence for the DHBTs is evidence of saturation charge storage suppression. A simple theory for the measured waveforms has been derived, in which the duration, T_s, of the negative transient is entirely attributed to minority carrier storage

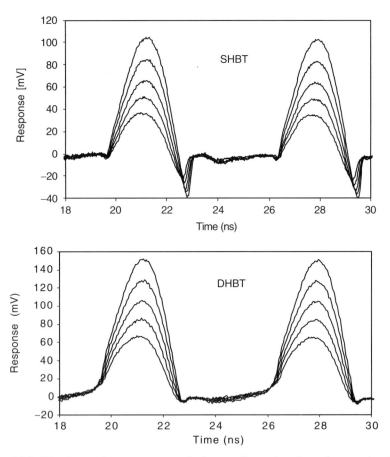

Figure 2.24. Waveforms of reverse recovery for base–collector junctions of conventional and DHBTs with sinusoidal excitation.

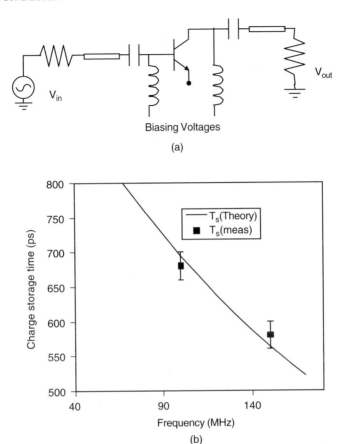

Figure 2.25. (a) Experimental system for the measurement of saturation charge storage in HBTs. (b) Measured saturation charge storage time using various measurement frequencies. The data may be fit with a recombination lifetime value of 1.9 ns.

(ignoring depletion capacitance effects). Theory shows that $T_s = \tan^{-1}(\omega\tau/2\omega\tau)$, where τ is the recombination lifetime in the collector, and ω is the angular frequency of the sinusoidal source. The data for HBTs with GaAs collectors (SHBTs) for varying values of ω are shown in Figure 2.25(b). The data can be fit with a recombination lifetime of 1.9 ns. Corresponding data for GaInP show charge storage below the level of detectability of the technique.

It should be noted that with SHBTs as well as Si homojunction bipolar transistors, a similar reduction of the saturation charge storage can be produced by Schottky clamping of the collector. However, this comes at a significant cost in terms of increased V_{CEsat}, since the turn-on voltage of the effective BC junction decreases dramatically. For example, V_{CEsat} increases to values of the order of 0.7 V when a GaAs Schottky diode is used together with a GaAs SHBT.

2.3.9 Thermal Resistance

A disadvantage associated with the use of GaInP in the collector region of the device is an increase in thermal resistance between the transistor and the heat sink (generally located on the backside of the wafer). The thermal conductivity of GaInP is expected to be at least five times lower than the corresponding thermal conductivity of GaAs (<0.1 W/(cm·K) at 300 K, compared with 0.45 W/(cm·K) for GaAs), as a result of alloy scattering of phonons. Ordering of the Ga and In on the group III sublattice may mitigate the increase in thermal resistance, however. The increased thermal resistance occurs in a relatively thin layer, a small fraction of the overall thermal path; nonetheless, it is a critical part of the thermal path, before the thermal current spreads over a significant distance. It is estimated that the thermal resistance of a typical 2×20 μm² emitter finger may increase by more than 500°C/W from the use of a GaInP collector (while a representative total thermal resistance value for an HBT with GaAs collector is near 1600°C/W—although the value can vary depending on thickness of the substrate and metallizations).

2.3.10 Novel HBT Designs with Tunneling B/C Structure

It is desirable to implement transistors that have the advantages of reduced V_{CEsat} and reduced saturation charge storage associated with DHBTs, but do not suffer from reduced on-resistance, lower f_t, or increased thermal resistance, which are found in traditional devices with GaInP collectors. In recent work, novel structures have been explored in which a barrier is inserted at the BC junction to block holes, but in which the majority of the collector remains GaAs. A structure of this type is shown in Figure 2.26. Here the traditional HBT band structure is modified by placement of a 50 Å layer of GaInP at the base–collector junction. The layer is sufficiently thin that electrons can tunnel through (and thus in principle the collector current should be the same as for a conventional HBT). The layer is sufficiently thick, however, that holes cannot tunnel appreciably—and thus the hole current leading to the low V_{BC} turn-on voltage for the BC junction, and the hole storage component in transistor saturation, are eliminated. Since beyond the thin barrier the

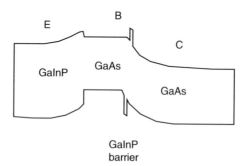

Figure 2.26. Band diagram of novel structure employing a tunnel barrier at the BC junction to suppress hole injection effects.

collector consists of GaAs, the key properties of saturation velocity, electron mobility, and thermal resistance are essentially unchanged from the case of the SHBT.

For the structure of Figure 2.26 to work effectively, it is necessary that the probability for holes to tunnel through the BC barrier be low, while the electron tunneling probability should be relatively high. With a barrier of GaInP, the majority of the bandgap offset lies in the valence band rather than the conduction band, which is favorable to having a large ratio of electron to hole tunneling probability. The low effective mass of electrons also assists in their tunneling (although light holes also benefit from low mass). Finally, in most conditions of operation of the HBTs, there is a potential drop between base and collector, which assists collection of electrons and impedes collection of holes. Figure 2.27 shows calculations of band diagrams and associated tunneling probability for ideal GaInP barriers in GaAs structures, which suggest that conditions desirable for HBT operation can be achieved.

It is known that real heterojunctions between GaInP and GaAs cannot be characterized by simple considerations of electron affinity and ideal abrupt interfaces. It is likely that exchange of P and As at the growing interface affects the junction characteristics significantly. For MOCVD growth, it is characteristic that junctions employing GaAs grown on top of GaInP have a greater tendency to block electron flow than corresponding junctions in which GaInP is grown on top of GaAs. Thus for the simplest implementation of the transistors described, a growth of the GaInP B/C barrier on top of the GaAs base would be preferable. This leads to the design of collector-up devices, as pictured in Figure 2.28. Such collector-up devices are advantageous from several standpoints. For example, the base–collector capacitance from the extrinsic device regions (where the base contacts are deposited on the base in a conventional transistor) is eliminated. This tends to dramatically increase f_{max}, increase output resistance, and reduce instability problems, since C_{BC} is a key parasitic that provides unintentional feedback from output to input. On the other hand, for collector-up HBTs to have adequate current gain, it is essential to block current flow from the emitter to the extrinsic regions of the base (under the base contacts), since these injected minority carriers are lost to recombination.

Mochizuki et al. [33] have successfully implemented a collector-up GaInP/GaAs transistor with interposed GaInP blocking layers, using implantation of boron to suppress injection in the extrinsic regions of the device. I–V characteristics are shown schematically in Figure 2.29. The device succeeds in reducing the V_{CEsat} value to near zero, and the on-resistance is kept low. The current gain in the structure is adequately high, around 40. This structure is very promising for power amplifier applications.

2.3.11 New Materials for Improved DHBT Collectors

The use of thin layers at the BC junctions to mitigate the problems of standard DHBTs can benefit significantly from new semiconductor materials presently under development. An attractive candidate is GaInPN, which differs from the conventional GaInP by the addition of small amounts of nitrogen. It has been shown that in

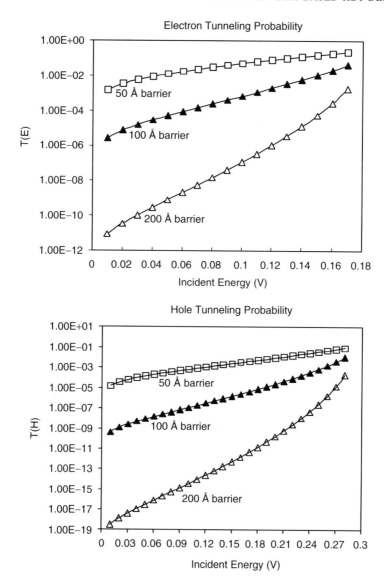

Figure 2.27. Calculated tunneling probability for electrons and holes at GaAs junctions with interposed GaInP layers.

most III–V compounds, the addition of nitrogen leads to a drop in the bandgap energy (although if one were to linearly interpolate between the bandgap of GaAs or GaInP and that of GaN—which has a very high bandgap energy of 3.4 eV—the result expected would be an increase in bandgap energy). The decrease in bandgap is attributed to the large electronegativity of nitrogen. It has been found to correspond

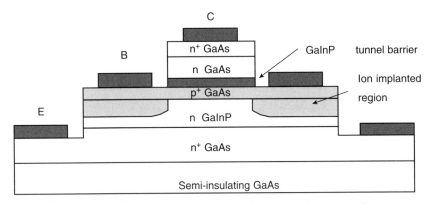

Figure 2.28. Structure of collector-up tunneling-collector HBT designed for low turn-on voltage.

almost entirely to a decrease in conduction band energy (rather than a rise of valence band energy). This result leads to the possibility of developing "tailor-made" materials for the HBT structures. In principle, it is possible to add N to GaInP thin barriers interposed between base and collector, in such a fashion that the conduction band energy is matched to that of GaAs, as shown in Figure 2.30. At the same time, a large valence band barrier could be sustained. In this scenario, electron flow will be unimpeded, and hole flow will be blocked. The resultant devices should show very low offset voltage, low on-resistance, and negligible saturation charge storage. At the same time, the body of the collector will consist of GaAs. No changes would be expected in f_t, series resistance, or thermal resistance of the collector from those of a SHBT.

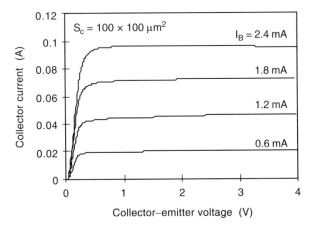

Figure 2.29. Experimental *I–V* characteristics obtained with collector-up tunneling-collector HBT.

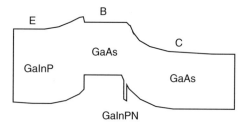

Figure 2.30. Band diagram of HBTs fabricated with thin layers of GaInPN interposed between base and collector.

2.3.12 DHBT Application to Amplifiers

The low values of V_{CEsat} and the absence of significant charge storage allows high frequency amplifiers to utilize load lines that extend to the lowest possible voltages, without severe phase nonlinearity. At the same time breakdown voltage can be increased (which is of particular concern when the power amplifier load is unintentionally mismatched). Thus amplifiers currently fabricated with SHBTs may benefit from the use of DHBTs. Additionally, to exploit the characteristics of DHBTs to maximum advantage, amplifiers with unconventional architecture may be utilized. Switching-mode circuits, in which transistors are driven between cutoff and saturation, are possible with DHBTs at significantly higher speeds than with SHBTs. Efficiency of switching-mode amplifiers is frequently above 80%. However, their output is typically too nonlinear for use with many modulation formats. Various possible architectures, including LINC amplifiers, envelope elimination and restoration amplifiers, and bandpass delta-sigma Class S amplifiers [34] offer opportunities to improve the linearity while preserving the high efficiency of switching-mode operation. A related architecture is based on the implementation of a dc–dc converter within the power amplifier itself, in which the dc–dc converter uses a switching-mode power transistor to optimize the amplifier efficiency [35]. GaInP DHBTs are very promising for the implementation of these circuits.

2.4 CONCLUSION AND OUTLOOK

In this chapter, we have addressed the most critical issues relevant to the HBT's emitter and collector structure designs and the overall performance optimization. With a novel on-ledge Schottky potentiometer, we are able to quantify the degree of emitter ledge depletion. With this in-place apparatus, we can easily monitor the effectiveness of the emitter ledge passivation and create emitter structures that will provide HBTs with high current gain, low $1/f$ noise, and excellent reliability. By biasing this potentiometer externally, we can even change the current gain and minimize the $1/f$ noise level and its dependence on input base currents. This opens new possibilities to produce variable gain amplifiers and analog signal mixers by using single HBT with an additional ledge electrode.

On the other hand, collector design is a major determinant of the HBT's speed and breakdown and offset voltages, which directly affects the power amplifier's power-added efficiency and linearity. For an optimized collector design, we have adopted new material systems and created new collector layer designs based on bandgap engineering. We found that GaInP DHBTs provide lower offset voltage, higher breakdown voltage, and potentially lower on-resistance than corresponding devices with GaAs collectors. They exhibit dramatically lowered saturation charge storage time. The devices are well-suited to the requirements for next-generation power amplifiers for wireless communications. Novel amplifier architectures employing transistors in switching mode are also possible to exploit the characteristics of these devices.

Over the past decade, significant advances have been made in GaAs-based HBT technology. Many of them are associated with the theme of making HBT integrated circuits a realistic possibility in RF/wireless system applications. Over the next decade, it is likely that significant changes in the HBT's capability will be further expanded and demonstrated. It is probable that the HBT evolution will continuously respond to the system drive to higher frequencies and to lower power dissipation, better linearity, lower noise, and higher efficiency in communications to satisfy the needs for battery-operated systems. HBT technologies will also advance in areas of reduced device dimensions and extended ability to tailor material properties, particularly bandgap or bandgap alignment, in all three dimensions to optimize the HBT's performance. As a result, HBT technology will remain a performance leader for advanced wireless communications systems.

REFERENCES

1. O. Nakajima, K. Nagata, H. Ito, X. Tshibashi, and T. Sugeta, "Emitter–base junction size effect on current gain H_{fe} of AlGaAs/GaAs heterojunction bipolar transistors," *Japanese Journal of Applied Physics*, Vol. 24, No. 8, L596–L598 (1985).
2. T. S. Henderson and P. K. Ikalainen, "Reliability of self-aligned, ledge passivated 7.5 GHz GaAs/AlGaAs HBT power amplifiers under RF bias stress at elevated temperatures," in *GaAs IC Symposium*, 1995, pp. 151–154.
3. M. Wetzel, M. C. Ho, P. Asbeck, P. Zampardi, C. Chang, C. Farley, and M. F. Chang, "Modeling emitter ledge behavior in AlGaAs/GaAs HBTs," *1997 MANTECH Digest*.
4. W. Liu, "Microwave and D. C. studies of npn and pnp AlGaAs/GaAs heterojunction bipolar transistors," Ph.D. dissertation, Stanford University, 1991.
5. H. H. Lin and S. C. Lee, "Super-gain AlGaAs/GaAs heterojunction bipolar transistors using an emitter edge-thinning design," *Applied Physics Letters*, Vol. 47, No. 8, 839–841 (1985).
6. L. H. Camnitz, "Improved current gain stability in C-doped HBTs with an AlGaAs surface passivation layers," in *Technical Digest, GaAs Reliability Workshop*, Monterey, 1991.

7. J. R. Waldrop and M. F. Chang, "Negative differential resistance of AlGaAs/GaAs heterojunction bipolar transistors: influence of emitter edge current," *IEEE Electron Device Letters*, Vol. 16, No. 1 (Jan. 1995).
8. W. Liu and J. Harris, "Parasitic conduction current in the passivation ledge of AlGaAs/GaAs heterojunction bipolar transistors," *Solid State Electronics*, Vol. 35, 891–895 (1992).
9. S. M. Sze, *Physics of Semiconductor Devices*, Wiley, New York, 1981, pp. 251–258.
10. S. M. Sze and W. Shockley, "Unit-cube expression for space-charge resistance," *Bell System Technical Journal*, Vol. 46, 837 (1967).
11. P. Ma, P. Zampardi, L. Zhang, and M. F. Chang, "Determining the effectiveness of HBT emitter ledge passivation by using an on-ledge Schottky diode potentiometer," *IEEE Electron Device Letters*, Vol. 20, 460–462 (Sept. 1999).
12. P. Ma, Y. Yang, P. Zampardi, R. T. Huang, and M. F. Chang, "Modulating HBT's current gain by using externally biased on-ledge Schottky diode," *IEEE Electron Device Letters* (Submitted).
13. W. Liu and J. S. Harris, "Diode ideality factor for surface recombination current in AlGaAs/GaAs heterojunction bipolar transistors," *IEEE Transactions on Electron Devices*, Vol. 39, No. 12, 2726–2732 (1992).
14. H. Sugahara, "GaAs HBT reliability," in *Current Trends in Heterojunction Bipolar Transistors, Selected Topics in Electronics and Systems*, Vol. 2, M. F. Chang (Ed.), World Scientific, Singapore, 1996, pp. 227–240.
15. A. Abidi, "Low power radio frequency ICs for portable communications," in *RF/Microwave Circuit Design for Wireless Communications*, L. Larson (Ed.), Artech House, Boston, 1996, pp. 43–98.
16. P. Ma and M. F. Chang, "Dependence of HBT's $1/f$ noise on the emitter ledge potential," *IEEE Electron Device Letters* (Submitted).
17. H. Sugahara et al., "Improved reliability of AlGaAs/GaAs HBTs with a strain-relaxed base," *IEEE GaAs IC Symposium Technical Digest*, San Jose, CA, 1993, pp. 115–118.
18. T. Henderson et al., "Characterization of bias-stressed carbon doped AlGaAs/GaAs power HBTs," *IEDM 1994 Technical Digest*, San Francisco, 1994, pp. 187–190,
19. M. J. Mondry and H. Kroemer, *IEEE Electron Device Letters*, Vol. 6, 175 (1985).
20. W. Liu, E. Beam III, T. Kim, and A. Khatibzadeh, in *Current Trends in Heterojunction Bipolar Transistors*, M. F. Chang (Ed.), World Scientific, Singapore, 1996, pp. 241–301 (and numerous references therein).
21. F. Ren, C. R. Abernathy, S. J. Pearton, and P. W. Wisk, *Electronics Letters*, Vol. 28, 1150 (1992).
22. T. Takahashi, S. Sasa, A. Kawano, T. Iwai, and T. Fujii, *Proceedings 1994 IEDM*, p. 191.
23. H. C. Kuo, J. M. Kuo, Y. C. Wang, C. H. Lin, H. Chen, and G. E. Stillman, *Journal of Electronic Materials*, Vol. 26, 944 (1997).
24. N. Pan, J. Elliott, M. Knowles, D. P. Vu, K. Kishimoto, J. K. Twynam, H. Sato, M. T. Fresina, and G. E. Stillman, *IEEE Electron Device Letters*, Vol. 19, 115 (1998).
25. B. J. Baliga, *Journal of Applied Physics*, Vol. 53, 1759 (1982).
26. S. L. Fu, T. P. Chin, M. C. Ho, C. W. Tu, and P. M. Asbeck, *Applied Physics Letters*, Vol. 66, 3507 (1995).

REFERENCES

27. J. P. R. David, D. R. Ghin, M. Hopkinson, M. A. Pate and P. N. Robson, *Proceedings 1995 International Symposium on Compound Semiconductors*, p. 569.
28. W. Liu, T. Henderson, E. Beam III, and S. K. Fan, *Electron Letters*, Vol. 29, 1885 (1993).
29. P. F. Chen, Y. M. Hsin, and P. M. Asbeck, *Proceedings 1997 IEEE International Symposium on Compound Semiconductors*, p. 443.
30. H. Leier, A. Marten, K. H. Bachem, W. Pletschen, and P. Tasker, *Electronics Letters*, Vol. 29, 868 (1993).
31. J. I. Song, C. Caneau, W. P.Hong, and K. B. Chough, *Electronics Letters*, Vol. 29, 1881 (1993).
32. R. M. Flitcroft, B. CLye, H. K. Yow, P. A. Houston, C. C. Button, and J. P. R. David, *Proceedings 1997 IEEE International Symposium on Compound Semiconductors*, p. 435.
33. K. Mochizuki, R. J. Welty, and P. M. Asbeck, "GaInP/GaAs collector-up tunneling collector heterojunction bipolar transistors with zero-offset and low knee-voltage characteristics," submitted for publication.
34. A. Jayaraman, P. F. Chen, G. Hanington, L. Larson, and P. Asbeck, *IEEE Microwave and Guided Wave Letters*, 8 (1998).
35. G. Hanington, P. F. Chen, and P. M. Asbeck, *Technical Digest 1998 IEEE MTT-S*.

3

InP-BASED DEVICES AND CIRCUITS

DIMITRIS PAVLIDIS, DONALD SAWDAI,* AND GEORGE I. HADDAD
Department of Electrical Engineering and Computer Science
The University of Michigan, Ann Arbor

3.1 INTRODUCTION

InP-based devices and circuits can be used in a variety of applications extending from the low-frequency spectrum of several hundred megahertz to the millimeter-wave range. The choice of device type is influenced not only by its maturity in terms of manufacturing but also by fundamental mechanisms that determine device performance. High electron mobility transistors (HEMTs) are best suited, for example, for low-noise applications up to very high frequencies, while heterojunction bipolar transistors (HBTs) are considered best for power amplifiers and low phase noise oscillators. This chapter presents basic considerations regarding the properties of HBT and HEMT devices, as well some of their circuit and system applications, and addresses the relative merits of each technology.

3.2 HBT FUNDAMENTALS

3.2.1 General NPN Heterojunction Bipolar Transistor Considerations

The HBT is based on the same principles of operation as the homojunction bipolar junction transistor (BJT). In the typical forward-active mode of operation, the emitter–base junction is forward biased, and the base–collector junction is reverse

---

*Currently with TRW, One Space Park, Mailstop D1/1302, Redondo Beach, CA 90278.

RF Technologies for Low Power Wireless Communications
Edited by Tatsuo Itoh, George Haddad, and James Harvey
ISBN 0-471-38267-1   Copyright © 2001 by John Wiley & Sons, Inc.

biased. Electrons are therefore injected from the emitter into the base region, where they diffuse across the base to the base–collector junction. The high electric field in the base–collector junction sweeps these electrons into the collector. While BJTs are made out of a single material, the HBT employs a semiconductor in the emitter that has a wider bandgap than the base semiconductor. The wide-gap emitter, the use of a thin, highly doped base that presents short base transit time and small base resistance, as well as the lightly doped emitter for minimum base–emitter capacitance allow HBTs to operate at much higher frequencies than BJTs.

A basic theoretical drift-diffusion analysis of the NPN homojunction transistor structure can be found in many semiconductor texts [1], and basic properties of HBTs have also been published [2]. The basic theoretical results for the dc current gain and high-frequency response are summarized in the following two sections, including some modifications to include effects of the heterojunction. The two subsequent sections describe the maximum voltage and current sustainable by an HBT design, which determine the maximum output power that it can generate. Note that while all notation is for NPN HBTs, the results hold to the same degree of accuracy for PNP HBTs.

3.2.1.1 dc Current Gain of HBTs Analytical and simulation-based approaches are available for evaluating the HBT operation. The theory for gain is less developed for HBTs with abrupt emitter–base (EB) junctions. The most significant effects are due to the spike in the conduction band resulting from the conduction band discontinuity ΔE_C. First, the spike adds a barrier to electron injection into the base, which increases the turn-on voltage V_{BE}. For example, measured values of $\Delta E_C/\Delta E_V$ for InP/InGaAs and InAlAs/InGaAs abrupt heterojunctions are 0.25/0.34 eV and 0.48/0.24 eV, respectively [3]. Measurements of InAlAs/InGaAs NPN HBTs demonstrate approximately a 0.4 V increase in V_{BE} when moving from a graded EB junction to an abrupt EB junction [4], which agrees with $\Delta E_C = 0.48$ V. Second, the spike allows electrons to cross the junction by both thermionic emission and tunneling, which increases the ideality factor n_c. Both of these effects decrease the injection efficiency. However, the spike also causes hot carriers to be injected into the base, which decreases the neutral base recombination.

3.2.1.2 Frequency Response of HBTs The two most common figures of merit for the high-frequency performance of HBTs are the cutoff frequency, f_T, and the maximum frequency of oscillation, f_{max}. At f_T, the small-signal current gain $|h_{21}| \equiv h_{fe} \equiv \partial I_C/\partial I_B$ drops to unity. At f_{max}, the small-signal power gain, as measured either by the maximum-stable-gain/maximum-available-gain G_{max} or by Mason's unilateral gain U, drops to unity. The total emitter-to-collector delay τ_{ec} is related to f_T as [5]

$$\tau_{ec} \equiv \frac{1}{2\pi f_T} = \tau_e + \tau_b + \tau_{pcd} + \tau_c$$

$$= \frac{kT}{qI_E}C_{BE} + \left(\frac{W_B^2}{2D_B} + \frac{W_B}{v_{bc}}\right) + \frac{W_C}{2v_c} + \left(R_C + R_E + \frac{kT}{qI_E}\right)C_{BC}, \quad (3.1)$$

where τ_e, τ_b, τ_{pcd}, and τ_c are the emitter charging, base transit, precollector delay, and collector charging times; D_B is the minority-carrier diffusion coefficient in the base; W_B is the neutral base width; W_C is the collector depletion width; v_{bc} is the velocity with which electrons are swept into the base–collector depletion region (which is close to but slightly smaller than the electron saturation velocity v_{sat}); v_c is a weighted average electron velocity in the collector; and the R and C values are parasitic resistances and junction capacitances. Usually the τ_{pcd} delay is on the order of one-half of the time it takes for the carriers to actually traverse the depletion region.

The conventional expression employed for f_{max} is

$$f_{max} \approx \sqrt{\frac{f_T}{8\pi R_B C_{BC}}}, \qquad (3.2)$$

where R_B and C_{BC} are the total base resistance and the base–collector capacitance. While a useful tool for analyzing the performance of HBTs, this expression only holds approximately due to distributed effects in the base–collector junction and to transit-time effects. As shown by Eq. (3.2), R_B and C_{BC} must be minimized in order to allow a high f_{max}. A standard rule-of-thumb for microwave amplifiers is to design HBTs such that $f_{max} \approx 2f_T$. According to Eq. (3.2), this implies that $f_T \approx [32\pi R_B C_{BC}]^{-1}$. On the other hand, $f_T > f_{max}$ is more appropriate for most digital switching applications, which implies that $f_T > [8\pi R_B C_{BC}]^{-1}$.

3.2.1.3 Breakdown Voltage of HBTs Since the base–collector junction is reverse biased, sufficiently large V_{CB} or V_{CE} will cause avalanche breakdown in the junction. If the HBT is operated in the common-base mode, then the breakdown BV_{CB0} occurs approximately when the peak electric field in the collector E_{max} reaches E_B, the maximum electric field that can be supported in the collector material. A thick collector with low doping is required for a high breakdown voltage. The breakdown voltage BV_{CE0} in the common-emitter configuration when biased via I_B is necessarily lower than BV_{CB0}. The Johnson figure of merit (FOM) for breakdown is given by

$$\text{FOM} = (\text{BV}) \times (f_T) \leq E_B v_{sat}/\pi \qquad (3.3)$$

and is constant for a given collector material. It shows that, for a given collector material, changing the collector thickness effects a direct trade-off between BV and f_T. Monte Carlo simulations of InP-based HBTs with 3000 Å collector region and a doping of 5×10^{16} cm^{-3} InGaAs (InP) collectors gave values of a similar figure-of-merit BV/($\tau_b + \tau_{pcd}$) of 0.7 (1.5). The nearly double figure of merit of the InP collector over the InGaAs collector indicates the higher breakdown and speed potential of InP-based double HBTs employing an InP collector.

3.2.1.4 Maximum Current Density in HBTs The characteristics of HBTs at high current densities are dependent on a number of effects, such as the Kirk effect, high-level injection, current crowding, self-heating, and voltage drops in parasitic resistances. While high-level injection decreases the current gain [6], it also decreases τ_b such that high-level injection does not cause a large degradation at microwave frequencies. Self-heating is the rise of the operating temperature of the HBT due to the power dissipated P_{diss}. Various heat-sink structures have been employed to decrease the thermal resistance between the HBT and the ambient environment. For example, both a thermal shunt and a thermal lens were used to reduce the thermal resistance of multifinger AlGaAs/GaAs HBTs by 2.5–3 times in order to allow operation at power densities of 10 mW/µm² [7]. Without such heat sinks, both the maximum I_C and V_{CE} may be limited by temperature rise. The maximum power density for HBTs on GaAs substrates has been estimated at 5–10 mW/µm² without thermal shunts and greater than 10 mW/µm² with thermal shunts [8]. Due to the higher thermal conductivity of InP substrates, InP-based HBTs will have slightly higher power densities than GaAs-based HBTs without the need for thermal shunts. Perhaps the most significant impact on HBT microwave performance at high current densities is the Kirk effect, also called base push-out. HBTs rarely show high current gain or high-frequency operation when I_C is greater than the Kirk effect threshhold I_K. In order to increase I_K, the collector doping must increase and the collector thickness must decrease; however, both of these effects also reduce the breakdown voltage.

3.2.1.5 Implications for HBT Optimization The optimization of the HBT layer structure and physical layout involves a number of trade-offs between dc current gain, f_T, f_{max}, BV, and maximum current density. This section introduces a typical mesa-isolated HBT design and some of the parameters that are often varied to optimize HBT performance.

An HBT cross section is shown in Figure 3.1. By self-aligning the contacts, the parasitic resistances R_B and R_C are minimized. As indicated earlier, f_T is dominated

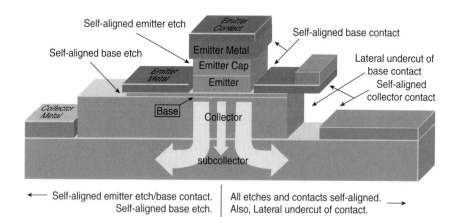

Figure 3.1. Two typical HBT cross sections (the asymmetry is for illustrative purposes only). The left side demonstrates a self-aligned emitter etch, self-aligned base contact, and a self-aligned base etch. The right side adds lateral undercut of the base contact and a self-aligned collector contact. The large arrows designate typical electron flow in the collector.

by τ_b and τ_{pcd}, and f_{max} is limited by both f_T and $R_B C_{BC}$. The primary trade-offs between f_T and f_{max} involve the base and collector thicknesses. Similarly, the collector thickness W_C and doping N_C also trade off the breakdown voltage and current handling capability. Increasing the base doping N_B will decrease R_B and hence increase f_{max}; however, any resulting decrease in the base mobility or lifetime will degrade the gain and f_T. For a well-designed HBT, the emitter thickness W_E and doping N_E have a minor impact on the charging times in f_T and on the injection characteristics that limit the current gain.

To a first order, performance is not affected by the emitter length except for scaling of current. However, decreasing the emitter width decreases the intrinsic R_B, C_{BE}, and C_{BC}, which increases high-frequency performance unless the extrinsic components of R_B and C_{BC} dominate. For very small emitter widths, the dc gain can drop due to excessive surface recombination. Finally, decreasing the extrinsic base width decreases the extrinsic C_{BC}, which increases the frequency performance.

3.2.2 InP-Based PNP HBTs

Research on InP-based HBTs has almost exclusively focused on NPN HBTs, since the electron mobility is several orders of magnitude higher than the hole mobility in InGaAs. PNP HBTs are also of interest, however, primarily for integration in circuits with NPN HBTs. Together, NPN and PNP HBTs can form a simple, efficient, and linear Class B power amplifier or output buffer. Such an AlGaAs/GaAs integrated NPN/PNP push–pull amplifier has demonstrated good power performance at 2.5 GHz [9]. PNP HBTs could also be used as active loads and current sources for NPN amplifier stages, which would provide higher gain per stage, reduced power consumption in the load, and reduced wafer area consumed by passive resistors. In addition, PNP and NPN HBTs could be used together for compact, high-speed complementary digital circuits, such as I^2L logic [10]. Overall, the integration of PNP with NPN HBTs offers simpler circuits with reduced component count and reduced power consumption.

Since both NPN and PNP HBTs are almost identical in theory, the design of PNP HBTs should be fairly similar to those of NPN HBTs. However, practical differences arise because the holes in semiconductors have greater effective mass and lower mobility than the electrons. For example, in InGaAs (the base and collector layers for most InP-based HBTs), the *average* reported electron mobility at room temperature varies from 11,600 cm^2/(V · s) for undoped layers to 4000 cm^2/(V · s) at $N_D = 10^{18}$ cm^{-3}. It is, however, much lower for holes: near 400 cm^2/(V · s) in undoped InGaAs.

The disparity between electrons and holes has several unfavorable implications for the performance of PNP HBTs as compared to the more popular NPN HBTs. For example, since the diffusion rate is proportional to the mobility, minority-carrier holes cross the PNP base more slowly, which increases the base transit time (τ_b) and reduces the gain by allowing more recombination in the neutral base. Moreover, the holes have slightly lower saturation velocity and do not experience velocity overshoot in the collector, both of which increase the collector transit time (τ_{pcd}).

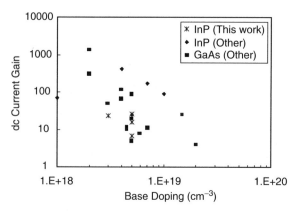

Figure 3.2. Published values for dc current gain versus base doping for PNP HBTs.

Overall, these effects result in reduced gain and slower performance for PNP HBTs when compared to NPN HBTs.

Conversely, PNP HBTs present one advantage over their NPN counterparts: the high-mobility majority-carrier electrons in the base greatly reduce the base access resistance and the base contact resistance. The lower base resistance primarily increases f_{max}. It also reduces the base spreading resistance and hence current crowding toward the edges of the emitter. For microwave PNPs, some of this low base resistance can be traded off to increase the HBT speed by making the base layer very thin (typically 300–500 Å).

The dc current gain from various published PNP HBTs is summarized in Figure 3.2. As can be seen, the dc current gain shows a strong decrease as the base doping increases. This general conclusion agrees with a specific study on GaAs-based PNPs, which demonstrated an increase in gain from 4 to 110 by decreasing the base doping from 2×10^{19} to 5×10^{18} cm^{-3}, even though the base thickness was increased from 300 to 500 Å [11]. Similarly, a study on InP-based PNPs demonstrated that gain increased from 90 to 420 by decreasing the base doping from 1×10^{19} to 4×10^{18} cm^{-3} [12].

Similarly, the microwave performance from various published PNP HBTs is summarized in Figure 3.3. Figure 3.3 demonstrates a decrease in f_T as the base thickness or the collector thickness increases, which indicates that the base and collector transit times are significant components of the total transit time τ_{ec} for all PNP HBTs. All figures demonstrate the range of results that have been obtained for PNP HBTs.

3.3 GENERAL SMALL-SIGNAL CONSIDERATIONS

3.3.1 InP-Based NPN HBTs

InP-based NPN HBTs have shown exceptional high-frequency performance. InP or InAlAs are the most commonly employed emitter materials, and the base is usually

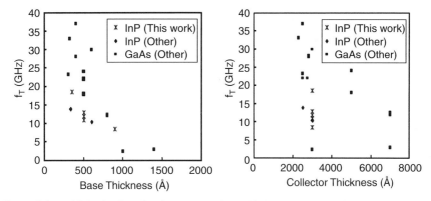

Figure 3.3. Published values for f_T versus (a) base thickness and (b) collector thickness for PNP HBTs.

InGaAs. Impressive microwave results have been published for InP-based NPN single HBTs with InGaAs collectors. A gas-source molecular beam epitaxy (MBE)-grown InP/InGaAs HBT with a 300 Å base carbon-doped at 3×10^{19} cm^{-3} and with a 2000 Å collector demonstrated a record f_T of 235 GHz [13]. The 1.6×10 µm^2 HBT employed a 0.5 µm base contact to reduce the extrinsic C_{BC}. In addition, the structure employed a 3×10^{19} cm^{-3} InP subcollector to reduce the thermal resistance by 50% compared to an InGaAs subcollector, which increased the breakdown voltage at high current levels. Another report on an InAlAs/InGaAs HBT grown by MBE demonstrated f_T and f_{max} of 78 GHz and 236 GHz, respectively [14]. The structure had a 700 Å, 5×10^{19} cm^{-3} Be:InGaAs base and a 6000 Å collector, which resulted in BV$_{CE0}$ = 7.6 V.

Impressive results have also been achieved with double HBTs employing an InP collector. One author reported $f_T = 144$ GHz and $f_{max} = 267$ GHz for a low-pressure MOVPE-grown InP/InGaAs/InP double heterojunction bipolar transistor (DHBT) with a 0.8×5 µm^2 emitter [15]. The 650 Å base was zinc-doped to 4×10^{19} cm^{-3}, and the 2700 Å collector enabled BV$_{CE0} \approx 8$ V. The structure employed a step-graded InGaAsP collector to prevent current blocking at the base–collector heterojunction. Finally, a high $f_T = 160$ GHz and a high $f_{max} = 162$ GHz were obtained simultaneously using a novel metallorganic chemical vapor deposition (MOCVD)-grown InP/InGaAs/InP structure [16]. The collector consisted of a 2000 Å undoped InGaAs region followed by a 1700 Å, 1×10^{17} cm^{-3} InP region, with a 300 Å p^+–i–n^+ sandwich between the layers to lower the heterojunction spike. This composite collector keeps electrons in the Γ valley in the InGaAs region to decrease τ_{pcd}, and the resulting BV$_{CE0}$ was 8 V. The base was zinc-doped to 2×10^{19} cm^{-3}, and the emitter size was 1.6×4.6 µm^2.

The best breakdown voltage for an InP-based HBT was BV$_{CE0}$ = 32 V for a gas-source MBE-grown InAlAs/InGaAs/InP DHBT [17]. This HBT used a novel 33-period InAlAs/InGaAs chirped superlattice to eliminate the heterojunction spike at the InGaAs/InP base–collector junction. The HBT had a 600 Å, 3×10^{19} cm^{-3}

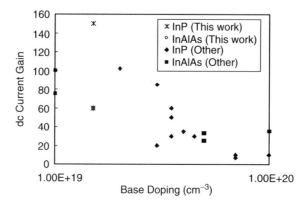

Figure 3.4. Published values for dc current gain versus base doping for InP-based NPN SHBTs with InP and InAlAs emitters.

beryllium-doped base and a 1 μm, 1.5×10^{16} cm^{-3} InP collector. A 12-finger 2×30 μm^2 power HBT using this design demonstrated a record power density for InP-based HBTs of 3.6 mW/μm^2 with power-added efficiency (PAE) of 56% at 9 GHz.

As mentioned earlier, InP-based NPN single HBTs (SHBTs) have displayed impressive frequency performance. Many reports have been published on such HBTs. Figure 3.4 summarizes a variety of published values for dc current gain as a function of the base doping. While most published HBT results have a base doping between 1×10^{19} and 1×10^{20} cm^{-3}, the decreasing trend of current gain with increased doping is pronounced. Figure 3.5(a) demonstrates a similar decreasing trend in f_T with respect to the base thickness. This strong trend indicates that τ_b is a large component of τ_{ec} for a well-designed NPN HBT. Figure 3.5(b) demonstrates the trade-off between f_T and BV_{CE0} as described by the Johnson figure of merit for HBTs: an increased breakdown requires a thicker InGaAs collector, which reduces f_T. Finally, thicker collectors reduce C_{BC} and result in an increase in f_{max}. Together, these plots exhibit the range of characteristics reported so far for InP-based single HBTs.

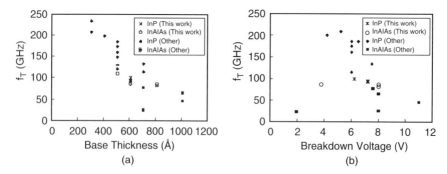

Figure 3.5. Published values of f_T versus (a) base thickness and (B) breakdown voltage BV_{CE0} for InP-based NPN SHBTs with InP and InAlAs emitters.

3.3.2 InP/InGaAs NPN HBT Layer Structure

A reference structure was employed for discussing the HBT small-signal characteristics. Its design was selected based on simulations of HBTs optimized for low base and collector transit times. The results shown here confirm that good current gain combined with state-of-the-art power performance can be obtained using the selected HBT design. The devices had a 600 Å base, which was designed for short transit time, and $\sim 1000\,\Omega/\square$ base sheet resistance. The moderate 5000 Å n^- collector design presented reduced electric fields in the collector, which allowed a fairly high HBT breakdown voltage of ~ 7 V without a large collector transit time. The moderately thick collector also helped to reduce the base–collector capacitance, further enhancing high-frequency performance. Unless otherwise indicated, all NPN HBT data presented here were obtained from the layer structure shown in Table 3.1. This wafer was grown by MOCVD in our laboratory and will be referred to as "Wafer A."

3.3.3 dc and Small-Signal Microwave Performance of InP/InGaAs NPN HBTs

This section presents the dc and small-signal S-parameter results from typical InP/InGaAs NPN HBTs with $2 \times 10\,\mu m^2$ emitter. The $2 \times 10\,\mu m^2$ emitter geometry gave reproducible results and was the basic "high-performance" HBT design used in this work. For large ($50 \times 50\,\mu m^2$) non-self-aligned HBTs, the breakdown voltage BV_{CE0} was 7.2 V at $I_C = 10\,\mu A$. The breakdown voltage achieved here is very good for InP/InGaAs single HBTs (see Fig. 3.5(b) for comparison). In general, the small-bandgap InGaAs collector produces low breakdown voltages due to the low electric field threshold for impact ionization, $\sim 2 \times 10^5$ V/cm [18]. The impact ionization rate α_n increases with temperature on the order of 0.8 cm^{-1}/°C, which causes the breakdown voltage of HBTs to decrease at high current levels where self-heating occurs [19]. The use of InP collectors with ~ 2.2 times the threshold field for impact

TABLE 3.1. Reference InP/InGaAs NPN SHBT Layer Structure (Wafer A) Used in This Work

Layer Name	Material[a]	Thickness (Å)	Type	Doping (cm^{-3})
Emitter cap	InGaAs	2000	n^+	2×10^{19}
	InP	700	N^+	2×10^{19}
Emitter	InP	1500	N	5×10^{17}
Spacer	InGaAs	100	i	—
Base	InGaAs	600	p^+	1.5×10^{19}
Collector	InGaAs	5000	n^-	5×10^{16}
Subcollector	InGaAs	5000	n^+	2×10^{19}
Buffer	InP	2000	I	—
Substrate	InP	—	LEC InP:Fe	

[a] All InGaAs layers are In$_{0.53}$Ga$_{0.47}$As.

ionization can be employed to increase the breakdown voltage. However, in such HBTs with an InGaAs base and InP collector, special care must be taken to eliminate the conduction band spike at the base–collector junction. Therefore this work employs thick (at least 5000 Å), lightly doped InGaAs collectors, which can create a larger depletion region in order to decrease the average electric field in the collector and hence increase the breakdown voltage.

For a typical microwave HBT with a 2×10 μm^2 emitter, the common-emitter I–V characteristics are shown in Figure 3.6, demonstrating collector current densities of $\sim 1.2 \times 10^5$ A/cm^2 and large-signal current gain of approximately 60.

The Gummel plot for this HBT is shown in Figure 3.7, from which the ideality factors $n_C = 1.2$ and $n_B = 1.6$ were extracted. Such ideality factors are reasonable, with published values of n_C from 1.02 to 1.5 and n_B from 1.02 to 2.0 for InP-based HBTs. The nonunity collector ideality factor is probably due to thermionic emission or tunneling of electrons at the abrupt emitter–base heterojunction. The higher base ideality factor indicates that significant recombination is occurring either in the emitter–base space-charge region, in the depleted spacer, or along the external surfaces of the emitter mesa. Increases in the base ideality factor have been linked to recombination in undoped spacers when the base dopant does not out-diffuse enough to convert the entire spacer into a p-type layer [20]. Therefore it is likely that the base dopants in this work diffuse less than 100 Å and that a thinner spacer may be more appropriate.

The small-signal S-parameter data were measured by an HP 8510B network analyzer from 0.5 to 25.5 GHz. The cutoff frequency f_T and the unity power gain frequency f_{max} were found to be 95 GHz and 55 GHz, respectively, at $I_C = 12.6$ mA and $V_{CE} = 1.5$ V. The bias dependence of f_T and f_{max} are shown in Figure 3.8. As can

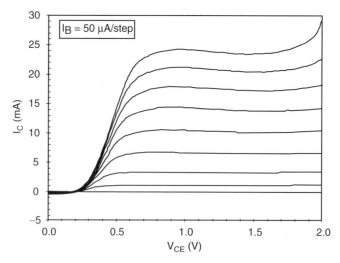

Figure 3.6. Common-emitter $I_C - V_{CE}$ characteristics for 2×10 μm^2 HBT from the reference Wafer A.

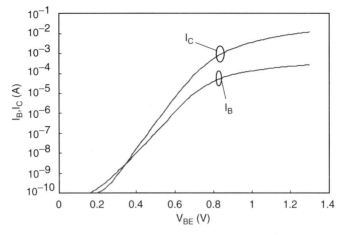

Figure 3.7. Forward Gummel plot for $2 \times 10 \ \mu m^2$ HBT.

be seen, the f_{max} increased to 58 GHz when V_{CE} was increased to 2.0 V. This increase of f_{max} with V_{CE} corresponds to an increase of $\sim 17\%$ in the base–collector depletion width, which slightly lowers C_{BC} by 15%. However, the increase in the depletion width for $V_{CE} = 2.0$ V caused a corresponding increase in the collector delay τ_{pcd} and hence f_T dropped slightly. Although the peak f_T occurred at $I_C = 12.6$ mA

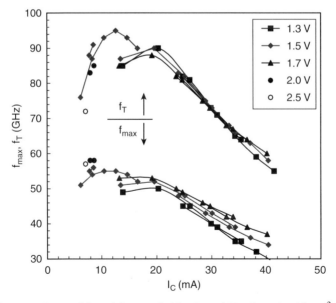

Figure 3.8. Dependence of f_T and f_{max} on dc bias I_C and V_{CE} for a $2 \times 10 \ \mu m^2$ HBT. Four identical HBTs were measured, two at low current densities and two at high current densities.

($J_C = 6.3 \times 10^4$ A/cm^2), the HBTs could be operated above $I_C = 40$ mA ($J_C = 2 \times 10^5$ A/cm^2), where base push-out and high-level injection degraded f_T to 60 GHz and f_{max} to 37 GHz at $V_{CE} = 1.7$ V.

3.3.4 Epilayer Dependence of HBT Characteristics

HBTs can be grown by different techniques such as MOCVD and MBE. Special p^-n^- collectors can be employed for lower collector transit time without sacrificing breakdown voltage. Monte Carlo simulations indicate that this p^-n^- collector reduces $\tau_b + \tau_{pcd}$ by 26–38% compared to a conventional n^- collector due to the lower electric field in the p^- region, which keeps the electrons in the high-velocity Γ valley [21]. While a 3000 Å p^- collector further reduces $\tau_b + \tau_{pcd}$ by 37–73% compared to a conventional n^- collector, the peak electric field for this design moves to the collector–subcollector junction and reduces the breakdown voltage by 2 V. Conversely, the p^-n^- collector spreads the electric field across the collector region more evenly, which increases the breakdown voltage by 1 V compared to the conventional n^- collector. Therefore the p^-n^- collector theoretically has enhanced speed–breakdown characteristics.

InAlAs rather than InP emitters can also be used. The use of MBE is interesting from the point of view of allowing high base doping with no dopant diffusion problems by using carbon as the p dopant. The use of a graded InAlAs/InGaAs emitter–base junction allows lowering of the turn-on voltage. By employing a thicker beryllium-doped base and low-doped collector, one can reduce R_B and C_{BC}, respectively, both of which enhance f_{max}. Compared to abrupt InP/InGaAs emitter–base junctions with $\Delta E_C/\Delta E_V \approx 0.25$ V/0.34 V, abrupt InAlAs/InGaAs junctions with $\Delta E_C/\Delta E_V \approx 0.48$ V/0.24 V have a 0.23 V larger turn-on voltage due to the larger ΔE_C [4]. However, a graded InAlAs/InGaAs emitter–base junction effectively eliminates the conduction band spike and reduces the turn-on voltage by $\sim$0.4 V ($\sim$0.15 V) compared to an abrupt InAlAs/InGaAs (InP/InGaAs) junction.

3.3.5 dc and Microwave HBT Characteristics

The characterization of wafers with different HBT designs allows some interesting conclusions to be drawn. Wafers with abrupt emitter–base heterojunctions give similar values for the collector ideality factor $n_C \approx 1.2$–1.3, which indicates that the electron injection into the base is significantly impacted by either tunneling or thermionic emission at the conduction band spike formed at the heterojunction. The lower $n_C \approx 1.1$ for graded emitter–base junction devices is due to its smooth conduction band at the heterojunction, which allows diffusion to dominate electron injection into the base. The removal of the conduction band spike also decreases the on-state V_{BE} (defined at highest f_T) to 0.76 V for the graded emitter–base junction, as compared to 0.92–0.96 V for the abrupt emitter–base junction designs. Similarly, the offset voltage in the common-emitter characteristics reduces from 0.20 V for abrupt junctions to 0.12 V for the graded junction. Monte Carlo simulations have indicated that abrupt InAlAs/InGaAs emitter–base heterojunctions can inject carriers into the base at high velocity, which can reduce the base transit time across a 1000 Å base from 2.06 ps for a graded EB junction to 0.85 ps for an abrupt EB junction [22]. Therefore, while a graded EB junction reduces the turn-on voltage, it may also penalize f_T.

Wafers with InP emitters demonstrate higher base ideality factor n_B than wafers with InAlAs emitters. In all cases, n_B is significantly higher than n_C, which implies significant carrier recombination in either the emitter depletion region or at the emitter mesa surface. The higher n_B for the InP emitters versus InAlAs emitters is therefore understandable, since the bandgap of InP is smaller, which tends to increase carrier recombination rates.

A higher collector doping allows a higher I_C before base push-out occurs, which gives a higher I_C at peak f_T. Both f_T and f_{max} can degrade due to a higher base contact resistance and higher C_{BC}. Devices with a thinner collector and higher doping, as well as higher base contact resistance, give a smaller f_{max} due to large C_{BC} and R_B. A thin, moderately doped collector allows a slight improvement in f_T although the resulting increase in C_{BC} can degrade f_{max}. The advantage expected for $p^- n^-$ HBTs in reduced transit time without sacrificing breakdown voltage can be masked by $R_B C_{BC}$; the power gain can in such a case be dominated by $R_B C_{BC}$ rather than by the collector transit time.

A higher base doping reduces the base mobility, which can account for a slight decrease in f_T. However, the higher base doping also decreases R_B. A higher collector doping leads to higher C_{BC}. Lateral undercut of the base contacts, which reduces the effective area of the base–collector junction, however, can reduce C_{BC} when compared to non-undercut HBTs. The use of both a thicker collector and lateral undercut of the base contact results in the smallest C_{BC}. If the base is also thicker and more highly doped this also decreases R_B. Together, a lower $R_B C_{BC}$ can significantly increase f_{max}. However, the thicker base and collector layers slightly increase the base and collector transit times, which can result in a decrease in f_T. HBTs with 2×10 μm^2 emitters without very advanced optimization can demonstrate f_T and f_{max} values up to 80 and 110 GHz, respectively. Smaller HBTs with 1×10 μm^2 emitters and a 1.5 μm base contact can demonstrate an even higher f_{max} of 130 GHz.

3.3.6 InP-Based PNP HBT Designs

A PNP layer structure with reasonable gain ($h_{fe} = 12$) is shown in Table 3.2. This structure was fixed as the standard PNP layers for the PNP studies in this work, and the results from this wafer were representative of all PNP HBTs fabricated in this work. As compared to typical NPN structures, the PNP design has a higher emitter doping and a lower base doping, both of which increase the dc gain of the HBT. The base thickness was originally reduced to 350 Å for increased dc gain and reduced base transit time; however, the final base thickness was fixed at 500 Å in order to ease fabrication during the nonselective emitter etch. The emitter–base spacer was retained from the NPN structure, where it compensated for beryllium out-diffusion from the highly doped base. However, in the PNP the purpose of the spacer was to

TABLE 3.2. Standard PNP Layer Structure Used in This Work (Wafer R)[a]

Layer	Type	Thickness (Å)	Doping (cm^{-3})
Emitter cap	p^+ InGaAs	2000	2×10^{19}
	P^+ InAlAs	700	1×10^{19}
Emitter	P InAlAs	1500	8×10^{17}
Spacer	i InGaAs	100	—
Base	n^+ InGaAs	500	5×10^{18}
Collector	p^- InGaAs	3000	3×10^{16}
Subcollector	p^+ InGaAs	5000	1×10^{19}
Buffer	InGaAs/InAlAs superlattice	~1000	—
Substrate	Semi-insulating (001) InP		

[a]Results from this wafer were representative of all studied PNP HBTs.

reduce the turn-on voltage of the HBTs for low-power amplifiers. Finally, the thickness of the collector was reduced to 3000 Å in order to decrease the collector transit time of the holes.

The designs of two MBE-grown wafers studied here are shown in Table 3.3. The wafers, R and S, were grown in the same MBE system at The University of Michigan. Wafer R used the standard layer structure from Table 3.2, and Wafer S had a linear doping gradient in the base, which was designed to decrease the base transit time.

3.3.7 PNP HBT dc and High Frequency Characteristics

A summary of characteristics for HBTs with 5×10 μm^2 hexagonal emitters fabricated from each PNP wafer is shown in Table 3.4. Several interesting points can be made by comparing this PNP data to NPN data. First, the gain of the PNPs is much lower than that of the NPNs, which is expected due to the poorer base transit properties of holes. The collector ideality factor is almost unity for the PNPs, as compared to 1.2–1.3 for NPNs with similar abrupt emitter–base heterojunctions and 100 Å spacers. This lower ideality factor indicates that the heavier holes are extremely unlikely to tunnel through the valence band spike at the heterojunction. Similarly, the base ideality factor is slightly higher for the PNPs, which implies greater recombination in the emitter–base junction. Of course, f_T and f_{max} are much lower for the PNPs due to the lower mobility and saturation velocity of holes. While

TABLE 3.3. Design Values for PNP HBT Layer Structures Fabricated Using the Standard HBT Layout and Process

Wafer	Source	Details (differences from structure in Table 3.2)
R	U-M	(Standard layer structure in Table 3.2)
S	U-M	Base doping = linearly graded from 5×10^{18} cm^{-3} (at emitter end) to 1×10^{18} cm^{-3} (at collector end)

GENERAL SMALL-SIGNAL CONSIDERATIONS 93

TABLE 3.4. Characteristics of 5 × 10 μm² HBTs from the Wafers of Table 3.3

Wafer	h_{fe}	n_C	n_B	Maximum f_T (GHz)	Maximum f_{max} (GHz)	V_{EC} @ f_T (V)	I_C @ f_T (mA)
R	12.5	1.00	1.60	11.1	30.6	4.0	11.7
S	4.2	1.03	1.65	12.9	25.3	4.5	6.5

both NPN and PNP HBTs have similar breakdown voltages, the NPNs can only be biased up to $V_{CE} \approx 2.5$ V at moderate and high current levels before device failure. However, the PNPs can safely be biased up to $V_{EC} \approx 4.5$ V at similar current densities, probably since the ionization rate, α_p, for holes is approximately one-half that of α_n for electrons in InGaAs [18]. Finally, the collector current density at peak f_T is much lower for PNP HBTs with similar collector doping levels. This can be attributed to base push-out (Kirk effect), which occurs at lower J_C for PNP HBTs due to the lower velocity of holes in the collector.

The 5 × 10 μm² HBT from Wafer S was very similar to that from Wafer R. The forward I–V data exhibited a smaller Early voltage than Wafer R, approximately 1 V versus 4 V, due to the lighter base doping at the collector edge. The small Early voltage caused the gain of Wafer S to increase past that of Wafer R at higher V_{EC}. The peak gain occurred at lower current levels than Wafer R, perhaps due to either variations in geometries during fabrication or earlier base push-out from slight variations in the collector doping. Similarly, the maximum frequency performance also occurred at 45% lower collector current. The 15% increase in f_T is due to the decrease in τ_b caused by the drift electric field in the base. However, the doping gradient that created this drift field also significantly increased R_B, resulting in an overall 13% decrease in f_{max}.

The forward I–V curve for a 5 × 10 μm² HBT from Wafer R is shown in Figure 3.9. The offset voltage was 0.20 V. As can be seen from the forward I–V curve, the gain increased significantly ($\beta > 20$ and $h_{fe} > 30$) at higher V_{EC}. However, the gain compressed rapidly for $J_C > 50$ kA/cm² due to base push-out, which agrees with the maximum collector current density prior to base push-out from simulations using similar collector doping. Base push-out has a very significant effect on both the gain and the high-frequency performance of PNP HBTs due to the short diffusion length of the holes. Preliminary results from pulsed measurements indicate that self-heating significantly also degrades h_{fe} for $V_{EC} > 2$ V.

Microwave characteristics were measured on-wafer from 0.5 to 25.5 GHz using coplanar probes and an HP8510 network analyzer. $|h_{21}|^2$, G_{max}, and U were calculated from the S-parameters and extrapolated (when necessary) at 20 dB/decade to find f_T, $f_{max}^{G_{max}}$, and f_{max}^U, respectively. For the 5 × 10 μm² HBT from Wafer R, the peak frequency performance was $f_T = 11$ GHz, $f_{max}^U = 27$ GHz, and $f_{max}^{G_{max}} = 31$ GHz at $V_{EC} = 4.0$ V and $I_C = 11.69$ mA. The bias dependence of the frequency performance is shown in Figure 3.10. Note that the bias point for maximum frequency performance is very close to that of maximum gain from the Gummel plot,

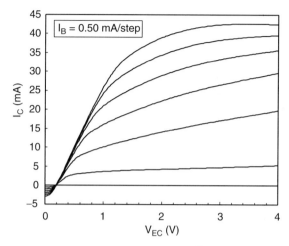

Figure 3.9. Forward I–V characteristics of 5×10 μm^2 PNP HBTs from Wafer R with $I_B = 0.5$ mA/step.

which indicates that hole transport rather than RC parasitics limits the high-frequency performance. The effect of base push-out can be seen in the rapid decrease of f_T and f_{max} at higher collector currents.

At similar bias conditions for a 1×10 μm^2 HBT from Wafer R, the best frequency performance was $f_T = 12$ GHz and $f_{max} = 35$ GHz. This is the highest reported f_{max} for any InP-based PNP HBT.

An analysis of the HBT characteristics demonstrated that the hole mobility and diffusion constant in the base are typical for 5×10^{18} cm^{-3} InGaAs. However, the hole lifetime (and therefore the diffusion length) in the base is approximately 20 times worse than reported values. Together, these values explain why these PNP HBTs demonstrate low dc current gain simultaneously with record high-frequency performance. If possible, base layers should be grown with longer hole lifetime. In order to better understand design trade-offs, this study of the quality of the base layer should be repeated for several different base doping levels.

3.4 LARGE-SIGNAL PROPERTIES

3.4.1 Single and Double NPN Heterostructure Bipolar Transistors

Typically, the microwave power characteristics of InP/InGaAs single HBTs have not been addressed due to their relatively inferior dc characteristics when compared to double HBTs, which implies early breakdown and thus limited power performance. On the other hand, single heterojunction bipolar transistors (SHBTs) are very attractive for higher frequency applications due to the absence of the heterojunction spike at the base-collector (BC) interface. Moreover, the homojunction BC structure

LARGE-SIGNAL PROPERTIES

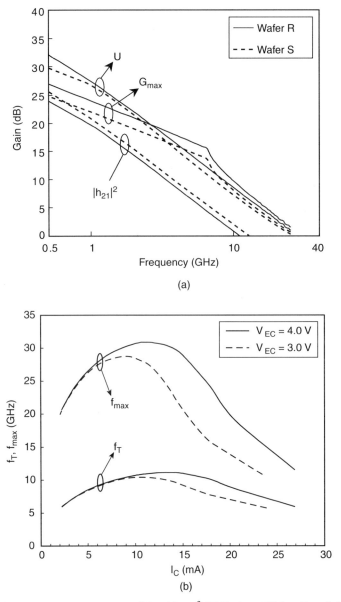

Figure 3.10. (a) Frequency response of 5×10 μm^2 HBTs from Wafers R and S. (b) Bias dependence of high-frequency performance for 5×10 μm^2 HBT from Wafer R.

offers direct compatibility for HBT integration with PIN diodes, since the latter can be realized by using the BC–subcollector region. Such integration is needed not only for optoelectronic integrated circuits (OEICs) but also for microwave monolithic integrated circuits (MMICs) with switching capabilities.

SHBTs and DHBTs were characterized at high-frequency large-signal operation using an X-band on-wafer measurement system in our laboratory using FOCUS electromechanical tuners.

3.4.1.1 Issues for Power Amplification in HBTs In order to reduce the supply voltage for longer battery life, single-transistor output stages are common in power amplifiers for hand-held units. Due to the linearity requirements set by wireless modulation schemes such as quadrature phase shift keying (QPSK), these output stages usually employ Class A or Class AB operation at power levels that do not exceed the 1 dB gain compression point ($P_{\text{O-1dB}}$).

Figure 3.11 shows a schematic representation of the I_C–V_{CE} characteristics of an HBT together with the key parameters defining its power performance. In the first order, the maximum output power and efficiency of an HBT under Class A are limited by the breakdown voltage BV_{CE0}, the knee voltage V_k, the maximum collector current I_{Cmax}, and the gain G. While it does not precisely scale with the HBT emitter area, I_{Cmax} increases with emitter area and can be approximated by $A \times J_{Cmax}$ for this analysis, where J_{Cmax} is the maximum collector current density and is limited by the HBT design. Under these assumptions and neglecting HBT nonlinearity, maximum output power is generated by sweeping between V_k at I_{Cmax} and BV_{CE0} at $I_C = 0$, resulting in an optimal (real part of the) load impedance,

$$R_L = \frac{BV_{CE0} - V_k}{AJ_{Cmax}}. \tag{3.4}$$

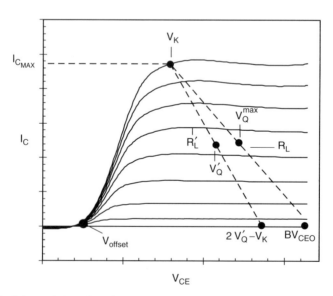

Figure 3.11. Schematic load lines for an HBT biased in class A for maximum possible output RF power P_L (V_Q^{max}, R_L) and typical RF power (V_Q', R_L').

LARGE-SIGNAL PROPERTIES   97

The peak linear output power (P_L) and power added efficiency (PAE) can then be estimated as

$$P_L = \frac{AJ_{Cmax}(BV_{CE0} - V_k)}{8} = 0.125 \frac{(BV_{CE0} - V_k)^2}{R_L} \qquad (3.5)$$

and

$$\text{PAE} = \eta_C \left(1 - \frac{1}{G}\right) \leq 0.5 \frac{BV_{CE0} - V_k}{BV_{CE0} + V_k}\left(1 - \frac{1}{G}\right). \qquad (3.6)$$

Other operation points V'_Q corresponding to different load impedance R'_L (see Fig. 3.11) may also be used, leading to smaller peak voltage $2V'_Q - V_K$. However, the maximum linear output power will also reduce by the corresponding voltage swing $2(V'_Q - V_K)$. Analogous expressions can be derived for Class B operation, where the prefactor in Eq. (3.5) is increased from 0.125 to 0.149, and the PAE expression has a similar form and peaks at 78% rather than 50%. Typically, the circuit technology limits the minimum R_L that can be synthesized for output matching, which then determines the maximum usable device area from Eq. (3.4). The peak output power and PAE are then limited by Eqs. (3.5) and (3.6).

Since power HBTs need high P_L and PAE, this brief analysis shows the required HBT characteristics: high G, high BV_{CE0}, and low V_k. High J_{Cmax} is also desired to reduce the required HBT area, which is limited by space, nonuniform heating, and signal phasing concerns. Note that the R_L presented here, which generates the most output power, is not directly related to the matched load impedance Z_L for the highest gain. Therefore the actual load impedance presented to the HBT must be chosen to trade off G, P_L, and PAE.

InP-based single and double HBTs each offer advantages for power amplifiers. Best reported values of some relevant parameters are shown in Table 3.5. The higher BV_{CE0} primarily allows DHBTs to generate more output power than SHBTs, and the lower thermal resistance to the substrate due to the InP collector allows for lower and more uniform junction temperatures. The higher saturation velocity of the InP collector also allows for proportionally higher J_{Cmax} before the onset of the Kirk effect at a given collector doping. With emitter–base and base–collector compositional grading, the offset voltage and V_K are theoretically slightly smaller for DHBTs. The higher BV_{CE0} together with this low V_K allows slightly higher efficiencies for DHBTs as compared to SHBTs. For example, by using Eqs. (3.5) and (3.6) for Class A operation together with the data in Table 3.5 and a system minimum R_L of 5 Ω, large DHBTs should be able to generate 25.2 W output power at 44% PAE, while similar SHBTs should generate 1.2 W output power at 42% PAE. Therefore DHBTs are the best choice for high-power applications.

However, the spike in the conduction band of DHBTs may limit J_{Cmax}, may increase V_K, and may also limit the gain at low J_C, which can introduce additional nonlinearities to the output characteristics. Since SHBTs are also easier to design and fabricate than DHBTs, they may be more cost effective and a better choice for

TABLE 3.5. Best Reported Results for InP-Based Microwave Single and Double HBTs as of early 1999

Parameters	InP SHBTs	InP DHBTs
Highest f_T (GHz)	235	160
Highest f_{max} (GHz)	236	267
Offset voltage (V)	0.20	0.10
V_K (V) @ 10^4 A/cm^2	0.30	0.25
Turn-on voltage (V) @ 10^4 A/cm^2	0.7	0.7
BV_{CE0} (V)	8.0	32
P_{out} (mW/μm^2) @ 10 GHz	1.37	3.6
Gain (dB) measured @ 10 GHz	11	10
v_{sat} of Collector[a] (cm/s)	6×10^6	9×10^6
Thermal conductivity of collector[a] (W/(cm·K))	0.05	0.68

[a]The bottom two rows show material parameters for InGaAs (SHBT) and InP (DHBT) collectors.

power amplifiers that require only moderate output power levels. A detailed experimental comparison of the power performance of SHBTs and DHBTs has been presented elsewhere [23].

To illustrate the analysis presented here, the power characteristics of three SHBTs were measured on-wafer at 8 GHz. All HBTs used bias, source impedance, and load impedance optimized for maximum output power under large-signal operation near the 1 dB gain compression point. Then the input power was swept while maintaining constant V_{CE} and V_{BE} bias, and the output power and the HBT currents were measured.

The optimal measured bias points and source/load impedances are shown in Table 3.6. These HBTs were representative of all devices measured. Note that none

TABLE 3.6. Optimal Bias Points and Source/Load Impedances for Maximum Output Power Under Large-signal Operation for Several NPN HBTs from Wafer A[a]

Parameters	HBT 1	HBT 2	HBT 3
A_E (μm^2)	2×10	3×10	Two-finger 3×10
V_{CE} (V)	1.8	1.8	2.0
V_{BE} (V)	0.87	0.82	0.82
I_{C0} (mA)	7.34	9.39	15.12
I_{B0} (mA)	0.22	0.27	0.52
Optimum Γ_S	0.209 ∠114°	0.336 ∠135°	0.327 ∠155°
Optimum Γ_L	0.202 ∠143°	0.222 ∠165°	0.298 ∠−175°
Peak gain (dB)	12.4	11.4	11.9
P_{0-1dB} (dBm)	10.4	11.2	14.4
Peak PAE (%)	30	29	35

[a]Large-signal performance under these conditions is also listed.

of the HBTs could be operated at V_{CE} above $\sim\frac{1}{3} BV_{CE0}$ without causing the HBTs to burn out at moderate-to-high collector currents, which forced the choice of low V_{CE} in Table 3.6. Also note that as P_{in} increased, self-biasing caused I_C to increase much above the I_{C0} values listed in the table.

Table 3.6 also shows the peak gain, output power at 1 dB gain compression (P_{0-1dB}), and peak associated PAE under these bias points and source/load impedances. These values and the real part of Z_L roughly correspond to predictions from the simplified expressions in Eqs. (3.4)–(3.6) when BV_{CE0} is appropriately reduced according to the peak V_{CE} applied. For example, the predicted Class A maximum output power for HBT 1 is 8.8 dBm. Similarly, the predicted Class A PAE for HBT 1 is 31%. Analysis of these equations indicates that the PAE is mostly limited by the difference ($BV_{CE0} - V_K$): if V_K were reduced to 0, the PAE would be 47–49% (where the theoretical maximum is 50% for Class A). More realistic HBT optimizations could produce $V_K \approx 0.3$ V, which would result in a PAE of 39% for HBT 1. Further analysis indicated that larger HBT emitters (with correspondingly smaller load impedances) can be used to generate more output power, provided that gain does not degrade as the HBT area increases.

3.4.1.2 Other Power Characteristics of InP HBTs The constant V_{BE} bias demonstrated better gain, output power, and PAE than the constant I_B bias for all HBTs measured at high input power levels. In general, the optimized load impedances for maximum gain, output power, and PAE did not coincide for the NPN HBTs measured.

Additional measurements revealed even higher output power, output power density, and PAE when the impedance was optimized for maximum output power. HBTs with a four-finger (2×10) μm^2 emitter biased in Class A showed the best output power density of 1.37 mW/μm^2 by generating 20.4 dBm output power at 10 GHz. Furthermore, a maximum power added efficiency of 33.9% was measured for this transistor at the same source and load terminations. The power density of 1.37 mW/μm^2 obtained from this four-finger device demonstrates very promising characteristics from unthinned InP/InGaAs SHBTs measured directly on 370 μm thick InP substrates.

Due to their inherently low breakdown voltage, the power performance of InP-based SHBTs has not previously been investigated. While the breakdown voltage increases with collector thickness, thicknesses beyond 1 μm are difficult to fully deplete at minimal achievable values of collector doping. In addition, very thick collectors significantly decrease f_T due to their increased collector delay. The highest reported BV_{CE0} for a InP-based SHBT is 11 V for a 7000 Å, 5×10^{15} cm^{-3} InGaAs collector, which resulted in $f_T = 45$ GHz [24].

Usually for power applications, DHBTs with InP collectors are used, which increases the breakdown voltage and hence the output voltage swing. While BV_{CE0} values as high as 32 V [17] have been reported with these DHBTs, the base–collector heterojunction must be designed carefully to suppress collector current blocking due to conduction band spikes. The best-published X-band power results for InP-based DHBTs demonstrated a power density as high as 3.6 mW/μm^2 with PAE of 56% at

9 GHz [17] by using a linearly graded InGaAs/InAlAs chirped superlattice (CSL) at the base–collector junction. These results indicate that the SHBTs presented here demonstrate very good power performance considering their low BV_{CE0} imposed by their InGaAs collector. A comparative study of the SHBTs from this work and several DHBT designs has also been published [23].

3.4.2 InP-Based PNP HBTs

3.4.2.1 Power Characterization of PNP HBTs

On-wafer power characterization of Wafers R and S was performed at 10 GHz using the load-pull system described earlier. The 5×10 μm^2 HBT from Wafer R was biased using constant V_{EB} and V_{EC}, while the source and load impedances were optimized for maximum gain at $P_{in} = -20$ dBm, resulting in $\Gamma_S = 0.740 \angle -179°$ and $\Gamma_L = 0.596 \angle 26°$. The power characteristics at these impedances, which are shown in Figure 3.12, demonstrate a small-signal gain of 10 dB, peak power-added efficiency of 24%, and a maximum output power density of 0.49 mW/μm^2. These characteristics are very similar to the InP-based NPN SHBTs of Sawdai et al. [23]. NPN HBTs with the same geometry showed slightly higher gain (+1 dB) and efficiency (+5%), while the PNP HBTs produced more output power (+3 dBm) before saturating. While power-handling capability has not previously been reported for InP-based PNP HBTs, GaAs-based PNP HBTs have demonstrated output power up to 0.63 mW/μm^2 [25], which is comparable to the HBTs presented in this work.

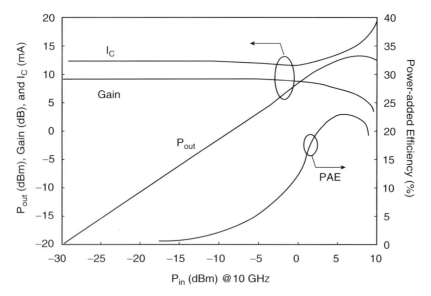

Figure 3.12. Power characteristics of 5×10 μm^2 HBT from Wafer R under constant V_{EB} bias with $V_{EC} = 4.0$ V. Maximum $P_{out} = 0.49$ mW/μm^2 and PAE = 24%.

Load-pull characteristics of the PNP HBT at $P_{in} = 3.17$ dBm and at 1 dB of gain compression demonstrated that the optimal load impedances for linearity, gain, and efficiency are approximately the same, indicating that circuit designs need not choose impedances to trade off these characteristics. In contrast, previous studies of NPN HBTs demonstrated that any single load impedance reduces either the gain by 4 dB, the maximum output power by 2 dB, or the efficiency by 14% from the best-matched values.

3.4.2.2 Discussion of Power Characteristics

The characteristics presented earlier give some insight to the limitations of power performance in PNP InP-based HBTs. In general, the power and efficiency performance is dominated by the microwave gain (or equivalently f_{max}) and the bias conditions. The expressions for power output and power-added efficiency were given in Eqs. (3.5) and (3.6).

It can be seen from these expressions that at maximum PAE, a large two-finger 5×10 μm^2 common-emitter HBT from Wafer R had a collector efficiency $\eta_C = 29.4\%$ but only 4.31 dB gain, which limited the PAE to 18.5%. This low gain was due to three factors: the small input impedance made matching difficult, the frequency of operation was close to f_{max}, and the HBT was operating under 2.8 dB gain compression at maximum PAE.

In order to increase P_{0-1dB}, HBTs must have their optimal f_{max} at higher I_C, which increases the available output signal swing. This implies that larger emitter area, delayed base push-out to higher J_C, and increased breakdown voltage are essential to power performance. While increasing the collector doping delays base push-out, it also decreases the breakdown voltage. However, more advanced structures such as a hole launcher in the collector also delay base push-out and may not decrease BV$_{EC0}$ as much as increased collector doping [26]. The higher gain and breakdown voltage for common-base HBTs clearly give a performance benefit over common-emitter HBTs. Typically, BV$_{BC0}$ for common-base HBTs is twice that of BV$_{EC0}$ for common-emitter HBTs, allowing for much higher bias voltages in the common-base configuration. When combining higher bias V_{BC} with its higher gain at frequencies close to f_T, common-base HBTs should be able to produce more output power at higher efficiencies than was presented here.

Although the doping gradient in the base of Wafer S decreased the base transit time, the base resistance also increased, resulting in lower f_{max}, less high-frequency gain, and worse power performance than Wafer R. Increasing the base doping in Wafer S should decrease the base resistance while maintaining the improved f_T, which would increase f_{max} and the power performance.

Finally, since the high-frequency characteristics of these PNP HBTs were dominated by transit times rather than by parasitic charging times, their intrinsic power performance was only slightly dependent on device geometry or area. However, the total device area still had a large influence on the input and output impedance of the HBT, making it difficult to fully impedance match very small or very large HBTs.

Device modeling and analysis also suggested several potential improvements to the PNP HBT design for increased high-frequency power performance. In general, f_T

was dominated by the base transit time, so either a doping gradient or a compositional gradient in the base can create an electric field to reduce the base transit time. The drift-diffusion simulations indicated that removing the emitter–base spacer would reduce the base transit time while increasing the dc gain, since a significant number of holes recombine in the spacer. These modifications will cause an associated increase in f_{max} and in the gain at 10 GHz, which should significantly improve the X-band power performance and efficiency. Slightly increasing the base doping and thickening the collector will reduce $R_B C_{BC}$, which in turn will also increase f_{max}. Finally, the power-added efficiency was partially limited by the parasitic emitter and collector resistances, indicating that a thinner emitter cap and a thicker subcollector should be used.

3.5 COMPARISON OF HBT, PHEMT, AND MESFET PROPERTIES

3.5.1 Introduction

GaAs-based devices and circuits have a well-defined place in commercial and defense applications as evidenced by their use in a variety of products and systems. Applications extend from the low-frequency spectrum of several hundred megahertz to the millimeter-wave range. The choice of a device is influenced by its maturity in terms of manufacturing but also by other criteria, which are related to fundamental mechanisms that determine the performance. Although metal semiconductor field effect transistors (MESFETs) have long been considered the most mature components, HEMT and HBT technology made significant advances due to the lessons learned from both III–V and Si manufacturing approaches. We present here basic considerations regarding the properties of HBTs, HEMTs/PHEMTs, and MESFETs, and address the relative merits of each technology.

3.5.2 Low-Noise Applications

For low-noise amplifier applications, the pseudomorphic high electron mobility transistor (PHEMT) is generally recognized as the best choice followed by the MESFET. The main source of noise in FETs is thermal diffusion, which is a result of random variations in carrier speed in the device channel. The latter leads to current variations and thus noise. Of particular importance is the presence of capacitive coupling between the gate and the channel, which results in the overall noise being determined by subtracting part of the gate noise from the drain noise. This is a unique property of FETs, which leads to very-low-noise performance.

Best noise performance is obtained by minimizing the source access resistance and maximizing the current gain cutoff frequency, f_t. The latter necessitates device design for maximum transconductance g_m and minimum gate capacitance C_{gs}, conditions that can to some extent also be controlled by proper bias choice. As a general rule, minimum noise figure (F_{min}) is obtained under $\sim I_{ds}/10$ conditions. There is, however, a difference in the bias range necessary for this purpose in

MESFETs and HEMTs; HEMTs appear to have a broader range of I_{ds} values than MESFETs over which F_{min} is achieved. This provides a larger margin in LNA circuit design. Since high gain is in general desired from the amplifier, and bias for F_{min} does not often coincide with bias for maximum gain, a trade-off is often made in gain versus noise. This turns out to be less severe in HEMTs due to their broader range of bias for F_{min}. Large gain also requires device designs with a heterojunction or other type of buffer below the channel to minimize carrier injection and reduce the output conductance.

The possibility of high base doping opens new opportunities for HBTs, which could be of interest for wide bandwidth, low-noise operation. An analysis of noise characteristics also shows that PHEMTs offer smaller bias sensitivity of noise performance than MESFETs. InP-based HEMTs offer the ultimate solution to low-noise operation. They are also the best choice for optoelectronic applications, which require compatibility with InP-based optical devices and operation speeds of 40 Gbit/s and above. Figure 3.13 summarizes the noise performance of various device types as a function of frequency. One observes excellent performance from InP-based HEMTs up to millimeter-wave frequencies, that is, a NF of 1.2 dB at 94 GHz using lattice-matched or strained designs.

Low-frequency noise is in general smaller in HBTs. With proper circuit design for reduced influence of nonlinearities and thus small noise upconversion, HBT oscillators can be designed with very small phase-noise performance [27].

3.5.3 Power Applications

Power amplification is characterized by parameters such as power compression, third order intermodulation distortion (IMD3), and power-added efficiency (PAE),

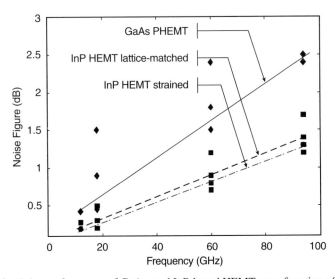

Figure 3.13. Noise performance of GaAs- and InP-based HEMTs as a function of frequency.

which are not normally examined in low-noise amplifiers. The most mature technology for this purpose, MESFETs, is not necessarily the best choice for power, especially when the operation frequency is high. This is due to the requirement of small gate length imposed by the high frequency, which translates to smaller channel thickness. This imposes higher doping in order to conserve the I_{ds} performance and thus smaller breakdown voltage and power. Despite these limitations, reasonable power characteristics of 0.53 W/mm have been obtained at 18 GHz from a MESFET with 600 μm gate periphery [28]. More recently, a 0.3 μm long gate MESFET with a GaAs channel on Al_2O_3 demonstrated a record PAE of 89% at 8 GHz with 9.6 dB gain, 0.12W/mm output power, and a small V_{ds} bias of 3 V [29]. The Al_2O_3 insulating buffer was obtained by wet oxidation of $Al_{0.98}Ga_{0.02}As$ on low-temperature AlGaAs.

Figure 3.14 summarizes the power performance of HEMTs, MESFETs, and HBTs. The use of heterojunctions opens new possibilities in optimizing device performance. AlGaAs/GaAs HEMTs allow higher frequency of operation than MESFETs. These devices however, are limited in terms of sheet carrier density n_s (10^{12} cm^{-3}) and thus current and power due to their small conduction band discontinuity ΔE_c. Despite this limitation, 100 W of power has been obtained at 2.1 GHz using a 86.4 mm gate HEMT [30]. Improved performance can be obtained by adding heterojunction channels or increasing ΔE_c. The former leads to higher overall n_s without at the same time compromising breakdown voltage since the doping of individual channels remains the same. The latter can be realized using the so-called PHEMT approach where the GaAs channel is replaced by $In_xGa_{1-x}As$, where x is usually about 0.2. The channel thickness is in this case of the order of 150 Å in order to ensure reasonably good material properties. PHEMTs have n_s values of at least 4×10^{12} cm^{-3} and therefore currents exceeding 1 A/mm can be achieved. A power

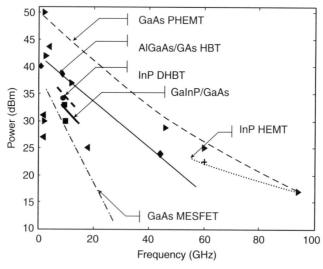

Figure 3.14. Power performance of HEMTs, MESFETs, and HBTs as a function of frequency.

exceeding 0.44 W/mm has been obtained with this approach at 44.5 GHz using a 1800 μm gate device [31]. Further n_s and thus I_{ds} improvements can be obtained by fabricating HEMTs on InP substrates. This technology offers the additional advantage of higher thermal conductivity for InP. InP substrates and the related technology is, however, less mature than GaAs. Most popular among all devices for power applications is the PHEMT. From the technological point of view, further improvements in PHEMT performance can be envisaged by employing double and asymmetric recess. Moreover, for InP-based HEMTs, which are in general limited in breakdown due to their low-bandgap InGaAs channel, some improvement can be envisaged by employing InGaAs/InP composite channels. The use of heterojunction barriers can help by reducing impact-ionization-generated holes from leaking to the gate.

Power backoff is normally required for better linearity performance of PHEMTs. This appears to be a much less demanding requirement in the case of HBTs, which appear to combine good linearity performance at an input power level that is appropriate for good PAE.

Power applications and development of small-size, high-efficiency chips have benefited from the availability of HBTs. These devices, although less mature in technology than PHEMTs, offer higher breakdown voltage, better threshold voltage uniformity, and exponential rather than power-law transfer characteristics. Of major importance is the consideration of thermal issues due to self- and adjacent-device heating. This can lead to a significant power performance difference in continuous-wave (CW) versus pulsed operation, which is not so pronounced in devices such as field effect transistors (FETs). Large-signal modeling of HBTs including thermal effects has led to very satisfactory results and agreement with experiment [32]. Moreover, thermally stable cascode (TSC) HBT designs have been demonstrated that manifest lower temperature increase than conventional HBTs while at the same time offering higher gain [32]. Other approaches include thermal shunts for heat removal from the top of the chip, but these add in general to the cost of technology.

Good power performance has been achieved with GaAs HBTs such as 10 mW/μm^2 at 10 GHz and 4 mW/μm^2 at 25 GHz [33]. InP DHBTs have also been reported [34] with 3.6 mW/μm^2 at 9 GHz, while InP SHBTs fabricated in our laboratory led to record performance for this type with 1.4 mW/μm^2 and a PAE of 43%. The InP-based design offers an attractive solution in terms of gain and high frequency of operation. These devices are also the best choice for high-linearity operation due to the cancellation of currents at high harmonics, which takes place at the internal nodes of the HBT.

3.5.4 High-Frequency Operation

If high-frequency operation is required, PHEMTs offer the best choice due to their high gain up to millimeter wavelengths and their good noise and power performance. Figure 3.15 summarizes the frequency performance of various types of FETs. Although applications at very high (millimeter-wave) frequencies cover a smaller part of the market than systems at microwave frequencies, the potential exists for use

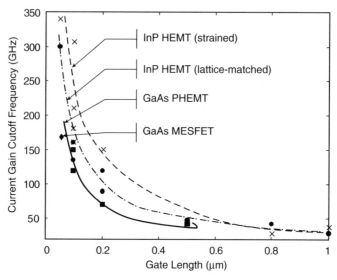

Figure 3.15. High-frequency performance of HEMTs and MESFETs as a function of frequency.

of PHEMTs in commercial systems such as, for example, the automotive anticollision radar system and local multipoint distribution systems (LMDSs). Space applications are also in demand of PHEMT technology at continuously higher frequencies.

InP-based HEMT technology offers the best choice for millimeter-wave systems. It offers excellent low-noise operation up to millimeter-wave frequencies and allows high gain values. HBTs based on GaAs and InP have also shown good high-frequency characteristics, and a record performance ($f_{max} = 250$ GHz) has recently been reported by NEC. Their maximum frequency of operation, however, is lower than that of HEMTs.

3.6 COMPLEMENTARY HBT PUSH–PULL AMPLIFIER

Power amplifiers for wireless communications systems must be able to generate a large output signal at high power-added efficiency (PAE) for viable transmission with long battery life for hand-held units. While GaAs-based devices are already in production for power amplifiers, the excellent high-frequency characteristics and low power consumption of InP-based HBTs make them good candidates for hand-held wireless applications. In addition, the output power capability of InP-based HBTs is also promising, with power levels up to 1.4 mW/μm^2 at 10 GHz already reported using NPN SHBT designs [24]. Double heterojunction designs offer even greater output power, such as a power density of 3.6 mW/μm^2 with PAE of 54% at 9 GHz [34].

In addition, most popular wireless communications systems use high-frequency carriers with time-varying envelopes, which adds a high-linearity requirement to the power amplifier. In order to create power amplifiers with better power performance and linearity than typical circuits using single transistors, this section presents NPN and PNP HBTs combined in a push–pull amplifier. These complementary push–pull circuits are more compact and much simpler to design than NPN-only push–pulls. The challenges in fabricating such a complementary amplifier include developing a high-frequency PNP HBT and integrating NPNs and PNPs on the same substrate.

In general, push–pull amplifiers typically produce linear output in Class AB or Class B, which can have efficiencies as high as 78%. To achieve the same linearity, single-transistor amplifiers must operate in Class A, with efficiency limited to 50%. In addition, since the output voltage swing is generated across two transistors in the push–pull amplifier, a somewhat greater output voltage swing may be possible. Similarly, the output current swing is approximately doubled, resulting in increased output power *without* a significant decrease in the input or output impedances, as would be the case if the area of a single-transistor amplifier were simply doubled.

3.6.1 Integration of NPN and PNP HBTs

Several monolithic and hybrid fabrication technologies are available for integrating NPN and PNP HBTs in order to build circuits. In the only report on complementary HBTs in the InAlAs/InGaAs material system, first PNP and then NPN HBT layers were grown uninterrupted on a mesa-patterned substrate, which resulted in a planar wafer with NPN and PNP HBTs after device fabrication [35].

Several more NPN/PNP integration technologies have been demonstrated in the AlGaAs/GaAs material system. Hybrid common-emitter push–pull amplifiers using 6-finger 2×20 μm^2 HBTs have demonstrated 6 dB of gain, 0.5 W of output power, and 42% PAE at 10 GHz [36]. Nonselective MOVPE was used to grow monolithic planar complementary HBTs: first the NPN layers were grown and etched back to form mesas, and then the PNP layers were grown over the whole wafer and etched back over the NPN regions. A push-pull amplifier on this wafer demonstrated 1 W output power with less than -50 dBc second harmonic output at 8 GHz [37]. Selective low-pressure MOVPE has also been used to create monolithic common-emitter push–pull amplifiers that demonstrate power combination at 10 GHz with power cancellation in the second harmonic [38]. Finally, selective MBE has been used to monolithically create common-collector push–pull amplifiers that produce 7.2 dB gain and 7.5 dBm output power at 2.5 GHz using 4-finger 3×10 μm^2 HBTs [9].

Of course, push–pull amplifiers could be designed without complementary devices at the expense of substantially increased circuit size and complexity. However, the epitaxial growth and device fabrication would be simpler for the noncomplementary circuits. For example, a hybrid push–pull amplifier was reported using two packaged GaAs MESFETs on coplanar substrate. By using coplanar-to-slot line baluns, 19 dBm output power was produced at 10 GHz [39]. Another approach employed Wilkinson combiners with either lowpass/highpass LC networks or 180°

delay lines to feed very large GaAs MESFETs operating in Class AB. Both circuits produced 3.2 W output power at 12.2 dB gain and 25% PAE in the 8–11 GHz band [40]. In another report, a FET amplifier using three single-ended stages and two push–pull stages, all coupled by transmission lines and dielectric resonators, produced 44 dBm output power at 77% PAE operating at 11.2 GHz with a 50 MHz bandwidth. Although operating in Class C ($\theta = 50°$), a high IP3 of 56 dBm was measured at 5 MHz separation [41]. Finally, Lange couplers were used to feed two GaAs FETs 180° apart in a push–pull configuration, which resulted in a 10-dB improvement in IP2 over the 2.5–6.5 GHz band [42].

3.6.2 Basic Push–Pull Simulations

In order to investigate the benefits of using PNP HBTs together with NPN HBTs in various integrated circuits, the large-signal characteristics of several NPN-only, PNP-only, and complementary amplifiers were simulated.

Most of the circuit simulations presented here are variations of the classic "Class B" (Fig. 3.16(a)) and "Class AB" (Fig. 3.16(b)) common-collector push–pull amplifiers [43]. Both circuits are fed from a single power supply such that both the NPN and PNP are biased for maximum output power without exceeding the maximum voltages that they can withstand (2.5 V for NPN and 4.5 V for PNP). The input and output capacitors serve as dc blocks to remove any voltage offset from the input and output, which enables the single power supply.

The advantage of the push–pull amplifier can be seen in the harmonic content of the Class B signal in Figure 3.17. At the beginning of P_{out} saturation, the second harmonic content of the output of the NPN and PNP amplifiers are both -7 dBc. The even-order harmonic cancellation in the push–pull amplifier reduces its second harmonic content to -21 dBc. However, since the odd-order harmonics do not inherently cancel in push–pull amplifiers, its third harmonic shows little variation from the NPN and PNP amplifiers at similar levels of self-biasing.

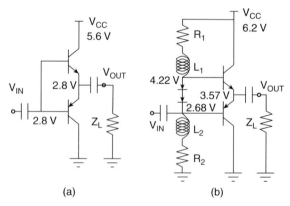

Figure 3.16. Circuit diagram of simulated (a) Class B and (b) Class AB common-collector push–pull amplifiers with single bias supply.

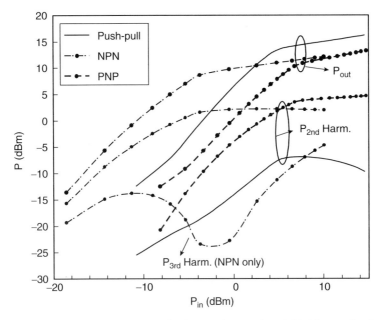

Figure 3.17. Simulated power output at fundamental, second, and third harmonics for Class B common-collector push–pull, NPN-only and PNP-only amplifiers at 1.9 GHZ.

The Class AB push–pull amplifier also demonstrates second harmonic cancellation: −25 dBc at P_{out} saturation, versus −16 dBc for the NPN and PNP amplifiers. The third harmonic is also reduced from −20 dBc for the NPN and PNP amplifiers to −30 dBc for the push–pull amplifier at the same power levels.

In the Class B circuit, the linearity is limited by the crossover distortion when both HBTs are off, which consumes approximately 65° of the output cycle (see Fig. 3.18). While charge storage transients [44] can be seen for both HBTs when moving between the on and off half-cycles, the transients are more pronounced for the PNP due to its slower operation. The second harmonic content is slightly reduced (−4 dB) when moving from Class B to Class AB, resulting from the increased symmetry between the NPN and PNP large-signal output characteristics under Class AB bias (see Fig. 3.18). Simultaneously, the third harmonic content is greatly reduced (−14 dB) due to the elimination of the crossover distortion, which can clearly be seen from the output current characteristics. Note that the higher gain of the NPN causes it to saturate at lower output power than the PNP, which appears as clipping of the peak current in the NPN waveform.

While both circuits may have applications as power amplifiers in various handheld units, the elimination of crossover distortion in the Class AB circuit of Figure 3.16(b) makes it more appropriate for applications with strict linearity requirements. For example, many current digital modulation schemes such as CDMA employ modulation constellations, which result in time-varying envelopes. Since a nonlinear amplifier such as the Class B circuit in such a system would distort the transmitted

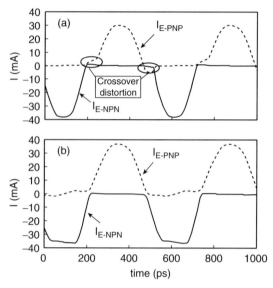

Figure 3.18. Output current components for (a) Class B and (b) Class AB common-collector push–pull amplifiers driven to maximum PAE at 1.9 GHz. Crossover distortion is evident in the Class B output current.

waveform and result in an unacceptable bit-error rate, such modulation schemes require a linear amplifier such as the Class AB circuit.

3.6.3 Issues for Push–Pull Amplifiers

The above simulations demonstrate several advantages of the push–pull amplifiers over SHBT amplifiers. First, the nonlinear distortion is reduced due to harmonic cancellation, although this reduction is limited by the difference in the gain characteristics of the NPN and PNP HBTs. The maximum output power increased by 3 dB, which is primarily due to the increased current swing across the two HBTs. Additional output power may be achievable in the case where a higher supply voltage can be used in the push–pull amplifier than in SHBT amplifiers without breakdown. While the push–pull amplifier produces more output power than either HBT is capable of producing, each HBT in the circuit operates safely by only amplifying approximately one-half of the input signal.

The simulated push–pull amplifiers had only moderate PAE when compared to the theoretical maximum of 78% or higher for the "Class B" amplifier. The primary limitation on PAE was the large knee voltage for the HBTs used in the simulations: 0.5 V and 2.1 V at $I_C = 30$ mA for the NPN and PNP, respectively, which together reduced the available output voltage swing from the 5.6 V dc supply to 3.1 V (see dynamic load lines in Fig. 3.19). Small-signal modeling of the HBT data indicated that reducing the collector and emitter parasitic resistances would reduce the knee voltage, especially for the PNP. Also, using a graded emitter–base junction or using a

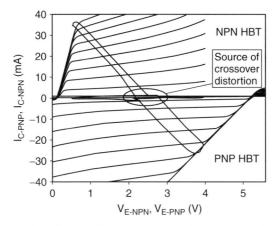

Figure 3.19. Dynamic load lines of NPN (top) and PNP (bottom) HBTs for Class B common-collector push–pull amplifier driven to maximum PAE at 1.9 GHz with $V_{CC} = 5.6$ V. Crossover distortion occurs when both HBTs are in the off state, as shown in the figure.

wide-bandgap collector can reduce the offset voltage, which would also reduce the knee voltage. A graded emitter–base junction would also reduce the turn-on voltage [4], which would cause the common-collector amplifier to saturate at higher input power, decrease the crossover distortion, and reduce the dc power consumption. The other limitation on efficiency is the decreased gain at small bias, which reduces the overall gain and linearity of the Class B amplifier, especially for small input power. Enhanced emitter–base junction designs and fabrication technologies can increase the gain at low bias, which would enable the Class B amplifier to improve its linearity and gain while maintaining a higher PAE than the Class AB amplifier.

Similarly, the common-collector topology simulated up to this point is typically used for current-driving output buffers in power amplifiers. On the other hand, common-emitter amplifiers are more commonly used for gain stages due to their higher voltage gain. Therefore the possibility of a common-emitter push–pull amplifier is also of interest.

3.6.4 Push–Pull Amplifier Simulations with Optimized InP-Based PNP HBT

Since the PNP HBT model used earlier was based on fabricated devices that have not been extensively optimized in layer structure or in lateral layout, the earlier simulations do not demonstrate the full potential of complementary InP-based HBTs. To investigate some additional potential of InP-based push–pull amplifiers, an optimized PNP HBT model was developed by reducing the parasitic resistances of the collector (by 66%) and of the emitter (by 40%). These reductions could easily be achieved by making the collector ohmic contact self-aligned to the base metallization, by increasing the subcollector thickness to 1 μm, and by decreasing the emitter cap and the undepleted low-doped emitter thicknesses. While further enhancements, such as a graded emitter–base junction, should allow for even higher efficiencies, the optimization of parasitic resistances presented here demonstrates simple techniques that can increase circuit performance. Highlights of simulations of the optimized PNP characteristics are a reduced knee voltage (1.1 V at $I_C = 30$ mA) and a reduced turn-on voltage (by 0.3 V).

Under the same bias conditions, the Class B push–pull amplifier using the optimized PNP HBT demonstrated a peak PAE of 52%, which is a 12% improvement over the same amplifier using the nonoptimized PNP. Most of this improvement is due to the decreased knee voltage of the optimized PNP. Although the amplifier gain was unchanged, the decreased turn-on and knee voltages allowed the amplifier with the optimized PNP to saturate at 1 dBm higher output power (Fig. 3.20). Since the optimized PNP dc characteristics were more similar to the NPN, the second harmonic content also decreased by -5 dBc at saturation. However, the resulting $10°$ decrease in crossover distortion was not enough to significantly impact the third harmonic content of the amplifier.

The results were very similar when switching to an optimized PNP HBT in the Class AB push–pull amplifier: the PAE increased to 44% (an 8% improvement), and the second harmonic content decreased by -4 dBc at saturation. Note that, in this

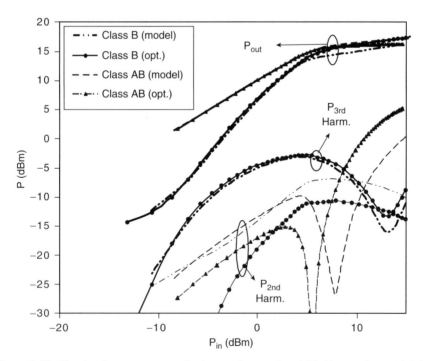

Figure 3.20. Simulated power output at fundamental, second, and third harmonics at 1.9 GHz for common-collector push–pull amplifiers using either modeled or optimized PNP HBTs at Class B and Class AB.

circuit, the bias supply was decreased from 6.2 to 5.2 V, in order to produce the same gain and saturated output power as the Class AB amplifier with the nonoptimized PNP. Again, the third harmonic content was unchanged.

3.6.5 Push–Pull Experiment

In order to demonstrate experimentally the benefits of complementary technologies, a complementary common-emitter push–pull amplifier was designed and fabricated using the NPN and PNP HBTs described earlier in this chapter.

A passive coplanar circuit was designed and fabricated on 10 mil alumina substrates to feed NPN and PNP common-emitter HBTs with separate signals via a common GSGSG input probe (see Fig. 3.21). The HBTs were thinned to 200 µm, cleaved, and then mounted in the circuits on the alumina substrates. Gold bond wires connected the HBT chips to the electroplated gold interconnects on the circuit.

The circuit characteristics were measured at 8 GHz for four sets of bias points: Bias A with both NPN and PNP in Class A ($I_{C0} = 20.3$ mA and 11.6 mA, respectively); Bias B with slightly lower NPN current ($I_{C0} = 14.2$ mA and 11.5 mA, respectively); Bias C with NPN in Class AB and PNP in Class A ($I_{C0} = 5.66$ mA and 11.5 mA, respectively); and Bias D with both NPN and PNP in Class AB ($I_{C0} = 2.03$ mA and 2.41 mA, respectively). The gains for the individual HBTs and for the circuit at all bias points are shown in Figure 3.22 for both small-signal and large-signal inputs. In general, its higher-frequency performance enables the NPN HBTs to produce more gain at 8 GHz than the PNP HBTs. Consequently, the gains of the push–pull circuit at Bias A and Bias B fall between the respective NPN and the PNP gain. However, for both HBTs, the microwave gain reduces for decreasing bias. Therefore Bias C and Bias D were chosen so that the NPN and PNP gains were identical at small-signal and very similar at large-signal inputs.

Typical input versus output power curves (at Bias C) are shown in Figure 3.23, which also shows the power output at the second harmonic for single-tone excitation

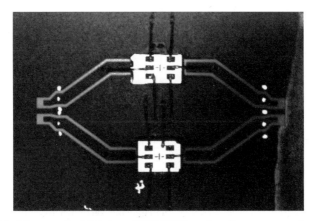

Figure 3.21. Micrograph of fabricated NPN/PNP push–pull common-emitter amplifier.

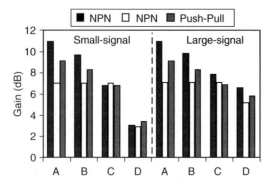

Figure 3.22. Measured gain of NPN HBT, PNP HBT, and push–pull amplifiers at 8 GHz for four bias points.

and the third-order intermodulation (IM3) power output for two-tone excitation at 500 kHz separation. Note that, for the circuit, this graph shows the *total* power input to the circuit and *total* power output from the circuit; therefore each HBT in the circuit is receiving half of the input power. However, the data plotted for the individual HBTs were measured for independent SHBT amplifiers. The simulations indicated that if the HBTs were perfectly matched, both HBTs would produce the same gain and saturate at the same P_{in}, which would result in 3 dB increase in output power for the push–pull circuit. Since the gains were fairly matched in Bias C and

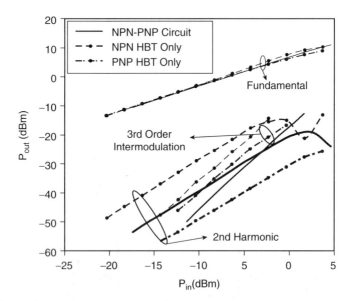

Figure 3.23. Measured power output at fundamental, second harmonic, and third-order intermodulation for NPN HBT, PNP HBT, and push–pull amplifiers at 8 GHz for Bias C.

Bias D, the circuit produced over 2 dB more output power at 1 dB of gain compression than the individual HBTs under the same bias conditions. However, under Bias A at 1 dB of gain compression, the circuit produced only ~1.4 dB more output power than the individual HBTs, due to the gain mismatch between the NPN and the PNP HBTs: when the circuit was at 1 dB compression ($P_{in} = +0.9$ dBm), each HBT received -2.1 dBm input power, causing the high-gain NPN to be over saturated and the low-gain PNP to be undersaturated.

Figure 3.23 also demonstrates increased linearity for the push–pull amplifier. Over most of the power levels measured, the circuit demonstrated ~9 dBc less IM3 and ~11 dBc less second harmonic content than the NPN HBT. At this bias, the circuit also demonstrated less IM3 than the PNP HBT. Linearity improvements are slightly less for the circuit at Bias A: at the highest power levels measured, the circuit demonstrated less IM3 (by ~7 dBc) and smaller second harmonic content (by ~9 dBc) when compared to the NPN HBT. These numbers indicate an advantage for the push–pull circuit even under Class A operation. The slightly better linearity performance at Bias C is due to the better-matched gain between the NPN and PNP HBTs at that bias.

For more insight, the push–pull amplifier output spectra were decomposed into NPN and PNP components using data from the individual HBTs (see Fig. 3.24). Note that since the phase shifter was set for peak output power of the circuit at the fundamental frequency, the fundamental output from each HBT recombines perfectly in phase. Under this condition, Figure 3.24 demonstrates that the second harmonics of the two HBTs recombine out of phase, as predicted analytically for all

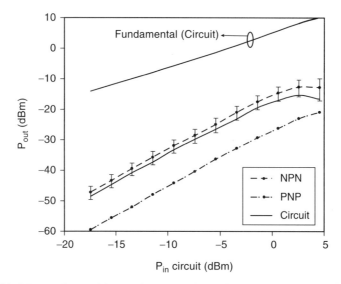

Figure 3.24. Measured second harmonic characteristics for NPN and PNP HBTs inside the push-pull amplifier at 8 GHz for Bias D. The vertical bars represent the limits for constructive and destructive interference.

even harmonics in a push–pull amplifier. However, since the second-harmonic content of the PNP HBT is much smaller than that of the NPN HBT, these two components only partially cancel, resulting in a significant second-harmonic component for the push–pull amplifier. Therefore the NPN HBT limits the second-order linearity of the push–pull amplifier. For NPN and PNP HBTs with perfectly matched linearity characteristics, the resulting push–pull amplifier should have no second-harmonic component.

Finally, the power-added efficiencies of the SHBTs and the push–pull circuit were compared at the various bias points. Note that the input power to the HBTs was not pushed to peak efficiency in order to avoid device burnout. Typical *measured* efficiencies were around 25% for the NPN HBT and 15% for the PNP HBT and the circuit. In the PNP HBT, parasitic collector and emitter resistances and low gain at microwave frequencies caused the reduced efficiency. This high dc power consumption of the PNP in turn dictated the low efficiency of the push–pull circuit. The measured efficiency characteristics of the push–pull amplifier were independent of bias, since the decrease in gain at Class AB bias points compensated for the reduced dc power consumed in the HBTs. Ideally, HBTs with more uniform gain at lower current levels would display increased efficiency at Class AB than at Class A.

3.7 SYSTEM CONSIDERATIONS

3.7.1 Applications

There is a large variety of applications for all three device types discussed. GaAs MESFETs are used, for example, in a wide range of analog–digital cellular applications. Digital cordless telephones also employ power MESFETs. The handset size, however, imposes a need for low voltage, single power supply, and high PAE. Biasing in MESFETs used in such applications is therefore arranged by employing a source resistor, which leads to the required negative gate bias. This approach results, however, in high-frequency gain degradation and PAE reduction. The alternative approach is HBTs, which can be operated using a single power supply and have a low knee voltage in their *I–V* characteristics. The change from analog to digital cellular phone systems using, for example, a $\pi/4$-shifted QPSK signal, imposes a requirement of much higher output power for even lower supply voltage. The adjacent channel leakage power should also be maintained low in this application with less than -48 dBc at 50 kHz off-center frequency. Performance within these specifications and with 1 W output power at 950 MHz using 1.5 V supply has been demonstrated with PHEMTs [45]. High PAE is also important for long battery life in wireless applications, and in satellite applications, where the available power is limited. The demand for high PAE devices in satellite systems becomes increasingly higher as phased arrays are introduced to produce multiple beam spots and improve communications performance.

InP-based HEMTs and other InP-based components such as PIN diodes are particularly suitable for millimeter-wave applications such as mobile satellite communications systems and collision avoidance systems (CASs).

HBTs are attractive at cellular radio frequencies for power amplification. This can range from the watt level for handsets to 150 W or more for base stations. The advantages offered are high gain per amplification stage and competitive if not better linearity and PAE. An important advantage of HBTs is that they offer good linearity properties without necessarily backing up the power, as is often done in FETs. This leads to good linearity properties combined with high PAE. On the other hand, the smaller operating voltage imposed by handset applications leads to difficulty in achieving high PAE. The V_{CE} offset voltage of the HBT and the collector–emitter resistance determines the on-state voltage, which limits the output signal. Reduction of V_{BE} is also often sought despite the fact that a high value can be useful for power down of an amplifier. The use of GaAs DHBTs and InP-based technology can be attractive for this purpose. High-speed, low-power applications using small-area devices also benefit from GaInP designs due to their large ΔE_v, which increases injection efficiency and thus enables higher base doping and speed, while the low surface recombination velocity of InGaP leads to reduced base leakage. The use of small emitter area HBTs is the most effective way for reducing power consumption while keeping the current density high. Although the latter guarantees high-speed performance, the small emitter size compensates the speed improvement by increased emitter resistance R_E. Graded emitter designs have been proposed for this purpose [46].

HBTs have also been introduced in fiber optic transmission systems operating at 10 Gbits/s rates, and higher speed systems are currently envisaged. Although the actual operation speeds can be satisfied using GaAs technology, InP HBTs are the preferred solution for future generations that are envisaged up to 80 Gbits/s.

3.7.2 GaAs-Based Complementary Push–Pull Amplifier

While GaAs-based technology is more mature than that of InP, power amplifiers for wireless communications can better benefit from many of the characteristics of InP-based HBTs. Some of the relevant HBT characteristics are presented in Table 3.7. Note that, while one PNP GaAs HBT demonstrated f_{max} of 66 GHz with an associated unilateral gain U of ~ 17 dB at 10 GHz, the power performance was not reported [47]. Similarly, the f_T and f_{max} were not reported for the only PNP GaAs HBTs with published power performance [48].

In general, the improved frequency performance of InP-based NPN HBTs over GaAs-based NPN HBTs produces more gain at microwave frequencies. The (typically) smaller offset voltage, lower contact resistance, and lower sheet resistance of the emitter cap and subcollector layers reduce the saturation "knee" voltage of InP-based HBTs. The smaller knee and turn-on voltages increase the amplifier efficiency and allow the use of low-voltage batteries. Moreover, with the change from analog to digital cellular phone systems with modulation such as π/4-shifted QPSK, higher output power is required with even lower supply voltages, therefore making InP HBTs more attractive than their GaAs counterparts. Finally, the higher thermal conductance of the InP substrate allows for more power dissipation in any given HBT design. While InP-based SHBTs do suffer from low breakdown voltages, this

TABLE 3.7. Typical Characteristics (Except f_T and F_{max}, Which Are the Best-Reported Characteristics) of the InP-Based and GaAs-Based NPN HBTs and PNP HBTs

Parameter	NPN HBTs		PNP HBTs	
	InP	GaAs	InP	GaAs
Highest f_T (GHz)	235	171	14	37
Highest f_{max} (GHz)	236	255	35	66
Offset voltage (V)	0.2	0.4	0.3	0.5
Turn-on voltage (V) @ 10^4 A/cm^2	0.75	1.4	0.83	1.6
Single HBT BV_{CE0} (V)	7	20	6	7
P_{out} (mW/μm^2) @ 10 GHz	1.37	3.0	0.49	0.63
Gain (dB) measured @ 10 GHz	11	14	10	8
Thermal conductivity of substrate (W/(cm·K))	0.7	0.5	0.7	0.5

limitation can be overcome in DHBTs with InP collectors. Also note that by using a graded emitter–base junction, which is achievable in both material systems but is much easier to employ in GaAs-based HBTs, both the offset and turn-on voltages can be reduced significantly.

InP-based HBTs offer an additional advantage for push–pull operation: the turn-on voltage V_{BE} is approximately 0.75 V, versus 1.4 V for GaAs-based HBTs. This smaller turn-on voltage reduces both crossover distortion and overall power consumption.

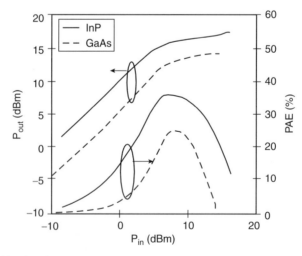

Figure 3.25. Simulated power output and PAE at 1.9 GHz for Class AB common-collector push-pull amplifiers using typical HBTs in either InP-based or GaAs-based technologies.

In order to more directly compare InP- to GaAs-based push–pull amplifiers, the Class AB common-collector amplifier from Section 3.6.2 was also simulated using large-signal HBT models derived from published characteristics of typical NPN and PNP AlGaAs/GaAs HBTs [47–49]. When using the same $V_{CC} = 6.2$ V for both circuits, the peak gain of the GaAs-based circuit (7.7 dB) was 2.6 dB less than the InP-based circuit (see Fig. 3.25), primarily due to the better frequency performance of the InP NPN HBT. Similarly, the larger turn-on voltag and knee voltage of the GaAs HBTs reduced the output power at 1 dB of gain compression to 12.7 dBm, which is 3.1 dB lower than for the InP push–pull circuit. Together with the larger turn-on voltage, the reduced P_{out} limited the GaAs-based PAE to 25%, as compared to the InP-based PAE of 36%. Of course, the P_{out} could be increased more than that of the InP-based amplifier by increasing V_{CC}, due to the higher breakdown voltage of the GaAs HBTs; however, the PAE would still be lower. While these results using typical HBTs do not necessarily reflect the full potential of each technology, they do indicate power advantages for InP-based HBTs over GaAs-based HBTs in push–pull amplifiers for applications with low-voltage power supplies, such as hand-held units.

3.8 CONCLUSIONS

The characteristics of high-speed InP-based NPN and PNP heterojunction bipolar transistors and the application of these HBTs to simple push–pull circuits have been discussed. Experimental characteristics are shown for NPN HBTs with f_T up to 100 GHz and f_{max} up to 130 GHz with $BV_{CE0} = 8.0$ V, which represents very good performance of InP-based SHBTs. Similarly, state-of-the-art PNP HBTs showed f_T up to 19 GHz and f_{max} up to 35 GHz with $BV_{ECO} = 6.8$ V. These NPN and PNP HBTs were combined in a simple push–pull amplifier that demonstrated power combination in the fundamental and cancellation in the second harmonic at 8 GHz.

Several key features of the HBT process enhance its performance by minimizing parasitic resistances and capacitances. For example, the base contacts can be made self-aligned to the emitters, which reduces the base access resistance. Similarly, the collector resistance can be reduced by collector contacts self-aligned to the base contacts. The trench etch under the microbridges reduces the extrinsic emitter–base and base–collector capacitances. By including a mesa etch using the base contact patterns as the etch mask, the extrinsic base area and hence C_{BC} are minimized. Finally, laterally undercutting the base contacts during the collector etch reduces C_{BC} even further. Under large-signal operation, NPN HBTs produced up to 1.37 mW/μm^2 of output power.

InP-based single and double HBTs each offer advantages for power amplifers. The higher BV_{CE0} primarily allows DHBTs to generate more output power than SHBTs, and the lower thermal resistance to the substrate due to the InP collector allows for lower and more uniform junction temperatures. The higher sturation velocity of the InP collector also allows for proportionally higher J_{Cmax} before the onset of the Kirck effect at a given collector doping. With emitter–base and base–collector compositional grading, the offset voltage and V_K are theoretically slightly smaller for DHBTs. The higher BV_{CE0} together with this low V_K allows slightly higher efficiencies for DHBTs and compared to SHBTs. Therefore DHBTs are the best choice for high-power applications. However, the spike in the conduction band of DHBTs may limit J_{Cmax}, may increase V_K, and may also limit the gain at low J_C, which can introduce additional nonlinearities to the output characteristics. Since SHBTs are also easier to design and fabricate than DHBTs, they may be more cost effective and a better choice for power amplifiers that only require moderate output power levels.

PNP HBTs exhibited record high-frequency performance for InP-based PNPs, with f_T of 19 GHz for a 350 Å base and f_{max} of 35 GHz for a 500 Å base. While inherently slower than their NPN counterparts, the low access and spreading resistance of the n-type base alleviates the penalty to f_{max}. Two-dimensional drift-diffusion simulations indicated that the current gain is limited by Auger recombination in the neutral base, and the high-frequency performance is limited by the hole transit time. Therefore the most significant performance improvement can be obtained using a very thin base (~ 300 Å) with low base doping and a built-in drift electric field to accelerate the holes toward the collector. Simulations also indicated that removing the emitter–base spacer should further increase the dc gain and high-frequency performance at the expense of increased turn-on voltage.

The use of a modest doping gradient in the base ($5 \rightarrow 1 \times 10^{18}$ cm^{-3}) of PNP HBTs demonstrated an increase in f_T by 16% over PNPs with uniform bases, and further improvements can be expected through the use of a compositional gradient in the base. Moving to an HBT layout with a self-aligned collector contact significantly reduced the collector resistance, resulting in a 0.18 V decrease in the knee voltage, a 7% higher f_T, and a 1.14 times increase in PAE. Power performance of the PNPs at 10 GHz was comparable to the NPN HBTs, providing up to 10 dB of gain, 0.49 mW/μm² of output power, and 24% power-added efficiency. While the experimental InP-based PNP results are encouraging, simulations of various PNP designs and reports of GaAs-based PNPs with f_{max} up to 66 GHz both suggest room for improvement. Analysis indicated that the primary performance limitations on these HBTs were the hole lifetime in the base, hole recombination in the emitter–base space-charge region (including in the spacer and at the surface), the hole transit time across the base and collector, and the parasitic resistances.

While the current PNP HBT characteristics are not ideal, they were sufficient to demonstrate a complementary push–pull amplifier operating at 8 GHz. Both simulations and experiments demonstrate an improvement to the linearity characteristics and maximum output power of NPN HBT amplifiers by including a PNP HBT in a push–pull configuration. The experimental results confirmed push–pull action with cancellation of the second harmonic and addition of the IM3 distortion when the fundamental recombined in-phase. However, the mismatch in the NPN and PNP HBT characteristics limited the linearity improvements. These results demonstrate that PNP HBTs are sufficiently advanced and have a practical application in high-frequency microwave electronics.

REFERENCES

1. B. Streetman, *Solid State Electronic Devices* 3rd ed., Prentice-Hall, Englewood Cliffs, NJ, 1990.
2. S. M. Sze (ed.), *Modern Semiconductor Device Physics*, Wiley, New York, 1998.
3. S. Datta, S. Shi, K. Roenker, M. Cahay, and W. Stanchina, "Simulation and design of InAlAs/InGaAS pnp heterojunction bipolar transistors," *IEEE Transactions on Electron Devices*, Vol. 45, 1634–1642 (1998).
4. J. Cowles, R. Metzger, A. Gutierrez-Aitken, A. Brown, D. Streit, A. Oki, T. Kim, and A. Doolittle, "Double heterojunction bipolar transistors with InP epitaxial layers grown by solid-source MBC," in *9th International Conference on Indium Phosphide and Related Materials*, pp. 548–550, 1997.
5. M. Das, "High-frequency performance limitations of millimeter-wave heterojunction bipolar transistors," *IEEE Transactions on Electron Devices*, Vol. 35, 604–614 (1988).
6. R. Muller and T. Kamins, *Device Electronics for Integrated Circuits*, 2nd ed., Wiley, New York, 1986.
7. B. Bayraktaroglu, R. Fitch, J. Barrette, R. Schere, L. Kehias, and C. Huang, "Design and fabrication of thermally-stable AlGaAs/GaAs microwave power HBTs," in *Proceedings IEEE/Cornell Conference on Advanced Concepts in High Speed Semiconductor Devices and Circuits*, pp. 83–92, 1993.
8. B. Bayraktarouglu, "HBT power devices and circuits," *Solid-State Electronics*, Vol. 41, 1657–1665 (1997).
9. K. Kobayashi, D. Umemoto, J. Velebir, A. Oki, and D. Streit, "Integrated complementary HBT microwave push–pull and Darlington amplifiers with P-N-P active loads," *IEEE Journal of Solid-State Circuits*, Vol. 28, No. 10, 1011–1017 (Oct. 1993).
10. P. Enquist, D. Slater, and J. Swart, "Complementary AlGaAs/GaAS HBT I^2L (CHI^2L) technology," *IEEE Electron Device Letters*, Vol. 13, 180–182 (1992).
11. W. Liu, D. Hill, D. Costa, and J. Harris, "High-performance microwave AlGaAs-InGaAS pnp HBT with high-dc current gain," *IEEE Microwave and Guided Wave Letters*, Vol. 2, 331–333 (1992).
12. L. M. Lunardi, S. Chandrasekhar, and R. A. Hamm, "High-speed, high-current-gain P-N-P InP/InGaAs heterojunction bipolar transistors," *IEEE Electron Device Letters*, Vol. 14, 19–21 (1993).
13. H. Masuda, K. Ouchi, A. Terano, H. Suzuki, K. Watanabe, T. Oka, H. Matsubara, and T. Tanoue, "Device technology of InP/InGaAs HBTs for 40-Gb/s optical transmission application," *GaAs IC Symposium Technical Digest*, 139–142 (1997).
14. H.-F. Chau and Y.-C. Kao, "High f_{max} InAlAs/InGaAs heterojunction bipolar transistors," *IEEE IEDM Technical Digest*, 783–787 (1993).
15. S. Yamahat, K. Kurishima, H. Nakajima, T. Kobayashi, and Y. Matsuoka, "Ultra-high f_{max} and f_r InP/InGaAs double-heterojunction bipolar transistors with step-graded InGaAsP collector," in *GaAS IC Symposium Technical Digest*, pp. 345–348, 1994.
16. Y. Matsuoka, H. Nakajima, K. Kurishima, T. Kobayashi, M. Yoneyama, and E. Sano, "Novel In/InGaAs double-heterojunction bipolar transistors suitable for high-speed IC's and OEIC's," in *6th International Conference on InP and Related Materials*, pp. 555–558, 1994.

17. C. Nguyen, T. Liu, M. Chen, H.-C. Sun, and D. Rensch, "AlInAs/GaInAs/InP double heterojunction bipolar transistor with a novel base–collector design for power applications," *IEEE Electron Device Letters*, Vol. 17, 133–135 (1996).
18. S. M. Sze, *Physics of Semiconductor Devices*, 2nd ed., Wiley, New York, 1981.
19. A. Neviani, G. Meneghesso, E. Zanoni, M. Hafizi, and C. Canali, "Positive temperature dependence of the electron impact ionization coefficient in $In_{0.53}Ga_{0.47}As/InP$ HBT's," *IEEE Electron Device Letters*, Vol. 18, 619–621 (1997).
20. B. Hong, J.-I. Song, C. Plamstrøm, B. Van der Gaag, K.-B. Chough, and J. Hayes, "DC, RF, and noise characteristics of carbon-doped base InP/InGaAs heterojunction bipolar transistors," *IEEE Transactions on Electron Devices*, Vol. 41, 19–25 (1994).
21. H.-F. Chau, D. Pavlidis, G.-I. Ng, K. Tomizawa, D. Baker, C. Meaton, and J. Tothill, "Improved breakdown-speed tradeoff of InP/InGaAs single heterojunction bipolar transistor using a novel p^-n^- collector structure," in *5th International Conference on InP and Related Materials*, pp. 25–28, 1993.
22. J. Hu, D. Pavlidis, and K. Tomizawa, "Monte Carlo studies of the effect of emitter junction grading on the electron transport in InAlAs/InGaAs heterojuction bipolar transistors," *IEEE Transactions on Electron Devices*, Vol. 39, 1273–1281 (1992).
23. D. Sawdai, K. Yang, S. Hsu, D. Pavlidis, and G. Haddad, "Power performance of InP-based single- and double-heterojunction bipolar transistors," *IEEE Transations on Microwave Theory and Techniques*, Vol. 47 (Aug. 1999).
24. L. Tran, D. Streit, K. Kobayashi, J. Velebir, S. Bui, and A. Oki, "InAlAs/InGaAs HBT with exponentially graded base doping and graded InGaAlAs emitter–base junction," in *4th International Conference on Indium Phosphide and Related Materials*, pp. 438–441, 1992.
25. D. Umemoto, J. Velebir, K. Kobayashi, A. Oki, and D. Streit, "Integrated *npn/pnp* GaAs/AlGaAs HBTs grown by selective MBE," *Electronics Letters*, Vol. 27, 1517–1518 (Aug. 15, 1991).
26. Y. Tateno, H. Yamada, S. Ohara, S. Kato, H. Ohnishi, T. Fujii, and J. Fukaya, "3.5 V, 1 W High efficiency AlGaAs/GaAs HBTs with collector launcher structure," in *1994 International Electron Devices Meeting*, pp. 195–198, 1994.
27. M. Tutt, D. Pavlidis, A. Khatibzadeh, and B. Bayraktaroglu, "Investigation of HBT oscillator noise through $1/f$ noise upconversion studies," in *IEEE MTT-S International Microwave Symposium Digest*, pp. 727–730, June 1992.
28. G. Gaquière, B. Bonte, D. Théron, Y. Crosnier, P. Arséne-Henri, and T. Pacou, "Breakdown analysis of an asymmetrical double recessed power MESFET," *IEEE Transaction on Electron Devices*, Vol. 42, 209–214 (Feb. 1995).
29. T. Jenkins, L. Kehias, P. Parikh, J. Ibbetson, U. Mishra, D. Docter, M. Le, K. Kiziloglu, D. Grider, and J. Pusl, "Record power-added efficiency using GaAs on insulator MESFET technology," in *GaAs IC Symposium Technical Digest*, pp. 259–262, 1998.
30. S. Goto, K. Fujii, H. Morishige, S. Suzuki, S. Sakamoto, N. Yoshida, N. Tanino, and K. Sato, "A 100W S-band AlGaAs/GaAs Hetero-structure FET for Base stations of wireless personal communication," in *GaAs IC Symposium Technical Digest*, pp. 77–80, Nov. 1998.
31. P. M. Smith, C. T. Creamer, W. F. Kopp, D. W. Ferguson, P. Ho, and J. R. Wilhite, "A high-power Q-band PHEMT for communication terminal applications," in *IEEE MTT-S International Microwave Symposium Digest*, pp. 809–812, 1994.

32. S. Hsu, B. Bayraktaroglu, and D. Pavlidis, "Comparison of conventional and thermally-stable cascode (TSC) AlGaAs/GaAs HBTs for microwave power applications," in *Proceedings of the Topical Workshop on Heterostructure Microelectronics for Information Systems Applications (TWHM-ISA '98)*, Paper S3-6, pp. 36–37, Aug./Sept. 1998.

33. S. Tanaka, S. Murakami, Y. Amamiya, H. Shimawaki, N. Furuhata, N. Goto, K. Honjo, Y. Ishida, Y. Saito, K. Yamamoto, M. Yajima, R. Temino, and Y. Hisada, "High-power, high-efficiency cell design for 26 GHz HBT power amplifier," in *IEEE MTT-S International Microwave Symposium Digest*, pp. 843–846, 1996.

34. C. Nguyen, T. Liu, M. Chen, and R. Virk, "Bandgap engineered InP-based power double heterojunction bipolar transistors," in *9th International Conference on Indium Phosphide and Related Materials*, pp. 15–19, 1997.

35. W. Stanchina, R. Metzger, J. M. Pierce, J. Jensen, L. McCray, R. Wong-Quen, and F. Williams, "Monolithic fabrication of NPN and PNP AlInAs/GaInAs HBTs," in *5th International Conference on Indium Phosphide and Related Materials*, pp. 569–571, 1993.

36. H. Tserng, D. Hill, and T. Kim, "A 0.5-W complementary AlGaAs-GaAs HBT push–pull amplifier at 10 GHz," *IEEE MW & Guided Wave Letters*, Vol. 3, No. 2, 45–47 (Feb. 1993).

37. D. Hill, H. Tserng, and T. Kim, "65/90 GHz Complementary HBT technology," *Electronics Letters*, Vol. 30, 597–598 (Mar. 31, 1994).

38. D. Slater Jr., P. Enquist, and J. Hutchby, "Harmonic cancellation in monolithic AlGaAs/GaAs npn/pnp HBT push–pull pairs,, " in *Proceedings IEEE/Cornell Conference on Advanced Concepts in High Speed Semiconductor Devices and Circuits*, pp. 305–314, 1991.

39. P.-C. Hsu, C. Nguyen, and M. Kintis, "Uniplanar broad-band push–pull FET amplifiers," *IEEE Transactions on Microwave Theory and Techniques*, Vol. 45, 2150–2152 (1997).

40. H. Henry, R. Freitag, R. Brooks, A. Burk, and M. Murphy, "A compact 3W X-band GaAs MMIC amplifier based on a novel multi-push–pull circuit concept," in *GaAs IC Symposium Technical Digest*, pp. 327–330, 1991.

41. S. Toyoda, "Push–pull power amplifiers in the X-band," in *IEEE Microwave Theory and Techniques Symposium*, pp. 1433–1436, 1997.

42. M. Tsai, "A wide-band push–pull amplifier upgrades IP2," in *IEEE Microwave Theory and Techniques Symposium*, pp. 511–514, 1990.

43. D. Sawdai and D. Pavlidis, "InP-based complementary HBT amplifiers for use in communication systems," *Journal of Solid State Electronics*, Vol. 43, 1507–1512 (Aug. 1999).

44. P. Chen, Y. Hsin, and P. Asbeck, "Saturation charge storage measurements in GaInP/GaAs/GaAs and GaInP/GaAs/GaInP HBTs," in *24th International Symposium on Compound Semiconductors*, pp. 443–446, 1997.

45. N. Iwata, K. Yamaguchi, and M. Kuzuhara, "Double-doped power heterojunction FET for 1.5 V digital cellular applications," in *Topical Workshop on Heterostructure Microelectronics*, Sapporo, Japan, pp. 66–67, Aug. 1996.

46. T. Niwa, Y. Amamiya, M. Mamada, H. Shimawaki, H. Sakaki, J. C. Woo, N. Yokoyama, and Y. Hirayama, "High-f_T AlGaAs/InGaAs HBTs with reduced emitter resistance for

low-power-consumption, high-speed ICs," in *25th International Symposium on Compound Semiconductors*, Nara, Japan, Oct. 1998.

47. D. Slater, P. Enquist, J. Hutchby, A. Morris, and R. Trew, "pnp HBT with 66 GHz f_{max}," *IEEE Electron Device Letters*, Vol. 15, 91–93 (Mar. 1994).

48. D. Hill, T. S. Kim, and H. Q. Tserng, "X-band power AlGaAs/InGaAs p-n-p HBT's," *IEEE Electron Device Letters*, Vol. 14, 185–187 (Apr. 1993).

49. D. Hill, T. Kill, and H. Tserng, "AlGaAs/GaAs pnp HBT with 54 GHz f_{max} and application to high-performance complementary technology," in *51st Annul Device Research Conference*, Paper IIA-6, 1993.

4

Si/SiGe HBT TECHNOLOGY FOR LOW-POWER MOBILE COMMUNICATIONS SYSTEM APPLICATIONS

LARRY LARSON
Department of Electrical and Computer Engineering, University of California—San Diego

M. FRANK CHANG
Department of Electrical Engineering, University of California—Los Angeles

What is claimed is:
1) ...
2) A device as set forth in claim 1 in which one of the separated zones is of a semiconductor material having a wider energy gap than that of the material in the other zones.

— Claim 2 of U.S. Patent 2 569 347 to W. Shockley
Filed 26 June 1948
Issued 25 September 1951
Expired 24 September 1968.

4.1 INTRODUCTION

As the above patent from William Shockley in 1948 demonstrates, the concept of heterojunction techniques for improving the performance of semiconductor devices dates back to the earliest days of microelectronics. Although heterojunction technology has been a common feature of III–V-based electronic and optoelectronic devices during the past 25 years, the silicon/silicon germanium heterojunction bipolar transistor (Si/SiGe HBT) is one of the first of these class of devices to be combined with high-volume, low-cost silicon manufacturing.

Silicon integrated circuit technology is an ubiquitous feature of the modern electronics industry, constituting a greater than $200B/yr technological "tsunami." Silicon is by far the most common semiconductor material in use today. Why is this? Silicon has numerous practical advantages as a semiconductor material, including (1) ease of growth of large, low-cost, defect-free crystals; (2) availability of a stable high-quality dielectric—SiO_2—that can be grown nearly pinhole free with a resulting low surface-state density; (3) ease of doping and fabrication of ohmic contacts; and (4) excellent mechanical properties including strength and high thermal conductivity. It is these factors, and many others, that have made silicon technology so dominant.

However, until recently, the use of silicon microelectronics technology for applications in the microwave range—defined as frequencies above 300 MHz—was mostly hypothetical. This was because the comparatively low intrinsic speed of electrons in silicon, which is related to their effective mass and saturated drift velocity, was inferior to competing III–V technologies of GaAs and InP. As a result, until recently III–V technology was considered the most promising technology for applications in the lower microwave frequency range—precisely those areas where the greatest commercial interest exists for wireless mobile communications. The popularity of III–V devices at these frequencies, despite the well-known limitations of III–V technology including higher defect densities, poor thermal conductivity, and lack of a passivating native oxide, is a testimony to the premium placed on achieving the best possible performance in a low-power mobile wireless application.

Despite its intrinsic material disadvantages from a speed perspective, silicon technology has exhibited dramatic improvements recently due to continuing advances in lithography and in materials and processing technology. In fact, several microprocessors have recently been announced with clock speeds literally in the microwave range [1]! One of the most important advances in silicon technology—improving both the high-frequency and dc characteristics of the transistor—has been the addition of alloys of silicon and germanium to the base of a bipolar transistor, resulting in the Si/SiGe HBT. This device emerged from a variety of industrial and university research laboratories in the early 1990s and is now in high-volume commercial production for a wide range of high-frequency communications applications. The question then becomes: Can a silicon-based technology like Si/SiGe HBT exhibit performance compatible with the demands of the low-power wireless communications marketplace, and does it retain the low cost and manufacturability of its silicon pedigree?

This chapter will attempt to answer that question by summarizing the progress and prospects for this technology in the area of low-power mobile communications systems, with particular emphasis on circuit and device performance. It is organized in the following manner. Section 4.2 outlines the advantages and disadvantages of silicon technology for applications to high-frequency low-power communications circuits. Section 4.3 summarizes the key device design aspects of SiGe devices, particularly with respect to high-frequency applications. Section 4.4 provides an assessment of passive component technology in Si/SiGe technology. Section 4.5 provides a direct comparison of the performance of silicon-based HBTs to those based on III–V technologies. Finally, Section 4.6 provides an overview of reported circuit performance in Si/SiGe HBT technology, with emphasis on low-power communications applications.

4.2 SILICON TECHNOLOGY FOR ULTRA-HIGH-SPEED COMMUNICATIONS APPLICATIONS

The widespread availability of wireless communications, with its resulting "take-off" in terms of volume production, has been brought about by a combination of advances in integrated circuit technology, radio-frequency (RF) components, digital communications, and networking techniques. Of all these factors, the first two have been most instrumental in bringing the cost of the technology within reach of the average consumer. A radio transceiver that would have required an entire rack of equipment and several hundred watts of power fifteen years ago can now comfortably fit in the palm of a person's hand and operate from a single-cell battery for several hours.

Si/SiGe technology promises to satisfy the simultaneous requirements of outstanding high-frequency performance and low-cost silicon manufacturing. The silicon pedigree of Si/SiGe technology will provide for a straightforward path for "up-integration" of standard radio-frequency functions along with analog-to-digital conversion and digital signal processing onto a single high-performance integrated circuit. This migration of multiple functions onto a single die will have some profound implications for wireless transceiver architectures.

The high-frequency systems applications of a technology capable of achieving intrinsic transistor speeds in excess of 100 GHz are numerous and extend well into the millimeter-wave frequency region. In addition to the well-known cellular telephone and wireless local area network (LAN) market, they include applications such as automobile collision warning radar, wireless distribution of cable television, and millimeter-wave point-to-point radios.

Gray and Meyer [2] point out the multiple advantages of a highly integrated adaptive transceiver for future wireless communications systems. There is no fundamental reason why a single transceiver architecture could not in principle accommodate all of the multiple worldwide wireless standards, with their differing frequency bands and modulation and demodulation schemes. This would require an architecture where most of the intermediate frequency (IF) signal processing is performed in the digital domain, as well as choice of carrier frequency, modulation formats, carrier recovery, symbol timing recovery, power control, dc offset removal, and so on. A conceptual block diagram of such a highly integrated architecture is shown in Figure 4.1 [2]. In order to make this vision a reality, the semiconductor

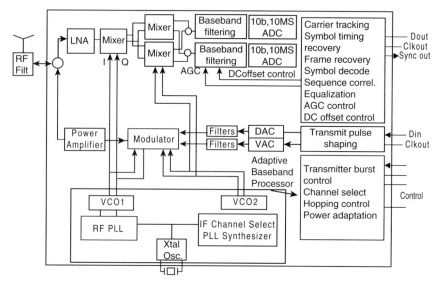

Figure 4.1. Block diagram of a possible future highly integrated adaptive transceiver [2]. Si/SiGe HBT BiCMOS technology is an ideal candidate for implementation of this architecture.

technology used to manufacture the devices must have the highest performance as well as the ability to integrate both digital and analog functions in a common substrate. Clearly, a highly integrated BiCMOS technology would be essential to the realization of this approach, and the use of Si/SiGe HBTs will allow the performance of such a unit to be comparable to that of more specialized implementations.

One example of such a highly integrated Si/SiGe HBT/BiCMOS integrated circuit technology from IBM is shown in Figure 4.2 [3]. In this case, the technology has been specifically designed to be as compatible with existing CMOS technology as possible, as well as including all of the typical devices required for RF and analog applications such as high-quality resistors and capacitors. This requires that the SiGe films be thermodynamically stable—so that the usual extremes of temperature associated with MOS device processing do not degrade the performance of the bipolar device. It also requires that the metallization and process tooling be compatible with existing implementations of CMOS technology.

The HBT has a self-aligned polysilicon emitter and is isolated using a combination of deep and shallow trench isolation—a technique borrowed from dynamic RAM manufacturing. Silicided contacts are used throughout to minimize extrinsic resistance in the contacts. The base layer is grown using an ultrahigh vacuum chemical vapor deposition (UHV/CVD) technique, and typical base widths are in the 500–1000 Å range, with peak germanium profiles from 8% to 15%. These devices are usually far less aggressive in terms of their base doping and thickness than comparable III–V-based HBTs, in order to maintain stability under extremes of thermal cycling later in the process. Although growth temperatures are in the 600 °C range, subsequent annealing of implants can occur at temperatures of up to 1000 °C.

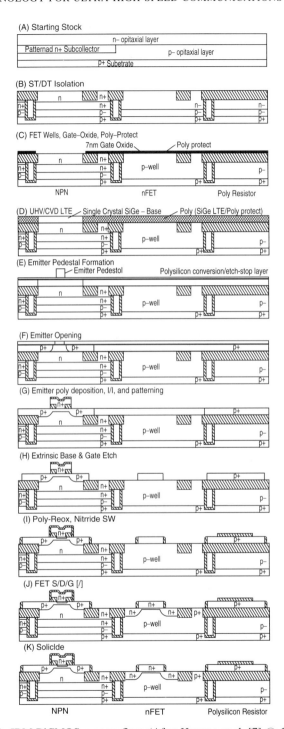

Figure 4.2. IBM BiCMOS process flow. (After Hareme et al. [7] © 1995 IEEE.)

TABLE 4.1. Elements in IBM's SiGe BiCMOS Process Technology [14]

Element	Characteristics
Standard SiGe HBT	47 GHz f_T at BV_{CE0} = 3.3 V
High-breakdown SiGe HBT	28 GHz f_T at BV_{CE0} = 5.3 V
Si CMOS	0.36 μm L_{eff} for 2.5 V V_{DD}
Gated lateral pnp	1.0 GHz f_T
Polysilicon resistor	342 Ω/□
Ion-implanted resistor	1600 Ω/□
Thin-oxide decoupling capacitor	1.5 fF/μm²
MIM precision capacitor	0.7 fF/μm²
Inductor (six-turn)	10 nH with Q-10 at 1.0 GHz
Schottky barrier diode	213 mV at 100 μA

Multilevel metallization is accomplished using chemical–mechanical polishing (CMP) of the tungsten via layers, with AlCu (and recently Cu alone [4]) metallization. The use of multilevel metallization, with a final thick level of metal, results in enhanced performance for important passive components such as inductors and capacitors. In essence, a standard CMOS technology has been upgraded with the addition of a high-performance HBT. A summary of the device characteristics of the IBM HBT/BiCMOS process are included in Table 4.1.

4.3 DEVICE DESIGN ASPECTS OF Si/SiGe HBTs FOR HIGH-SPEED OPERATION

The concept of "bandgap engineering" to improve the performance of semiconductor devices dates back to the earliest days of the bipolar transistor. In the case of Si/SiGe, the addition of germanium to the base of the transistor results in a variety of improvements to the performance of the device, resulting from the narrower bandgap of the alloy of silicon and germanium. Silicon has a bandgap of approximately 1.1 eV, and germanium has a bandgap of roughly 0.7 eV, so an alloy of the two is expected to have an intermediate value, controlled by the germanium concentration in the predominantly silicon film. An approximate value for the reduction of the bandgap energy as a function of germanium concentration and doping concentration is given by [5]

$$\Delta E_G(\text{meV}) \approx 28.6 + 27.4 \log_{10} \frac{N_A}{10^{18}} + 688x, \quad (4.1)$$

where x is the atomic fraction of germanium in the SiGe layer and N_A is the doping concentration.

Despite the potential improvement in performance, the synthesis of defect-free films of SiGe originally proved quite difficult, and high-performance semiconductor films were not developed in this material system until the mid-1980s. Part of the limitation arises from the fundamental mismatch in lattice constants between silicon and germanium (approximately 4%), so a SiGe film grown on an underlying silicon substrate exhibits compressive strain. Ideally, the concentration of Ge in the base layer would be quite high, but the compressive strain limits the maximum achievable film thickness. In order to maintain a defect-free film, the thickness of this layer must be strictly limited in order to maintain stability under subsequent thermal cycling. This limit is known as the Matthews and Blakeslee limit, and a curve of the limits of film thickness as a function of film strain appears in Figure 4.3 [6]. It is clear that even small amounts of germanium can quickly lead to significant strain in the resulting epitaxial film. Nevertheless, these small amounts of added germanium can lead to significant improvements in device performance.

The SiGe HBT offers three key advantages over conventional homojunction bipolar transistors: (1) a reduction in base transit time due to an enhanced built-in electric field in the base region; (2) an increase in transistor current gain, which allows for an increase in base doping and a decrease in base resistance; and (3) an increase in the Early voltage at a given cutoff frequency. A band diagram comparing heterojunction and homojunction transistor behaviors is shown in Figure 4.4. Note that the exact profile of the germanium concentration in the base affects the performance of the transistor in several differing ways, including current gain, speed, and output conductance. The introduction of germanium into the base offers an additional degree of freedom in the design of the device, improving the trade-offs between device scaling and speed that characterize homojunction bipolar transistor performance.

The outstanding high-frequency performance of Si/SiGe HBT technology, resulting from the improved base transit time, has been well established by a variety of groups [7,8]. In many applications, this speed performance advantage can be "traded-off" in a very satisfactory way for dramatically reduced power dissipation.

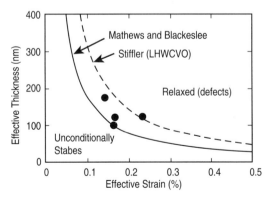

Figure 4.3. Effective thickness versus effective strain. The dashed curve is based on theoretical calculations by Stiffler et al. (*J. Appl Phys.*, p. 1416, 1991) using a methodology originally developed by Matthews and Blakeslee (solid curve). (After Cressler [14] © 1998 IEEE.)

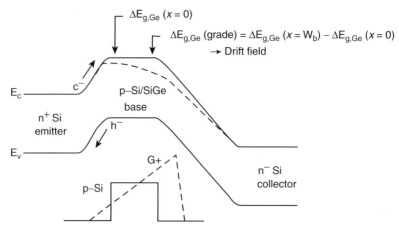

Figure 4.4. Si and SiGe HBT band diagram. (After Harame et al. [7] © 1995 IEEE.)

It is at these low power levels where Si/SiGe HBT technology has a distinct advantage compared to Si BJT or CMOS technology. The germanium content within the base of the SiGe HBT leads to a device of superior performance compared with a similarly structured silicon-only epitaxial-base transistor. In addition, the improvement in the performance of bipolar transistors must be seen in its historical perspective, where several complementary innovations lead to an overall dramatic improvement in performance. These innovations include the development of polysilicon emitters, epitaxial technology for growth of thin, heavily doped base layers, and trench isolation techniques that minimize parasitic capacitance [7].

In most applications of RF and communications circuits, the intrinsic speed of the transistor determines its performance for amplifier applications. The most commonly cited high-frequency figure of merit for a transistor is its unity current-gain cutoff frequency—or f_T—which is defined as

$$f_T = \frac{1}{2\pi}\left(\frac{1}{g_m}(C_{be} + C_{bc}) + \tau_e + \tau_b + \tau_c\right), \quad (4.2)$$

where g_m is the transconductance, C_{be} and C_{bc} are the base–emitter and base–collector junction capacitances, and τ_e, τ_b, and τ_c are the emitter, base, and collector delays, respectively. The transistor f_T is an excellent gauge of the "intrinsic" speed of the transistor and is not dramatically affected by the parasitic resistances used to connect to the device.

The high-frequency performance of Si/SiGe HBTs was demonstrated early in the development of the technology, with dramatic improvements in transistor f_T resulting from the improved transport properties in the base region. A particularly dramatic example of this improvement was demonstrated by Meyerson [9], where the peak reported f_T of silicon bipolar devices nearly doubled in the late 1980s as a

DEVICE DESIGN ASPECTS OF Si/SiGe HBTs FOR HIGH-SPEED OPERATION 133

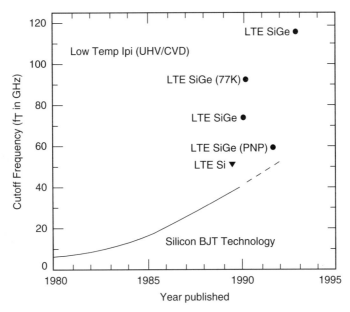

Figure 4.5. f_T trend chart for silicon-based bipolar transistor performance showing dramatic jump in performance with the advent of SiGe technology [9].

result of the fabrication of high-performance SiGe HBTs by a variety of laboratories. This dramatic increase is shown graphically in Figure 4.5 [9].

In a conventional bipolar junction transistor, one of the major limiting effects on device speed is the base transit time τ_b. The use of a graded germanium profile within the base—starting with a low germanium concentration at the base–emitter junction and ending with a high germanium concentration at the base–collector junction—results in a built-in electric field in the base region, reducing the base transit time considerably. The resulting reduction in base transit time is given by [10]

$$\frac{\tau_{b,\text{SiGe}}}{\tau_{b,\text{Si}}} = \frac{2}{\eta}\left(\frac{kT}{\Delta E_{g,\text{Base}}}\right) \times \left(1 - \frac{1 - e^{\Delta E_{g,\text{Base}}/kT}}{e^{\Delta E_{g,\text{Base}}/kT}}\right), \quad (4.3)$$

where $\Delta E_{g,\text{Base}}$ is the change in the bandgap across the base region, and η is the ratio of the electron diffusion constants (D_n) in the SiGe and Si base layers. This ratio is less than unity, so an increase in the grading of the energy bandgap across the base can lead to an improved base transit time.

Although it is generally regarded as a useful intrinsic figure of merit, the f_T can be somewhat misleading for high-frequency applications because it mostly ignores the effect of parasitic resistances. The use of bandgap engineering on the base region also improves the high-frequency performance of the transistor through the improvement in the unity power gain frequency—f_{max}. This is the frequency where the maximum power gain available from the device is unity, and is approximated by

$$f_{\max} \approx \sqrt{\frac{f_T}{8\pi C_{bc} r_b}}, \qquad (4.4)$$

where r_b is the series base resistance. The quantity $f_{\max}$ is a useful figure of merit because it sets an upper limit on the frequency where any power whatsoever can be extracted from the transistor. In contrast, the f_T of the device only measures the current gain.

This result illustrates the importance of base resistance in improving the high-frequency performance of the bipolar transistor. With a homojunction transistor, a desirable reduction in the base resistance inevitably leads to an undesirable reduction in transistor current gain, or β (I_C/I_B). This can be seen by examining the effect of base doping on current flow in the device. The current density in the Si/SiGe HBT can be approximated by [11]

$$J_{C,\text{SiGe}} \approx \frac{q D_{nb} n_{io}^2}{W_b N_b} \left(e^{q V_{be}/kT} - 1 \right) \lambda \eta \left(\frac{(\Delta E_{g,0}/kT) e^{\Delta E_{g,0}/kT}}{1 - e^{-\Delta E_{g,0}/kT}} \right), \qquad (4.5)$$

where W_b is the width of the base region, N_b is the doping density in the base, D_{nb} is the diffusion constant of electrons in the base region, n_{io} is the intrinsic carrier concentration in silicon, λ is the ratio of the density-of-states product in SiGe to Si ($N_C N_v(\text{SiGe})/N_C N_v(\text{Si})$) and $\Delta E_{g,0}$ is the change in bandgap energies at the emitter–base junction.

Note that the increase in base doping leads to a reduction in injected electron current at the base–emitter junction, but no change in the injected hole current from the base into the emitter. This lowers the current gain dramatically in a silicon homojunction transistor, but the use of a narrower bandgap base at the base–emitter junction dramatically reduces this effect, and the base resistance can be reduced without resorting to unacceptably low current gain. In fact, the ratio of current gains in the homojunction and heterojunction cases is given by [7]

$$\frac{\beta_{\text{SiGe}}}{\beta_{\text{Si}}} = \frac{J_{C,\text{SiGe}}}{J_{C,\text{Si}}} = \lambda \eta \left[\frac{(\Delta E_{g,0}/kT) e^{\Delta E_{g,0}/kT}}{1 - e^{-\Delta E_{g,0}/kT}} \right]. \qquad (4.6)$$

This ratio is considerably larger than unity, so even a small amount of bandgap reduction in the base can lead to a substantial improvement in the current gain. At the same time, this allows the base to be more heavily doped, reducing base resistance and improving $f_{\max}$ considerably. The base–emitter junction energy gap can also be specifically designed so that the β of the device exhibits a nearly zero temperature coefficient [12]—a distinct advantage for analog circuits and some power amplifier applications.

A good example of the improvements that can be achieved with the use of heterojunction technology can be seen from the transistor data presented in Figure 4.6, comparing transistor f_T and f_{max} as a function of collector current for epi-base SiGe HBTs and Si BJTs of *comparable base resistance* (approximately 12kΩ/□) [13]. The electrical characteristics of the two device types are compared in Table 4.2. In this case, the peak f_T for the Si BJT and SiGe HBT are 38 and 50 GHz, respectively. The peak f_{max} values of the Si BJT and SiGe HBT are 54 and 66 GHz, respectively. The epi-base SiGe HBT achieves higher f_T and f_{max} values than the epi-base Si BJT over its entire range of operation. In addition, for a given required f_T

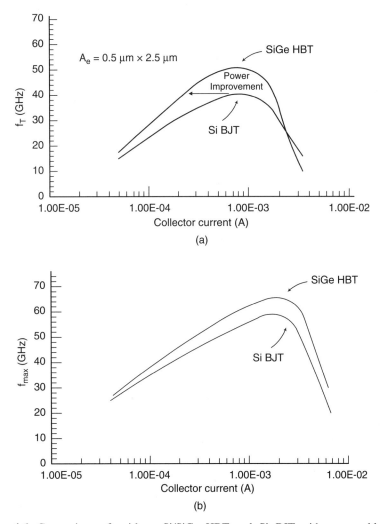

Figure 4.6. Comparison of epi-base Si/SiGe HBT and Si BJT with comparable base resistivity and geometry. (a) f_T versus collector current and (b) f_{max} versus collector current.

TABLE 4.2. Electrical Characteristics of Wafers Used to Compare Si BJT and SiGe HBT Performance

Type	Parameter ($A_E = 0.5 \times 2.5$ μm^2) Si BJT	SiGe HBT
R_{BI} (Ω/□)	12,121	12,099
β @ $V_{CE} = 0.72$ V	48	89
V_A (V)	20.3	33
β V_A (V)	974	2937
BV_{CEO} (V)	3.5	3.4
BV_{CBO} (V)	9.6	11
BV_{EBO} (V)	4.2	4.2
J_C @ 300 K (A/cm^2)	7.2×10^{-11}	1.5×10^{-10}
J_{OE} @ 300 K (A/cm^2)	2.0×10^{-12}	1.9×10^{-12}
R_E (Ω)	19.1	19.8
f_T (GHz)	38	50
f_{max} (GHz)	54	66
Low bias r_b (Ω)	130	130

or f_{max}, the SiGe HBT requires roughly one-third the collector current of an "equivalent" Si BJT for equivalently sized devices, dramatically lowering the power requirements in those circuits that are required to operate at very high frequencies.

Several other aspects of heterojunction technology can be used to improve the performance of the bipolar transistor. Specifically, the Early voltage of the transistor can be improved through the use of tailored germanium in the base, which reduces the output conductance of the transistor considerably. The output conductance is a measure of how much the neutral base of the device can be depleted with reverse bias on the collector–base junction, resulting in a rise in collector current through an effective reduction of base width. The Early voltage is defined as

$$V_A \approx \frac{J_c}{(dJ_c/dV_{cb})|_{V_{BE}}} = \frac{J_c}{(dJ_c/dW_b) \times (dW_b/dV_{cb})|_{V_{BE}}}, \quad (4.7)$$

which illustrates the importance of minimizing the base width variation due to changes in the collector–base voltage. An approximate expression for the improvement in Early voltage as a function of the device parameters is given by [14]

$$\frac{V_{A,SiGe}}{V_{A,Si}} = e^{\Delta E_{g,Base}/kT} \times \left(\frac{1 - e^{\Delta E_{g,Base}/kT}}{e^{\Delta E_{g,Base}/kT}} \right), \quad (4.8)$$

which is greater than unity, indicating the substantial improvement in Early voltage achievable with Si/SiGe HBT technology.

The resulting collector–emitter or collector–base breakdown voltage can also be improved—albeit indirectly—through the use of heterojunction technology. A high collector–base junction breakdown voltage (BV_{CBO}) is desirable for use in high-power/high-linearity applications. This parameter is the collector–base breakdown voltage measured with the base short-circuited and the emitter open-circuited. The breakdown voltage is limited by the collector doping and width, because the base is highly doped. In most cases, avalanche breakdown is the dominant phenomenon limiting the voltage across the junction. In this case, the electric field accelerates carriers with sufficient energy to the point where additional carriers are created, where collisions occur during transit. Electrons and holes generated in the base–collector space-charge region drift to the collector and base under the influence of the electric field. This effect is illustrated in Figure 4.7. A thick, relatively lightly doped collector increases the breakdown voltage at the expense of collector depletion transit time (τ_c). This increase in the collector transit time can degrade the f_T considerably, but the use of germanium in the base can recover some of the lost speed of the device, resulting in a very desirable speed versus breakdown voltage trade-off. In addition, the lightly doped collector minimizes the collector–base capacitance, further enhancing the f_{max} of the device.

In fact, for most circuit applications, the collector–emitter breakdown voltage (BV_{CEO}) is a more useful figure of merit. This breakdown voltage is measured where the emitter is short-circuited and the base is open-circuited and is often considerably lower than the collector–base breakdown voltage. This reduction in breakdown voltage can be understood by observing that the holes created by the avalanche process in the collector–base space-charge region drift into the quasineutral base. If the base current is zero—since it is open-circuited—then many of these holes are injected into the emitter, and β electrons are then injected back into the base. In other words, the intrinsic current gain of the transistor lowers the breakdown

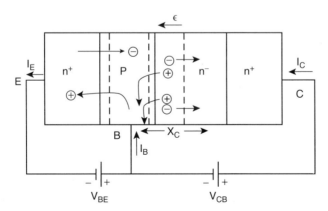

Figure 4.7. Electron and hole current flow in avalanche breakdown. (After J. S. Yuan, *SiGe, GaAs, and InP Heterojunction Bipolar Transistors*, Wiley Interscience, New York, 1999, p. 286.)

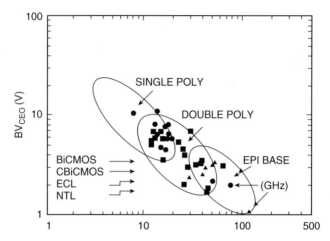

Figure 4.8. Measured breakdown voltage versus cutoff frequency (f_T in GHz) from a variety of recently reported structures. (After Nakamura and Nishizawa [15] © 1995 IEEE.)

voltage of the device. As a result, the collector–emitter breakdown voltage of the device is given by

$$BV_{CE0} = \frac{BV_{CB0}}{\beta^{1/n}}, \qquad (4.9)$$

where n is an empirical constant between 2 and 6. This illustrates one of the disadvantages of an *excessively high* current gain for a bipolar transistor if breakdown voltage is a major consideration. Figure 4.8 plots the measured BV_{CE0} as a function of f_T for a variety of scaled silicon bipolar transistors [15], and the effect of device scaling on breakdown voltage is clear.

An example of this approach appears in Greenberg et al. [16], where a 47 GHz f_T, 65 GHz f_{max} Si/SiGe HBT device with a 2.8V BV_{CE0} was modified through a change in the collector doping profile to exhibit a 6 V BV_{CE0}. The resulting f_T of the device degraded from 47 to 26 GHz, but the f_{max} remained virtually unchanged. As a result, the gain of the device remained well over 20 dB at 1.8 GHz. This is an example of the use of SiGe technology to enhance the breakdown voltage of the device, without excessively compromising RF performance. The major drawback of this approach is now the early onset of the so-called Kirk effect, which is the collector current density beyond which the f_T of the device begins to drop dramatically because of base "push-out" effects [17].

4.4 PASSIVE COMPONENT AND PACKAGING TECHNIQUES FOR SILICON-BASED LOW-POWER MICROWAVE SYSTEMS

Advanced packaging techniques will also be required in order to fully utilize Si/SiGe technology in systems applications above 10 GHz. Typical bond-wire and lead-frame inductances limit the high-frequency performance of packaged devices to less than 5 GHz [18]. Flip-chip bonding of X-band and Ku-band Si/SiGe microwave monolithic integrated circuits (MMICs) was recently demonstrated [19], with outstanding results, in a quasihybrid packaging environment.

One major potential limitation of silicon technology for very-high-frequency applications is the availability of low-loss transmission line structures for impedance-matching applications. Historically, transmission lines in silicon technology have suffered from extremely high losses resulting from the relatively high conductivity of typical silicon substrates. In addition, the ohmic losses of transmission lines implemented in silicon have tended to be high, because of the relatively thin Al-based metallization employed in most VLSI processes.

Despite these drawbacks, a number of approaches have been attempted to realize these structures in a monolithic silicon environment, including coplanar transmission lines on lightly doped substrates [20,21], thick deposited SiO_2 for realization of microstrip structures [22], and thick polyimide layers to separate the transmission lines from the silicon substrate [23]. These existing approaches have a number of drawbacks. Lightly doped silicon substrates are known to have a high degree of residual internal stress, which makes subsequent high-temperature processing very difficult [24]. The deposition of thick layers of high-quality SiO_2 is also limited by high levels of internal stress, limiting the critical thickness to less than 10 μm in most cases. Previous thick polyimide approaches have employed gold-based metallization [25], which is often incompatible with high-volume semiconductor processing. Recent results have demonstrated outstanding low-loss characteristics in silicon through bulk micromachining techniques [26]. All of these processes require an undesirable significant departure from standard silicon VLSI processing techniques.

Recent results have demonstrated a manufacturable, low-loss, transmission line process in silicon technology, using thick spun-on polyimide dielectrics [27]. A plot of the measured transmission line losses, in dB/wavelength for a variety of microstrip transmission line widths, is shown in Figure 4.9, and a plot of the

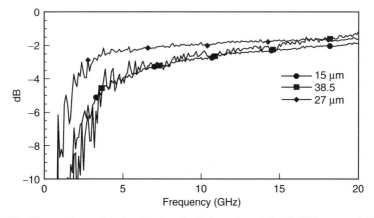

Figure 4.9. Measured transmission line loss in dB/wavelength for Si/SiGe transmission line structures fabricated with a thick dielectric layer [27].

measured uniformity of the characteristic impedance and effective permittivity across a 200 mm wafer is shown in Figure 4.10. Clearly, these results are adequate for many high-frequency amplifier applications, although the losses are substantially higher than what can be achieved in GaAs or hybrid monolithic integrated circuit (MIC) technology.

Although planar monolithic inductors have a long history, there has been renewed interest recently in their development for RFIC voltage controlled oscillators (VCOs) in particular. The realization of high-performance inductors—with performance

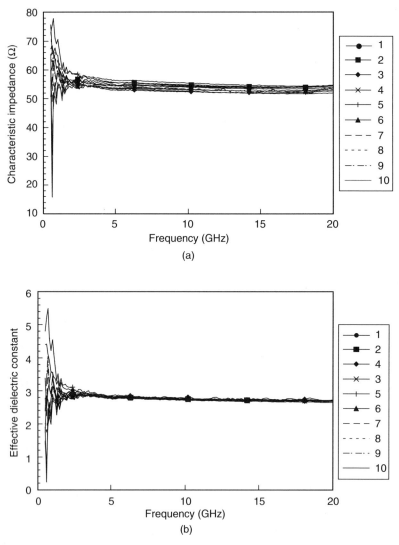

Figure 4.10. Measured variation of Si/SiGe HBT transmission line parameters across a 200 mm wafer: (a) characteristic impedance and (b) effective dielectric constant.

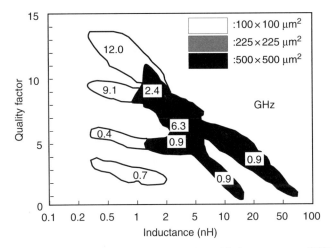

Figure 4.11. Inductor quality factor (Q) as a function of inductance on a Si/SiGe process. Note that the peak Q falls as the inductance rises. (After Burghartz et al. [33] © 1997 IEEE.)

comparable to hybrid implementations—is *fundamentally* limited by the fact that inductor Q is roughly proportional to the *area* of the inductor [28]; the inevitable area limitations of a monolithic integrated circuit render dramatic improvements in Q nearly impossible. Most current efforts at Q enhancement involve modest reductions in series resistance, or elimination of substrate loss effects. These efforts include the use of thick gold metallization [29], multiple metal layers in parallel [30], bulk micromachining techniques for the removal of resistive material underneath the inductor [31], and spun-on thick dielectrics [32] to physically separate the inductor from the lossy silicon substrate. Peak values of monolithic inductor Q in the 5–20 range have been achieved to date, but this is still well below what is achievable using off-chip hybrid components, which have typical peak Q values in the 50–500 range.

Si/SiGe technology has recently demonstrated outstanding inductor quality factors in a standard microelectronics "back-end" metallization process. This is another area where traditional microelectronic scaling has improved the performance of passive microwave circuits. Figure 4.11 plots the measured quality factor as a function of inductance at differing frequencies for inductors fabricated in the IBM SiGe HBT technology [33].

4.5 SILICON TECHNOLOGY VERSUS III–V TECHNOLOGIES

There is absolutely no question that III–V-based devices in GaAs or InP technology will exhibit superior f_T and f_{max} compared to a Si/SiGe device for a specified geometry. This is a result of the higher electron mobility of III–V devices (8500 cm^2/(V · s) in GaAs versus 1450 cm^2/(V · s) in silicon), and higher peak electron velocity

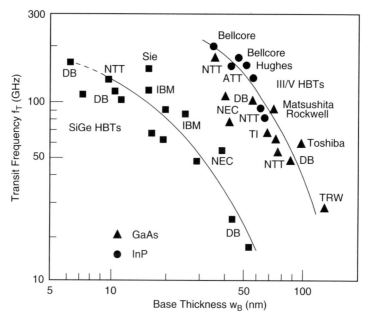

Figure 4.12. Comparison of HBT f_T as a function of base width for Si/SiGe and III–V technologies [34].

(2×10^7 cm/s in GaAs versus 10^7 cm/s in silicon) of these materials. These fundamental materials considerations have been borne out by experimental results, and an excellent comparison of the technologies was presented in Konig et al. [34]. A plot from that paper of relative speed (both f_T and f_{max}) as a function of base width is shown in Figures 4.12 and 4.13. Clearly, if maximum performance or speed is the *only* criteria, then III–V technology is the superior option.

III–V-based devices will also exhibit superior *breakdown voltage* properties at a given speed (either f_T and f_{max}) due to the well-known higher bandgap energies (GaAs has a bandgap of 1.4 eV compared to 1.1 eV for silicon). The impact ionization coefficients at a given electric field in silicon are dramatically higher in silicon than they are in GaAs or InP. The product of the intrinsic breakdown voltage and electron mobility—the well-known Johnson limit [35]—results in much higher gain for the GaAs power devices, even at a lower operating voltage. The result of this is that the product of transistor f_T and breakdown voltage is *material limited*. Silicon devices typically operate in a Johnson limit regime of f_T–breakdown product of approximately 100–200 GHz · V, compared with approximately 400–800 GHz · V for comparable III–V devices.

As with low-noise amplifiers, despite their high f_T, the performance of silicon-based power devices will lag that of GaAs-based power devices for the foreseeable future, even at lower power supply voltages. Clearly, in areas where the best performance is essential, III–V technology will demonstrate the best performance. However, its use may only be required at the critical antenna interfaces (the first

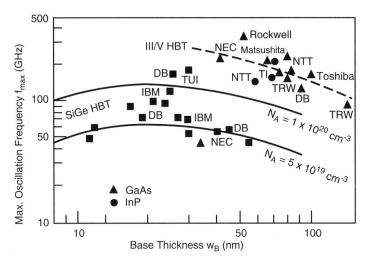

Figure 4.13. Comparison of HBT f_{max} as a function of base width for Si/SiGe and III–V technologies [34].

low-noise and final stage of power amplification), where the ultimate system performance is set, and relatively low levels of integration are necessary. The rest of the transceiver can plausibly be implemented in a highly integrated Si/SiGe BiCMOS circuit.

A comparison here to other areas of engineering science is appropriate. Titanium is a superior metal to steel in terms of its strength-to-weight ratio. Despite its obvious technical superiority, our cars are not constructed completely with titanium today; the more expensive metal is used *only* in those applications where the improvement in performance justifies the added cost. In the same vein, Si/SiGe technology can function as the workhorse "steel" of low-power communications applications, reserving III–V technology for those areas where its improved performance justifies the higher cost.

4.6 CIRCUIT DESIGN WITH Si/SiGe HBTs FOR MOBILE COMMUNICATIONS

Recent research results have demonstrated outstanding performance for Si/SiGe HBT circuits operating at frequencies above 10 GHz. They include frequency dividers operating to 28 GHz [36], Gilbert mixers operating to 12 GHz [37], power amplifiers with over 25 dBm of output power at 1.9 GHz [38], and VCOs with −103 dBc/Hz of phase noise at 7.5 GHz [39]. In addition, many commercial low-power communications products based on SiGe technology have been introduced into the marketplace recently, and it is anticipated that the volume of products will grow even more in the coming years. This section will highlight some of the recent circuit results in this technology.

4.6.1 Low-Noise Amplifiers

Low-noise amplifiers are one of the key performance bottlenecks in an RF system. They are required to contend with a variety of signals coming from the antenna—often of larger amplitude than the desired signal—and so both low noise and high linearity are simultaneously required. These requirements are often at odds with an additional requirement for low-power dissipation.

Low-noise amplifiers are usually required to amplify a signal coming from the antenna whose lowest level is very close to the thermodynamic limit of background noise—given by kTB, where k is Boltzmann's constant, T is the ambient temperature, and B is the bandwidth. In the case of a 50 kHz signal, this corresponds to a background noise power level of only 10^{-15} W at room temperature! So the intrinsic noise of the amplifier when connected to the antenna should not be much larger than the noise due to the antenna itself. The ratio of these two quantities is known as the noise figure (NF) of the amplifier. The minimum noise figure of a transistor is roughly proportional to the device base resistance and inversely proportional to the f_T. An approximate expression for the resulting minimum noise figure of a bipolar transistor—under optimum impedance matched conditions—is given by [40]

$$\mathrm{NF}_{\min} \approx 1 + \frac{f}{f_T}\sqrt{\frac{2I_{CQ}}{V_T} \times r_b \left[1 + \frac{1}{\beta_{dc}}\left(\frac{f_T}{f}\right)^2\right] + \frac{1}{\beta_{dc}}\left(\frac{f_T}{f}\right)^2}, \quad (4.10)$$

where I_{CQ} is the dc bias current. This expression illustrates the importance of minimizing the base resistance and maximizing the f_T of the device—in both of these areas, the Si/SiGe HBT performs exceptionally well, with the largest improvements occurring at lower collector bias currents. A plot of measured minimum noise figure for an advanced Si/SiGe HBT as a function of current appears in Figure 4.14. A minimum value of less than 0.5 dB at 2 GHz is possible to achieve with an appropriately scaled technology. This compares to nearly 1 dB for an equivalent silicon-only BJT.

This illustrates the importance of high transistor f_T at low dc currents for handheld wireless applications. Unfortunately, a typical transistor's f_T peaks at a relatively high current, and so the device is often biased at as low a current as possible—consistent with acceptable noise performance—in order to realize its lowest power operation. In fact, this relationship implies that there is a relatively straightforward trade-off between LNA gain, noise figure, and power dissipation for a given semiconductor technology. The higher the gain, for a given power dissipation and noise figure, the better the performance.

Figure 4.15 plots amplifier gain/dc power dissipation (in dB/mW) as a function of noise figure (in dB) for a variety of reported low-noise amplifiers in silicon and GaAs technology at 2 GHz. Most of the recently reported LNA results, fabricated in Si CMOS [41] or bipolar technologies [42,43], fall along a gain/($P_{dc} \times$ NF) line of approximately 0.4 (1/mW). By comparison, a recent SiGe HBT result [44] demonstrated a fully integrated LNA with 0.95 dB noise figure, 2 mW of power

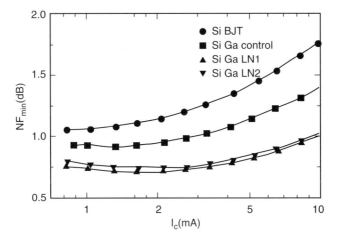

Figure 4.14. Minimum noise figure (NF$_{min}$) versus collector current at 2 GHz for fabricated Si BJT and alternative SiGe HBTs. There is a significant improvement over standard BJT performance. (After Niu et al. [53] © 1999 IEEE.)

dissipation, and 10.5 dB of gain at 2.4 GHz, for a figure of merit of approximately 5.5 (1/mW). The best reported GaAs LNAs have figures of merit of approximately 3.0 (1/mW) [45–47]. These results demonstrate the potential performance advantage of SiGe technologies at these frequencies, where dc power dissipation is a major consideration [48].

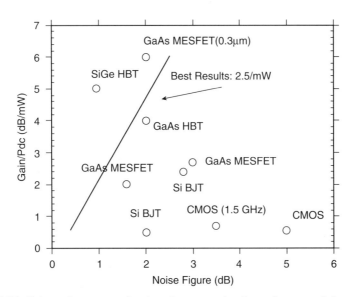

Figure 4.15. Gain to dc power ratio plotted versus noise figure for state-of-the-art 2 GHz LNAs [48]. Note that the SiGe HBT circuit provides the best result when power dissipation is a critical factor.

Linearity is an equally important figure of merit for front-end transistor amplifiers like low-noise amplifiers. The linearity of the LNA—as measured by its input intercept point (IIP3)—is at least as important as its noise figure, in most applications. If the nonlinearity of an amplifier can be modeled by a power series given by

$$S_0 = a_1 s_i + a_2 s_i^2 + a_3 s_i^3 + \cdots, \quad (4.11)$$

then the third-order input intercept point is given by

$$V_{\mathrm{IP}} = \sqrt{\frac{4\,a_3}{3\,a_1}}, \quad (4.12)$$

where a_1 is the first-order coefficient of the power-series expansion of the amplifier gain, and a_3 is the third-order coefficient of the power-series expansion of the amplifier gain. If we assume that the linearity of the circuit is dominated by the dc nonlinearity of the base–emitter junction, and the input to the amplifier is a voltage source, then the input intercept point for a standard bipolar transistor amplifier is approximately 100 mV—a rather low value for most applications [49].

However, the linearity of the amplifier can be improved through the use of series feedback in the emitter of the circuit. This improvement in linearity generally applies whether the feedback is inductive or resistive. If the feedback is resistive, then the noise figure is degraded by the addition of the thermal noise due to the resistor, and the resulting trade-off between noise and linearity is straightforward. If the feedback is inductive, then the feedback does not add any noise of its own, yet the linearity is improved. In the case of inductive feedback in the emitter of a BJT low-noise amplifier, it has been demonstrated that the minimum distortion occurs at a frequency given by [50]

$$f = \frac{1}{\sqrt{2C_{\mathrm{je}}(L_{\mathrm{b}} + L_{\mathrm{e}})}}, \quad (4.13)$$

which is also approximately the frequency where the minimum occurs in the noise figure. It was also demonstrated in Hull [50] that the simultaneous achievement in low noise and distortion is a substantial advantage of the common-emitter configuration compared with the common-base configuration. The exact expression for the distortion of a inductive feedback bipolar low-noise amplifier is extremely involved, but can be found in Fong and Meyer [51].

In this case, an often-used linearity figure of merit is the ratio of the third order input intercept point (IIP3) to the dc power dissipation. The IIP3 occurs at an input power level where the extrapolated output two-tone intermodulation products are equal in magnitude to those of the desired output signal. This is a crucial measurement, because it sets a lower limit on the received signal level that is indistinguishable from interference created by other unwanted signals. Field effect transistors (MOSFETs as well as GaAs MESFETs and PHEMTs) generally exhibit improved third order intermodulation distortion compared with bipolar devices, due to their near square-law current versus voltage behavior. On the other hand, bipolar transistor amplifiers have recently demonstrated outstanding linearity performance as well, apparently due to the cancellation of the resistive and capacitive nonlinearities in the base–emitter junction at certain frequencies [52,53]. As with the case of noise figure, the performance advantages of SiGe HBTs becomes significant if dc power dissipation is a critical parameter, although the improvement is less dramatic.

The use of improved circuit design approaches can improve the performance of low-noise amplifiers, and the "excess" high-frequency performance of the Si/SiGe HBT can be used to achieve improved linearity at lower microwave frequencies. An excellent example of this is the use of transformer coupled circuits—both low-noise amplifiers and mixers—which have demonstrated outstanding performance at very low power dissipation [54]. An example of this approach is shown in Figure 4.16 in the case of a combined RF low-noise amplifier and mixer designed to operate at 5.5 GHz. The low-noise amplifier transistors are driven differentially to reduce the effects of wirebond and package inductance. This also improves the quality factor of the inductors, whose Q increases significantly when driven in a differential configuration [55]. The transformer provides both shunt and series feedback to the circuit, providing for a simultaneous noise and gain match [56]. The turns ratio of the transformer provides additional voltage gain in the circuit, and the resulting performance is improved. The measured single side band (SSB) noise figure of the

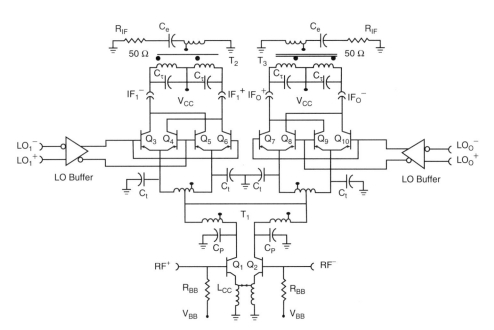

Figure 4.16. Schematic of SiGe HBT 5 GHz low-noise amplifier/downconverter. The low-noise amplifier is transformer coupled to the mixer stage, resulting in a compact circuit design and superior low-power performance. (After Long [56] © 1999 IEEE.)

receiver is 5.1 dB, with 15 dB of conversion gain; the power dissipation of the complete circuit is only 44 mW from a 2.2 V supply.

4.6.2 Power Amplifiers

Silicon technology is equally attractive for power amplifier applications, due to its low manufacturing cost and high thermal conductivity. However, the aforementioned Johnson limit tends to limit the peak performance achievable from the devices. As a result, for the past several years, AlGaAs/GaAs HBT power amplifiers have dominated the linear handset transmitter market due to their excellent linearity and power-added efficiency. However, GaAs-based integrated circuits are relatively expensive and must be thinned for optimum performance in power amplification. Compared with AlGaAs/GaAs HBTs, SiGe/Si HBTs are more attractive primarily due to their high substrate thermal conductivity (150 W/(m · °C)), comparable device performance ($f_f \approx 30$ GHz and $f_{max} \approx 50$ GHz), lower emitter/base turn-on voltage (~ 0.75 V), and substantially lower production cost. Unfortunately, SiGe/Si HBTs have their own disadvantages: the substrate is very conductive, adding significant parasitics to both active and passive components of the power amplifier. SiGe HBTs also have relatively low breakdown voltages ($BV_{CEO} \approx 5$ V; $BV_{CES} \approx 14.5$ V) and low Early voltage (~ 140 V), versus > 1000 V in GaAs HBTs. These characteristics are detrimental to the gain, linearity, and dynamic range of the power amplifier.

Despite these inherent limitations, the SiGe HBT has recently demonstrated outstanding power amplifier performance at microwave frequencies. Table 4.3 summarizes some typical load-pull power results at 900 MHz and 1.8 GHz [16]. In Class B operation, the results of a single 40 μm² emitter-area cell produces 0.8 mW/μm² power density and 69% power-added efficiency. Larger cells exhibit correspondingly greater output powers and comparable efficiencies. Figures 4.17 and 4.18 show load-pull data of output power and power-added efficiency for a 800 μm² Si/SiGe HBT power amplifier operated at 3.0 and 4.5 V bias. In both cases, the peak power-added efficiency was over 50%, at output powers of approximately 1 watt. This is very compatible with requirements for many cellular telephone and wireless applications, although still not as good as the best III–V results.

Recently, efforts have been made to produce SiGe/Si HBT power amplifiers for DECT and GSM handset transmission applications [57–60]. However, the output power of the DECT is relatively low (24 d Bm) and the linearity requirement of the

TABLE 4.3. Summary of Power Load-Pull Results for a Relatively Small (40 μm²) SiGe HBT Power Cell Operating at 1.8 GHz [16]

Class Operation	Bias Current	Maximum PAE	Output Power	Gain
B	2.0 mA	69%	15.2 dBm	24.9 dB
AB	6.5 mA	60%	15.2 dBm	28.9 dB
AB	12.5 mA	52%	15.2 dBm	29.0 dB
A	25.0 mA	42%	15.2 dBm	30.1 dB

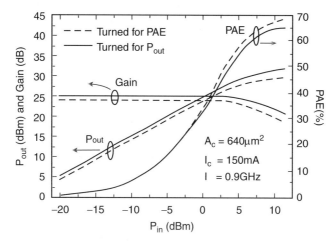

Figure 4.17. Load-pull data showing output power, gain, and power-added efficiency as a function of input power for a 640 μm² SiGe HBT power amplifier operating at 900 MHz. Results are shown for tuning for both maximum output power and maximum efficiency. (After Greenberg et al. [16] © 1997 IEEE.)

GSM is far less restrictive than that of CDMA applications. A dual-mode AMPS/CDMA handset power amplifier has one of the most demanding of requirements for the power amplifier. A two-stage Si/SiGe HBT power amplifier was recently demonstrated that met the performance goals for these applications [61]. A simplified schematic of the designed power amplifier is shown in Figure 4.19, which comprises driver and power amplification stages, input, interstage and output matching networks, and bias circuits for the driver and power stages, respectively.

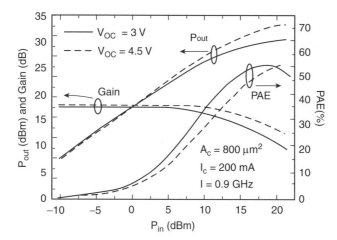

Figure 4.18. Load-pull data showing output power, gain, and power-added efficiency as a function of input power for a 800 μm² SiGe HBT power amplifier operating at 900 MHz at 3 V and 4.5 V dc bias. (After Greenberg et al. [16] © 1997 IEEE.)

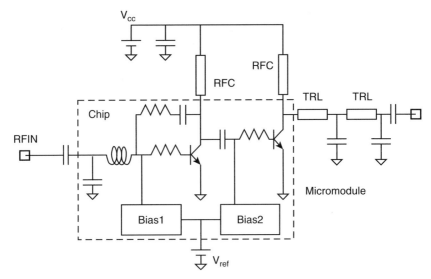

Figure 4.19. Simplified schematic of a cellular handset power amplifier. On-chip components are surrounded by the dashed line.

The total emitter areas of the driver and power HBTs were chosen to be 480 μm^2 and 3360 μm^2, respectively. Each HBT unit cell had an emitter size of 20 μm^2. To improve the HBT's thermal and electrical stability, a resistance of 170 Ω was added externally to ballast each individual base electrode. The base ballast resistors are effective in alleviating issues relevant to "thermal run away" (or second breakdown) of HBTs caused by non uniform current distribution among HBT unit cells due to process variations in base and emitter resistances [62].

An RC feedback network was used to linearize the first stage and bias the second stage at a low quiescent point to trade for a high PAE. An *LC* network and a one-pole filter are used for input and interstage matching, respectively. Both matching networks are implemented on-chip to facilitate the monolithic power amplifier design.

Figures 4.20 and 4.21 show the gain, PAE (%), and linearity (represented by first and second ACPRs) versus the output power over the frequency band (824–849 MHz). For CDMA operation, the amplifier satisfies linearity requirements at $V_{cc} = 3$ V with first ACPR better than -44.1 dBc and second ACPR better than -57.1 dBc with output power up to 28 dBm. The amplifier gain varies between 22 and 23 dB with power-added efficiencies of 36–37% at 28 dBm output power.

Figure 4.22 shows the gain, and PAE (%) versus the output power for AMPS operation. The amplifier satisfies the maximum output power requirement of 31 dBm at $V_{cc} = 3$ V with 21 dB gain and 49–51% PAE. As shown in Figure 4.23, the amplifier shows very low second and third harmonics, measured to be lower than -37 dBc and -55 dBc, respectively. The input return loss is always measured below -12 dB at any input level.

These results demonstrate that Si/SiGe HBT technology has the potential to demonstrate power amplifier performance that is competitive with III–V imple-

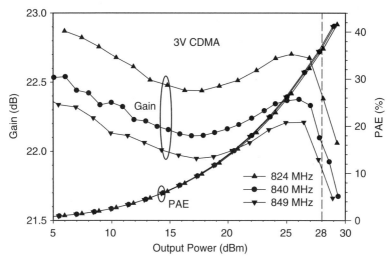

Figure 4.20. Gain and power-added efficiency versus CDMA power amplifier output power as a function of operating frequency.

mentations at the lower microwave frequencies common to modern wireless communications. At higher frequencies, the limitations imposed on the performance of the amplifier by the Johnson limit will tend to degrade the performance of the devices considerably compared to III–V technology.

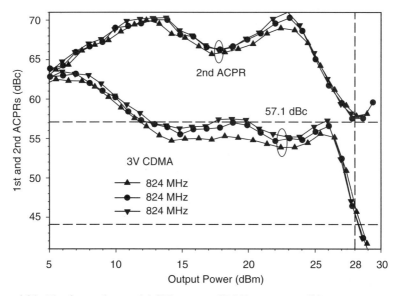

Figure 4.21. The first and second ACPRs versus CDMA power amplifier output power as a function of operating frequency.

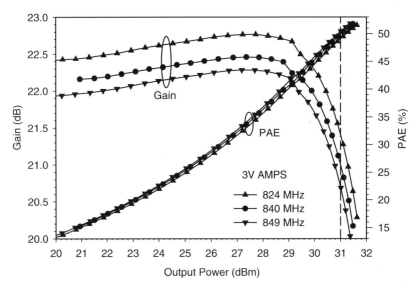

Figure 4.22. Gain and PAE of AMPS power amplifier versus output power as a function of operating frequency.

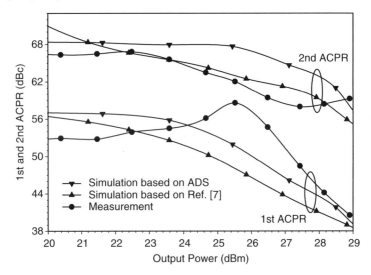

Figure 4.23. Comparison between the simulated and measured first and second ACPRs versus output power at $V_{cc} = 2.7\text{V}$ for CDMA operation at 840MHz.

4.7 CONCLUSIONS

Si/SiGe HBT technology has the potential to provide for a revolution in low-power, high-frequency transceiver design, which is in some ways comparable to the revolution in digital integrated circuit technology brought about by CMOS. Its unique combination of outstanding high-frequency performance, ultralow manufacturing cost, and high yield will provide abundant opportunities for new architectures and new systems in the near future. This chapter has summarized some of the key issues that are involved with this technology, from material growth, to device design, and finally circuit performance. It is clear that, although the performance of the SiGe HBT does not quite equal that of its counterpart, the III–V HBT, it is poised to become the "workhorse" technology for advanced mixed-signal and RF applications that are both cost and performance sensitive. In the coming years, we will expect to see greater use of this technology as it emerges into full-scale production from a variety of manufacturers.

ACKNOWLEDGMENTS

The author would like to acknowledge many useful discussions with Dr. D. Harame and Dr. B. Meyerson of IBM and Professor John Cressler of Auburn University and Peter Asbeck of the University of California–San Diego in preparation of this chapter.

REFERENCES

1. P. Hofstee et al., "A 1 GHz single-issue 64b PowerPC processor," in *Proceedings 2000 IEEE International Solid-State Circuits Conference*, pp. 92–93.
2. P. Gray and R. Meyer, "Future directions in silicon ICs for RF personal communications," in *Proceedings 1995 IEEE CICC*, pp. 83–90.
3. G. Freeman et al., "A 0.18 μm 90 GHz f_T SiGe HBT BiCMOS ASIC—compatible copper interconnect technology for RF and microwave applications," in *1999 International Electron Devices Meeting*, pp. 569–572.
4. J. N. Burghartz, D. C. Edelstein, K. A. Jenkiin, and Y. H. Kwark, "Spiral inductors and transmission lines in silicon technology using copper–damascene interconnects and low-loss substrates," *IEEE Transactions on Microwave Theory and Techniques*, Vol. 45, No.10, Pt. 2, 1961–1968 (Oct. 1997).
5. Z. Matutinovic-Krstelj, V. Venkataraman, E.J. Prinz, J. C. Sturm, and C. W. Magee, "Base resistance and effective bandgap reduction in n-p-n Si/SiGe HBTs with heavy base doping," *IEEE Transactions on Electron Devices*, Vol. 43, No.3, 457–466 (Mar. 1996).
6. S. S. Iyer, G. L. Patton, J. M. C. Stork, B. S. Meyerson, and D. L. Harame, "Heterojunction bipolar transistors using Si-Ge alloys," *IEEE Transactions on Electron Devices*, Vol. 36, No.10, 2043–2064 (Oct. 1989).
7. D. Harame et al., "Si/SiGe epitaxial-base transistors—Part I: materials, physics, and circuits," *IEEE Transactions on Electron Devices*, 455–468 (Mar. 1995).
8. D. Harame et al., "Optimization of SiGe HBT technology for analog and mixed-signal applications," *IEDM Technical Digest*, 874–876 (Dec. 1993).

9. B. Meyerson, "Issues in the commercialization of SiGe-based technology," in *Proceedings IEEE 1996 BCTM*, pp. 499–503.
10. H. Kroemer, "Two integral relations pertaining to the electron transport through a bipolar transistor with a nonuniform energy gap in the base region," *Solid-State Electronics*, Vol. 28, No. 11, 1101–1103 (Nov. 1985).
11. J. D. Cressler, J. H. Comfort, E. F. Crabbe, G. L. Patton, J. M. C. Stork, J. Y.-C. Sun, and B. S. Meyerson, "On the profile design and optimization of epitaxial Si- and SiGe-base bipolar technology for 77 K applications. I. Transistor DC design considerations," *IEEE Transactions on Electron Devices*, Vol. 40, No. 3, 525–541 (Mar. 1993)
12. J. D. Cressler, J. H. Comfort, E. F. Crabbe, G. L. Patton, J. M. C. Stork, J. Y.-C. Sun, and B. S. Meyerson, "Profile design issues and optimization of epitaxial Si and SiGe-base bipolar transistors and circuits for 77 K applications," in *1991 Symposium on VLSI Technology, Digest of Technical Papers*, pp. 69–70.
13. D. Harame, B. Meyerson, L. Larson, and P. Cunningham, "Epi-base BJT vs. epi-base HBT," private communication, 1995.
14. J. D. Cressler, "SiGe HBT technology: a new contender for Si-based RF and microwave circuit applications," *IEEE Transactions on Microwave Theory and Techniques*, Vol.46, No. 5, Pt. 2, 572–589 (May 1998).
15. T. Nakamura and H. Nishizawa, "Recent progress in bipolar transistor technology," *IEEE Transactions on Electron Devices*, Vol. ED-42, 390–398 (1995).
16. D. R. Greenberg, M. Rivier, P. Girard, E. Bergeault, J. Moniz, D. Ahlgren, G. Freeman, S. Subbanna, S. J. Jeng, K. Stein, D. Nguyen-Ngoc, K. Schonenberg, J. Malinowski, D. Colavito, D. L. Harame, and B. Meyerson, "Large-signal performance of high-BV_{CE0} graded epi-base SiGe HBTs at wireless frequencies," in *International Electron Devices Meeting 1997*, pp. 799–802.
17. C.T. Kirk, "A theory of transistor cutoff frequency (f_T) fall-off at high current density," *IEEE Transactions on Electron Devices*, Vol. ED-9, 164 (1962).
18. R. W. Jackson and S. Rakshit, "Microwave circuit modeling of an elevated paddle surface mount package," *IEEE Electrical Performance of Electronic Packaging*, 199–201 (1996).
19. M. Case, "SiGe MMICs and flip-chip MICs for low-cost microwave systems," *Microwave Journal*, 278–293 (May 1997).
20. A. C. Reyes et al., "High resistivity Si as a microwave substrate," in *Proceedings 46th Electronic Components and Technology Conference*, pp. 382–391, 1996.
21. A. C. Reyes et al., "Coplanar waveguides and microwave inductors on silicon substrates," *IEEE Transactions on Microwave Theory and Techniques Supplement*, Vol. 43, No. 9, Pt. 1, 2016–2022 (1995).
22. H. Sakai et al., "A novel millimeterwave IC on Si substrate using flip-chip bonding technology," *IEEE MTT-S International Microwave Symposium Digest*, Vol. 3, 1763–1766 (1994).
23. Bon-Kee Kim et al., "Monolithic planar RF inductor and waveguide structures on silicon with performance comparable to those in GaAs MMIC," *International Electron Devices Meeting Technical Digest*, pp. 717–720 (1995).
24. S. R. Taub and P. Young, "Temperature dependent performance on coplanar waveguide (CPW) on substrates of various materials," *IEEE MTT-S International Microwave Symposium Digest*, Vol. 2, 1049–1051 (1994).

25. M. Case, S. A. Maas, L. Larson, D. Rensch, et al., "An X-band monolithic active mixer in SiGe HBT technology," *IEEE MTT-S International Microwave Symposium Digest*, Vol. 2, 655–658 (1996).
26. J. Papapolymerou et al., "A micromachined high-Q X-band resonator," *IEEE Microwave and Guided Wave Letters*, Vol. 7, No. 6, 168–170 (1997).
27. L. Larson et al., "Low-loss microwave transmission lines implemented in a manufacturable Si/SiGe HBT process," *IEEE Transactions on Microwave Theory and Techniques* (Submitted).
28. G.Temes and J. LaPatra, *Introduction to Circuit Synthesis and Design,* McGraw-Hill, New York, 1977, see Chap. 7.
29. K. Ashby et al., "High-Q inductors for wireless applications in a complementary silicon bipolar process," *IEEE Journal of Solid-State Circuits*, Vol. 31, No. 1, 4–9 (1996).
30. J. Burghartz, "RF components implemented in an analog SiGe bipolar technology," in *1996 IEEE BCTM Digest*, Minneapolis, MN, pp. 138–141.
31. J. Chang et al., "Large suspended inductors on silicon and their use in a 2-μm CMOS RF amplifier," *IEEE Journal of Solid-State Circuits*, Vol. 14, No. 5, 246–248 (1993).
32. L. Larson et al., "Si/SiGe HBT technology for low-cost monolithic microwave integrated circuits," in *International Solid-State Circuits Conference*, San Fransisco, CA, 1996, pp. 80–81.
33. J. N. Burghartz, M. Soyuer, K. A. Jenkins, M. Kies, M. Dolan, K.J. Stein, J. Malinowski, and D.L. Harame, "Integrated RF components in a SiGe bipolar technology," *IEEE Journal of Solid-State Circuits*, Vol. 32, No.9, 1440–1445 (Sept. 1997).
34. U. Konig, A. Gruhle, and A. Schuppen, "SiGe devices and circuits: where are the advantages over III–V?," in *Proceedings 1995 IEEE GaAs IC Symposium*, pp. 14–18.
35. E. O. Johnson, "Physical limitation on frequency and power parameters of transistors," *IEEE International Convention Record*, Pt. 5, p. 27, 1965.
36. M. Case et al., "A 26 GHz digital frequency divider implemented in a manufacturable Si/SiGe HBT technology," in *Proceedings 1995 IEEE BCTM*, 1995.
37. J. Glenn et al., "12-GHz Gilbert mixers using a manufacturable Si/SiGe epitaxial-base bipolar technology," in *Proceedings 1995 IEEE BCTM*, pp. 186–189.
38. G. Henderson et al., "SiGe bipolar transistors for microwave power applications," in *1997 IEEE MTT-S International Microwave Symposium Digest*, pp. 1299–1302.
39. L. Larson et al., "A low-cost monolithic microwave integrated circuit technology in Si/SiGe HBT technology," in *Proceedings 1996 IEEE International Solid-State Circuits Conference*, pp. 88–89.
40. S. P. Voinigescu, M. C. Maliepaard, J. L. Showell, G.E. Babcock, D. Marchesan, M. Schroter, P. Schvan, and D. L. Harame, "A scalable high-frequency noise model for bipolar transistors with application to optimal transistor sizing for low-noise amplifier design," *IEEE Journal of Solid-State Circuits*, Vol. 32, No. 9, 1430–1439 (Sept. 1997).
41. A. Karanicolas, "A 2.7 V 900 MHz CMOS LNA and Mixer," in *International Solid-State Circuits Conference*, pp. 50–51 (1996).
42. J. Long and M. Copeland, "A 1.9 GHz low-voltage silicon bipolar receiver front-end for wireless personal communications systems," *IEEE Journal of Solid-State Circuits*, Vol. 30, No. 12, 1438–1448 (1995).

43. H. Takeuchi et al., "A Si wide-band MMIC amplifier family for L-S band consumer product applications," *1991 IEEE Microwave Theory and Techniques Symposium Digest*, pp. 1283–1284.
44. J. Long et al., "RF analog and digital circuits in SiGe technology," in *International Solid-State Circuits Conference*, pp. 82–83, 1996.
45. K. Cioffi, "Monolithic L-band amplifiers operating at milliwatt and sub-milliwatt dc power consumptions," in *1992 IEEE MMWMCS Digest*, pp. 9–12.
46. S. Hara et al., "Miniature low-noise variable MMIC amplifiers with low power consumption for L-band portable communications application," in *1991 IEEE Microwave Theory and Techniques Symposium Digest*, Atlanta, GA, pp. 67–70.
47. K. Kobayashi et al., "Ultra-low dc power GaAs HBT S- and C-band low-noise amplifiers for portable wireless communications," *IEEE Transactions on Microwave Theory and Techniques*, Vol. 43, No. 12, 3055–3061 (1995).
48. L. E. Larson, "Integrated circuit technology options for RFIC's—present status and future directions," *Proceedings of the IEEE 1997 Custom Integrated Circuits*, pp.169–176.
49. G. Vendelin, A. Pavio, and U. Rohde, *Microwave Circuit Design Using Linear and Nonlinear Techniques,* Wiley, New York, 1990.
50. C. D. Hull, "Analysis and optimization of monolithic RF downconversion receivers," Ph.D. dissertation, University of California, 1992.
51. Keng Leong Fong and R. G. Meyer, "High-frequency nonlinearity analysis of common-emitter and differential-pair transconductance stages," *IEEE Journal of Solid-State Circuits*, Vol. 33, No. 4, 548–555 (Apr. 1998).
52. S. Maas et al., "Intermodulation in heterojunction bipolar transistors," *IEEE Transactions on Microwave Theory and Techniques*, Vol. 40, No. 3, 442–447 (1992).
53. G. Niu, S. Zhang, J. D. Cressler, A. J. Joseph, J. S. Fairbanks, L. E. Larson, C. S. Webster, W. E. Ansley, and D. L. Harame, "SiGe profile design tradeoffs for RF circuit applications," in *International Electron Devices Meeting*, pp. 573–576 (1999).
54. J. Maligeorgos and J. Long, "A 2V 5.1–5.8 GHz image reject receiver with wide dynamic range," in *IEEE International Solid-State Circuits Conference*, pp. 322–323, 2000.
55. M. Danesh, J. R. Long, R. A. Hadaway, and D. L. Harame, "A Q-factor enhancement technique for MMIC inductors," in *1998 IEEE Radio Frequency Integrated Circuits (RFIC) Symposium, Digest of Papers*, pp. 217–220.
56. J. Long, "A 5.1–5.8 GHz low-power image reject downconverter," in *Proceedings of the IEEE 1999 Bipolar Circuits and Technology Meeting*, pp. 67–70.
57. A. Schuppen, S. Gerlach, H. Dietrich, D. Wandrei, U. Seiler, and U. Konig, "1-W SiGe power HBT's for mobile communication," *IEEE Electron Device Letters*, Vol. 19, No. 4, 341–343 (Apr. 1998).
58. J. N. Burghartz, J.-O. Plouchart, K. A. Jenkins, C. S. Webster, and M. Soyuer, "SiGe power HBT's for low-voltage, high-performance RF Applications," *IEEE Electron Device Letters*, Vol. 19, No. 4, 103–105 (Apr. 1998).
59. D. Harame, L. Larson, M. Case, S. Kovacic, S. Voinigescu, T. Tewksbury, D. Nguyen-Ngoc, K. Stein, J. Cressler, S.-J. Jeng, J. Malinowski, R. Groves, E. Eld, D. Sunderland, D. Rensch, M. Gilbert, K. Schonenberg, D. Ahlgren, S. Rosenbaum, J. Glenn, and B. Meyerson, "SiGe HBT Technology: Device and Application Issues," *IEDM*, 731–734 (1995).

60. F. Huin, C. Duvanaud, D. Masliah, J. M. Paillot, H. Mokrani, S. Gerlach, and K. Worner, "A low voltage integrated SiGe power amplifier for mobile applications," in *IMAPS '98, Wireless Communications Conference*, Session MP7, Part II.
61. P. Tseng, L. Zhang, G. Gao, and M. F. Chang, "A monolithic SiGe power amplifier for dual-mode (CDMA/AMPS) cellular handset application," in *Proceedings of the IEEE 1999 Bipolar Circuits and Technology Meeting*, pp. 153–156.
62. W. Liu et al., "The collapse of current gain in multi-finger hetrojunction bipolar transistors: its substrate dependence, instability criteria, and modeling," *IEEE Transactions on Electron Devices*, Vol. 41, No. 10, 1698–1707 (Oct. 1994).

5

FLICKER NOISE REDUCTION IN GaN FIELD-EFFECT TRANSISTORS

KANG L. WANG
Department of Electrical Engineering, University of California—Los Angeles

ALEXANDER BALANDIN
Department of Electrical Engineering, University of California—Riverside

Heterostructure field-effect transistors (HFETs) based on wide-bandgap compound semiconductors such as GaN have demonstrated potential for high-power density and high-frequency devise applications. One of the major factors, which so far restricted application of these devices for communications systems, is their noisiness. In most cases, high flicker noise level of GaN HFETs translates into unacceptable phase noise that limits performance of oscillators, mixers, and other electronic systems. In this chapter we review results of experimental investigation of the noise in GaN HFETs fabricated using different growth techniques and device processing methods. We examine the sources of noise in these transistors and present some methods, that allow us to reduce the noise in these devices. Effects of doping, deep levels, piezoelectric charges, potential barrier profile, and device dimensions are also discussed. Presented data of the low-temperature measurements reveal correlation of the flicker noise with generation–recombination noise in the doped channel GaN HFETs. The results reviewed in this chapter are important for the new generation of electronic technologies requiring a low phase noise level.

5.1 INTRODUCTION

Flicker or $1/f$ noise is the low-frequency phenomenon observed in nonequilibrium systems, for example, in systems in which electric current flows [1]. The name flicker noise stems from early vacuum-tube observations when a flicker-type fluctuation was recorded in the anode current. An intriguing property of flicker noise is the fact that its spectral density is inversely proportional to frequency. Theoretically, the noise power density would approach infinity as frequency approaches zero. Indeed, the $1/f$ dependence has been found to hold at frequencies as low as a few cycles per day [2,3]. In a more precise way, flicker noise should be described by the spectral density $1/f^\gamma$, where γ is a parameter typically in the range $0.8 < \gamma < 1.5$. The flicker noise, which manifests at low frequencies (usually 0.01–100 kHz) is an important figure of merit for GaN HFETs since this type of noise is the limiting phase noise factor for these transistors. The value of flicker noise is also a good indicator of the material quality.

It is important to know the value of flicker noise in solid state devices since this type of noise directly contributes to the phase noise of the device and it is the limiting noise figure for all kinds of high electron mobility transistors (HEMTs) and metal oxide semiconductor field effect transistors (MOSFETs). Especially when these devices are used as oscillators or mixers, the flicker noise limits the phase noise characteristics and degenerates performance of the electronic systems. In the past, many research efforts focused on the flicker noise in the Si CMOS, GaAs HEMT, and bipolar junction transistor (BJT). To date, the noise characteristics of the GaN-based structures have not been investigated in detail. It is expected that decrease of the characteristic device size and operation power will result in even more stringent requirements for the flicker noise level.

The $1/f$ noise is still a rather poorly understood phenomenon [4]. There is still no universally accepted $1/f$ noise theory. No single model has been able to explain all the diverse results obtained under different experimental conditions and from different devices. At this time, there are two competing models that are invoked to explain noise data in solid state electronic devices: the carrier density fluctuation model and the mobility fluctuation model. The former attributes noise to random trapping and de-trapping of free carriers by traps that have a particular distribution of time constants. The latter attributes noise to the mobility fluctuations due to carrier scattering on lattice (phonons), defects, impurities, and so on.

McWhorter [5] noted that a distribution of time constants leading to a $1/f$-type noise spectrum could arise naturally at a semiconductor–oxide interface from a spatially uniform distribution of tunneling depths to the trapping sites. The trapping and detrapping of carriers lead to fluctuations in the occupancy of the interface traps at the Fermi level, which modulate the local carrier concentration. According to the carrier trapping–detrapping model, the $1/f$-type noise spectral density is closely related to the Lorentzian peaks of the generation–recombination (G-R) noise. Simply put, the $1/f^\gamma$ arises from traps with a broad distribution in energy while the G-R noise is due to discrete energy states. The McWhorter model seems to work well for CMOS devices. This type of noise may be superimposed over the low-frequency noise due to carrier mobility fluctuations.

We should point out that there is a body of work pertaining to random telegraph noise in mesoscopic devices [6] caused by a single electron trap that supports the carrier density fluctuation model. On the other hand, the mobility fluctuation model attributes 1/f noise to spontaneous mobility fluctuation due to scattering of carriers. This model was successfully applied to a variety of material systems and structures ranging from long bulk resistors to short channel HEMTs. A related model attributes 1/f noise to random motion of impurities in a mesoscopic device smaller than the phase-breaking length of electrons [7]. This model is purely quantum mechanical and arises from quantum interference of electron waves scattered from a slowly moving impurity. It is relevant to cryogenic experiments where the phase-breaking length is large enough. Since in this chapter we are concerned with room-temperature and above phenomena only, this model is not relevant. In a number of experiments [8], particularly those pertinent to intrinsic semiconductor low-dimensional structures [9], the 1/f noise behavior has been successfully described by Handel's theory of 1/f noise [10,11]. Although, in material systems with a large number of defects (like GaN), a significant contribution to 1/f noise is expected to come from other mechanisms, such as the one associated with the carrier trapping–detrapping. In this chapter we do not make any attempts to develop a complete theory of 1/f noise in GaN systems or discriminate among existing ones. We rather provide a detailed account of empirical observations of flicker noise in GaN transistors, make suggestions regarding the dominant noise sources, and offer potentially successful methods of noise reduction.

Advances in GaN-related compound materials and heterojunction field effect transistors have led to demonstration of the high-power-density microwave operation of these devices. GaN HFETs exhibiting the cutoff frequency of 60 GHz and the maximum frequency exceeding 100 GHz have recently been reported by a group at UCLA [12]. Flicker noise in GaAs and Si transistors has been studied for several decades [5,10,13–20]. Although there is no single model that could explain all the diverse results obtained under different experimental conditions, the requirements for many material systems for reducing noise are known. However, it is not the case for GaN/AlGaN heterostructures. Very little is known about the physical origin of the low-frequency noise in GaN HFETs and the effect of the material quality on the noise level. It is also not clear which model (the mobility fluctuation or the number fluctuation through random carrier trapping–detrapping) best describes the 1/f noise in the GaN system [14–23]. There have been reported values of the low-frequency noise varying as much as four orders of magnitude. For example, Levinshtein et al. [22] and Kuksenkov et al. [23] reported values of the Hooge parameter higher than 10^{-2}. Although later, continuous progress in GaN material and device fabrication allowed different groups to obtain much lower values of the Hooge parameters. Levinshtein et al. [24] reported a value of $\alpha_H = (1.5-4) \times 10^{-4}$ for GaN HFETs grown on SiC, while Balandin et al. [13] reported a value of 4.9×10^{-5}–1.7×10^{-4} for GaN HFETs grown on sapphire. One possible explanation of the large level of flicker noise in large GaN resistors is the high density of imperfections and dislocations in this material.

It has previously been shown that GaN/AlGaN heterostructures have large piezoelectric coefficients that lead to strong electric polarization on (0001) faces of the wurtzite structures typically used to form GaN HFETs [25,26]. The latter results in appreciable charge densities, which are large enough to design HFETs without any channel doping. Due to this reason it is important to know how the channel doping influences the low-frequency noise in GaN devices. It is particularly interesting to know if Si atoms, which are used as dopants, may form different complexes of traps in GaN/AlGaN systems. It has recently been shown that the Si dopant forms two donor states in $Al_xGa_{1-x}N (0.5 < x < 0.6)$ [27].

In Section 5.3 we summarize results of the low-frequency noise measurements in the doped and undoped channel GaN HFETs. A higher aluminum content of the undoped channel devices leads to a higher piezoelectrically induced charge density, thus making up for the absence of doping and allowing for a meaningful comparison of the noise levels between the two types of devices. In order to clarify the physical origin of the low-frequency noise and understand the influence of material quality on device performance, we also present results of the low-temperature noise measurements (Section 5.4) and deep-level transient spectroscopy (DLTS) study of these devices (Section 5.5). In the next section, we concentrate on the experimental data obtained for GaN heterostructure field-effect transistors (HFETs) grown on both sapphire and SiC. We describe the noise spectral density dependence on biasing parameters, which is important for modeling of the device circuit performance. Without selecting any single theoretical description of the noise in GaN HFETs, we analyze several experimental situations and suggest models, that seem to describe them the best.

5.2 NOISE IN GaN HETEROSTRUCTURE FIELD-EFFECT TRANSISTORS

In this section we present results of experimental investigation of the flicker noise in $GaN/Al_{0.15}Ga_{0.85}N$ doped channel heterostructure field-effect transistors. Most of the results presented in this section have recently been reported by Balandin et al. [13,14]. The devices for this study were made from a commercial wafer grown by the standard metallorganic chemical vapor deposition (MOCVD) technique. During the device fabrication step special attention was paid to improvement of the ohmic contacts. The composition of various metal films and details of the rapid thermal annealing processes used for fabrication of GaN HFETs were reported by Cai et al. [28]. The top view of the GaN HFET and typical dimensions are shown in Figure 5.1. The layered structure was fabricated on a sapphire substrate. A 1.0 μm thick i-GaN buffer layer was followed by 50 nm thick n-GaN layer with the doping level of 5×10^{17} cm^{-3}, and 3 nm thick i-$Al_{0.15}Ga_{0.85}N$ undoped spacer layer. On top, there was 30 nm thick n-$Al_{0.15}Ga_{0.85}N$ layer with the doping level of 2×10^{18} cm^{-3}. The barrier and channel doping resulted in a sheet electron concentration of about 1.6×10^{13} cm^{-2}. Electron Hall mobility was determined to be 460 cm^2/(V·s) at room temperature.

Devices selected for this study had a fixed gate length $L_G = 0.25$ μm. A device with $L_{DS} = 3$ μm had the drain current $I_{DS} = 0.55$ A/mm at the gate bias $V_{GS} = -3.0$ V. The maximum transconductance for negative gate biases was determined to be $g_m = 102$ mS/mm at $V_{GS} = -5$ V. A device with $L_{DS} = 2$ μm

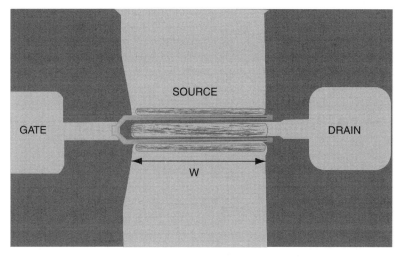

Figure 5.1. Top view of the examined device shown with the notations used in the text.

had the maximum transconductance for the negative gate biases $g_m = 133$ mS/mm at $V_{GS} = -3$ V for $L_G = 0.25$ μm. We have examined a large number of devices (more then 20 on each wafer) with the gate width $W = 2 \times 40$ μm, and four different source–drain separation distances $L_{SD} = 2$, 3, 4, and 5 μm. All examined devices were made on the same wafer. For these devices we obtained experimental dependence of the equivalent input referred noise power spectrum on frequency and gate and drain voltages. The measurements were carried out for both the linear region of the device operation corresponding to low drain–source voltage, V_{DS}, and the onset of the saturation region of operation (subsaturation) corresponding to $V_{DS} = 5$ V. Typical dc current–voltage characteristics of a GaN HFET are presented in Figure 5.2. In Figure 5.2(a) the gate bias V_{GS} is used as a parameter and it changes from 0 V with the step of -1 V. In Figure 5.2(b) the drain bias V_{DS} varies from 0 V with the step of $+1$ V.

The low-frequency noise measurement system that had been used for noise measurements reported by Balandin et al. [13,14] consisted of a low-noise amplifier with three stages: a dynamic signal analyzer and bias power supplies. The HP4142B modular source and monitor were used for current–voltage measurements. The amplifier was made from commercially available IC op amps and had an equivalent noise voltage on the order of 3 nV/Hz$^{1/2}$ and an equivalent noise current of 2 pA/Hz$^{1/2}$. A detailed description of the amplifier used for these measurements can be found in Balandin et al. [14]. The amplifier and GaN HFETs were enclosed in a shielded box during the measurements in order to prevent pick up of environmental noise.

Experimental noise spectra of two GaN HFETs for different gate bias V_{GS} and a fixed drain voltage $V_{DS} = 5$ V are shown in Figure 5.3. The threshold voltage for this device is $V_T = -7.5$ V. As one can see, the slope γ of the $1/f^\gamma$ dependence in all spectra is very close to 1, although it varies for different devices and gate bias values.

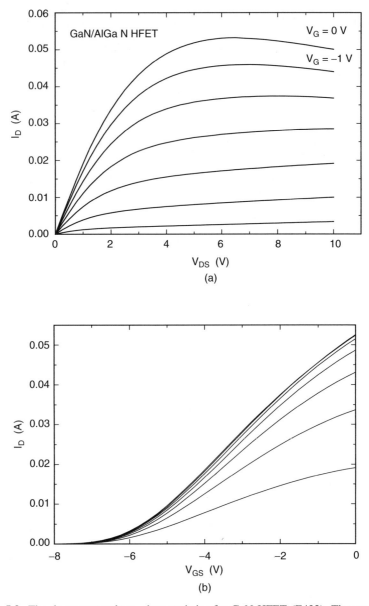

Figure 5.2. The dc current-voltage characteristics for GaN HFET (E432). The gate width $W = 80$ μm, the gate length $L_G = 0.25$ μm, and the drain–source separation distance $L_{DS} = 4$ μm. (a) The gate bias V_{GS} is used as a parameter and it changes from 0 V (top curve) with a step of -1 V. (b) The drain bias V_{DS} varies from 0 V (bottom curve) with a step of $+1$ V. (After Balandin et al. [14]. Copyright © 1999 IEEE; reprinted with permission)

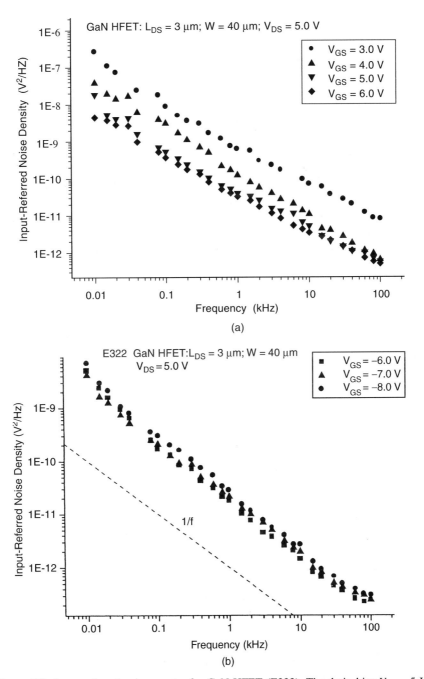

Figure 5.3. Input-referred noise spectra for GaN HFET (E332). The drain bias $V_{DS} = 5$ V. The gate dimensions are 0.25 μm by 80 μm, the drain–source separation distance $L_{DS} = 4$ μm. (a) Note a strong dependence of the noise spectral density on the gate bias until V_{GS} reaches the pinch-off conditions. (After Balandin et al. [14]. Copyright © 1999 IEEE; reprinted with permission)

No clear traces of the generation–recombination (G-R) bulges in the noise spectra of these devices at room temperature have been observed [14]. The equivalent input-referred noise spectral density shown in Figure 5.3 was obtained from the drain current noise spectral density using the relation $S_V = S_{ID}/g_m^2$, where S_V is the equivalent input-referred noise spectral density, S_{ID} is the drain current noise spectral density, and g_m is the transconductance of the device.

One can see from Figure 5.3 that, at relatively small absolute values of the gate bias $|V_{GS}|$, the noise spectral density S_V is gate bias dependent and decreases with a higher negative bias. At high absolute values of the gate bias (Fig. 5.3 b)) when the channel is almost pinched off, the noise density is about the same for different values of V_{GS}. This type of behavior was characteristic for all examined transistors. It is also interesting to note how close the experimental curves are to the $1/f$ function shown in the figure for comparison.

In order to have a quantitative characteristic of the overall noise in the device, we use the Hooge parameter α_H introduced via the commonly accepted equation [17, 18, 29–31]

$$\frac{S_{I_D}}{I_D^2} = \frac{\alpha_H}{Nf}, \tag{5.1}$$

where f is the frequency, I_D is the drain current, and N is the total number of carriers under the gate calculated from the drain–source current at which the noise was measured. The number of carriers in homogeneous samples is expressed as

$$N = \frac{L^2}{Rq\mu}. \tag{5.2}$$

Here μ is the mobility in the conducting channel, R is the resistance between two device terminals, L is the channel length, and q is the charge of an electron. Two quantities (R and μ) in this equation are determined experimentally. The resistance is found at a given V_{DS} during the noise measurements, while the mobility is determined for the layered structure using the Hall measurements. The gate leakage current for the devices was small (less then 1%), and hence its effect on their noise performance was neglected [13,14]. Using Eq. (5.1) and (5.2), one can express the Hooge parameter as follows:

$$\alpha_H = \frac{S_V g_m^2}{I_D V_D} \frac{L^2 f}{q\mu}, \tag{5.3}$$

where V_D is the voltage drop across the device terminals (source–drain). The values of α_H, calculated using the above equations, are very approximate numbers since the conducting channel of the device is not a homogeneous one. Meanwhile, consistent use of these equations for devices similar in design and utilization of the same type of mobility measurements gives a rather precise comparative characteristic of the noise level.

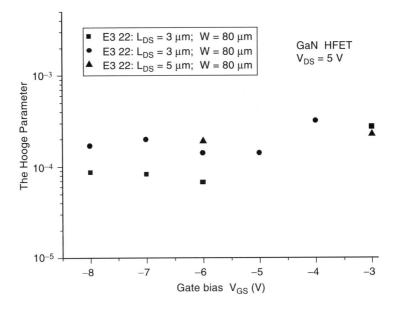

Figure 5.4. Hooge parameter α_H as a function of the gate bias. The results are shown for three devices biased at $V_{DS} = 5$ V. The gate dimensions are 0.25 μm by 80 μm. (After Balandin et al. [14]. Copyright © 1999 IEEE; reprinted with permission)

Finally, the Hooge parameter was calculated for different devices and bias points. The results for $V_{DS} = 5$ V are given in Figure 5.4. Despite some variations of the α_H values, they are all close to 10^{-4}. It is also seen that α_H almost does not depend on the gate bias. For all examined values of L_{DS}, the Hooge parameter was in the same range and did not show any clear trend. Based on this, it was concluded that the source–distance separation did not strongly affect the low-frequency noise performance of GaN HFETs. Very low values of the Hooge parameter determined for these devices indicate that GaN HFETs might be used in microwave applications, particularly with further progress in GaN material growth. For comparison, the measured α_H for the commercial GaAs MESFET NEC NE244 device is about 2×10^{-4}, and it is also not sensitive to the gate voltage [32]. In AlGaAs/GaAs MODFETs with gate dimensions 1 μm by 300 μm, the Hooge parameter was about 7.2×10^{-5} as reported by Duh and van der Ziel [32]. The value of the Hooge parameter determined for GaAs MODFETs by Peransin et al. [33] was 2×10^{-4}. All these numbers indicate that the overall noise level in the examined GaN HFETs is comparable to the one in conventional GaAs FETs. One should note that the Hooge parameter α_H in the presented analysis has been used as a figure of merit for purposes of comparison with other published results (similar to Refs. 29 and 30) and was not intended to suggest either the mobility fluctuation model or carrier density fluctuation noise model for the examined devices [14].

In an attempt to clarify the origin of the low-frequency noise in GaN devices, Balandin et al. [14] have extracted the exponent γ of the $1/f^\gamma$ noise power density for all examined devices and studied its gate bias dependence. The γ dependence on gate bias V_{GS} is shown in Figure 5.5 for four devices with different L_{DS}. All four devices are biased at $V_{DS} = 5$ V. One can see that γ is in the range of $1.0 > \gamma > 1.3$ and decreases with increasing (more negative) gate bias. Such a dependence of the magnitude of the $1/f^\gamma$ noise spectral density can be interpreted in terms of the modified carrier density fluctuation model [34,35], which is an extension of the well-established McWhorter formalism. The modified carrier density fluctuation model explains the liner dependence of the exponent γ on the gate bias by the nonuniformity of the trap distribution. The argument is that trap density across the bandgap varies with energy, and the band bending with increasing gate voltage will change the number of effective traps, and thus the time constants contributing to $1/f^\gamma$ noise. Since the quality of GaN/AlGaN heterojunctions remains rather poor, we expect a lot of imperfections (traps) nonuniformly distributed in both energy and space. Due to this reason, the above interpretation of the pronounced γ dependence on the gate bias seems rather realistic. We will address this issue in more detail in Section 5.4.

In order to model the noise response of an electronic system, one usually has to know the noise spectral density dependence on the applied gate bias. Here we present such a dependence for another set of GaN HFETs grown on sapphire

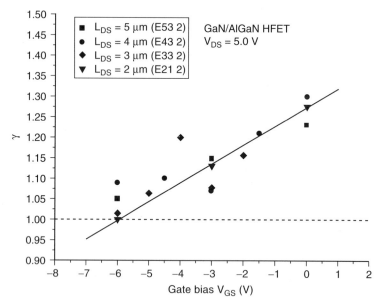

Figure 5.5. Dependence of the exponent γ of the $1/f^\gamma$ noise power density on the gate bias. The results are shown for four devices with different L_{DS}. All four devices are biased at $V_{DS} = 5$ V. The exponent γ approaches 1 with more negative bias. (After Balandin et al. [14]. Copyright © 1999 IEEE; reprinted with permission)

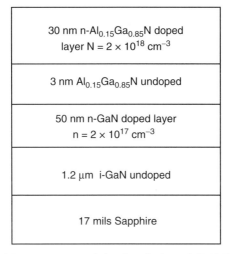

Figure 5.6. Uncapped layer structure of the doped channel GaN HFET with 15% of Al content in the barrier layer.

substrate. The data will be given for transistors with the larger gate length following the results of Balandin et al. [13] and Morozov et al. [36]. Devices that have been chosen for this investigation are GaN/Al$_{0.15}$Ga$_{0.85}$N doped channel HFETs fabricated by Wang et al. [37]. The MBE-grown layered structure used in this study is shown in Figure 5.6. The examined devices with the gate dimensions of 50 μm by 1 μm had been made from the same wafer in order to obtain consistent experimental dependence of the input-referred noise power spectrum on frequency and gate and drain voltages. The measurements again have been carried out for both the linear region of the device operation corresponding to low drain–source voltage, V_{DS}, and the saturation region of operation corresponding to $V_{DS} > 5$ V (see the current–voltage characteristics of these particular devices in Fig. 5.7).

Typical noise spectra of these devices for different gate bias V_{GS} values at a fixed drain voltage $V_{DS} = 5$ are shown in Figure 5.8. This drain voltage corresponds to the onset of the saturation region of operation for the device. The threshold voltage in both figures is $V_T = -4.6$ V. As one can see, the slope γ in these spectra is very close to 1. One can also notice from Fig. 5.8(a) that, for the positive gate bias, there is about an order of magnitude difference in noise figure for the spectra at $V_{GS} = 2$ V and at $V_{GS} = 0$ V. This difference becomes very small at negative V_{GS}. Figure 5.8(b) shows that the noise spectra change very little as the gate bias voltage varies from 0 to -3 V. Considering the error of the measurement, which was estimated to be 10% [13], one can conclude that there is no gate bias dependence for $V_{GS} < 0$ in the saturation regime (high V_{DS}). Since $V_{DS} = 5$ V is close to $-V_T$, one may assume that this change in the noise behavior is related to the pinch-off conditions, that is, $V_{DS} = V_{GS} - V_T$. In the linear regime of operation ($V_{DS} = 0.5$ V), Balandin et al. [13] reported a pronounced gate voltage dependence of the noise level for gate biases in

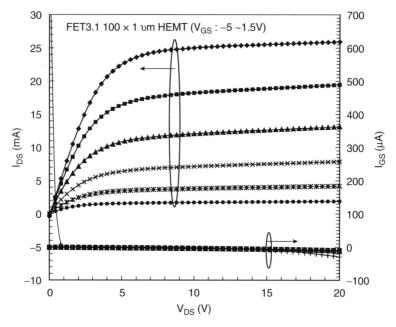

Figure 5.7. The dc current–voltage characteristics for GaN HFET (FET31) grown on sapphire.

the examined range $-1.5\text{ V} < V_{GS} < 1.5\text{ V}$. An example of such dependence is shown in Figure 5.9 for several values of the frequency of operation. This dependence was not plotted as a function of the effective gate voltage $V_G^* = V_{GS} - V_T$ since V_T itself was a function of V_{DS}. The inset in Figure 5.9 shows the V_T versus V_{DS} dependence.

The Hooge parameters for GaN/Al$_{0.15}$Ga$_{0.85}$N doped channel HFETs were determined using Eqs. (5.1) and (5.2). The resistance was found at a given V_{DS} during the noise measurements, while the mobility is determined for the layered structure using the Hall measurements. The gate leakage current for these devices was also rather small [13], and hence its impact on the device noise performance was neglected. The mobility was determined at room temperature to be $\mu = 650$ cm^2/(V·s) for one subset of the devices (FET31A), and $\mu = 320$ cm^2/(V·s) for another subset (FET31B). The values of the mobility extracted from the CV measurements were consistent with the Hall mobility measurement. The average values of the Hooge parameter have been calculated by applying Eq. (5.3) to the noise power spectral density at different frequencies. The Hooge parameter values for these high-quality transistors operating in the subsaturation region ($V_{DS} = 5$ V) turned out to be quite low. The lowest value attained at $V_G^* = 3$ V was $\alpha_H = 4.2 \times 10^{-5}$. With the increasing effective gate bias the Hooge parameter also increased. At $V_G^* = 5$ V the Hooge parameter was determined to be $\alpha_H = 1.7 \times 10^{-4}$.

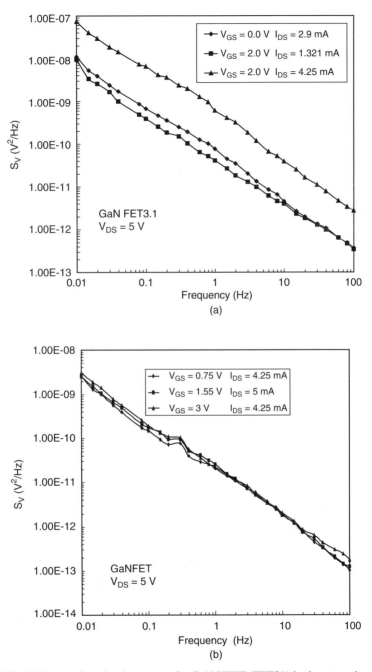

Figure 5.8. (a) Input-referred noise spectra for GaN HFET (FET31) in the saturation region of operation for different gate biases. (b) Input-referred noise spectra for negative values of the gate bias in the saturation region. The gate dimensions are 1 μm by 50 μm. (After Balandin et al. [13]. Copyright © 1999 IEEE; reprinted with permission)

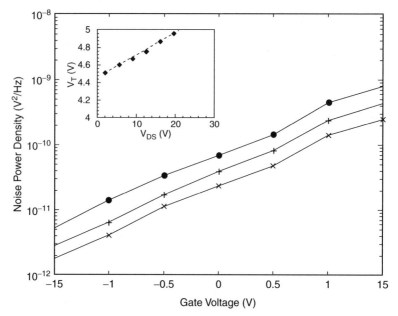

Figure 5.9. Noise power spectral density as a function of the gate bias in the linear regime of operation. The results are shown for three frequencies $f = 0.1$, 0.3, and 0.6 kHz, from the upper to lower curve, respectively. Inset shows the threshold voltage dependence on V_{DS}. (After Balandin et al. [14]. Copyright © 1999 IEEE; reprinted with permission)

5.3 EFFECT OF DOPING ON THE NOISE LEVEL IN GaN HFETs

It is known that GaN is a strong piezoelectric [25,26]. In a proper designed heterostructure, the lattice constant mismatch between GaN and $Al_xGa_{1-x}N$ layers is accommodated by the internal strain rather than by the formation of misfit dislocations. Because of the piezoelectric effect, this strain induces an electric field and significantly changes the carrier distribution near the interface [25,26,38]. The latter leads to an increase in the charge density in the two-dimensional (2D) channel [39]. Indeed, because of the piezoelectric effect, the strain induces an electric field that significantly changes carrier distribution near the interface [26]. The piezoelectric polarization P_z along the growth direction z associated with the effect is given by a simple formula,

$$P_z = 2d_{31}(c_{11} + c_{12} - (c_{13})^2/c_{33})\varepsilon_{xx}, \qquad (5.4)$$

where c_{lm} are the elastic stiffness coefficients, d_{31} is a piezoelectric tensor component, and ε_{xx} is the strain in the x direction. Here we assumed the (0001) growth direction, so that only one component of the piezoelectric tensor (d_{31}) is important.

The piezo-induced electric charge density at the interface between GaN and $Al_xGa_{1-x}N$ is given by

$$Q_z^{piezo} = \text{div } \mathbf{P} = qN_z, \quad (5.5)$$

where N_z is the charge density at the interface. The latter leads to an increase in the change density in the two-dimensional channel and manifests itself as an additional "piezo doping." The device channel or barrier regions can also be doped externally to increase the total number of carriers. In addition, an external doping can be used in order to increase the total number of carriers. Owing to this strong piezoelectric effect, one can design a GaN HEMT-like device without intentional external doping. Thus it is very important to investigate how external doping, or its absence, affects the noise level in GaN transistors. In this section we outline results of this investigation reported by Balandin et al. [21,36,40,41].

For comparative study, two $GaN/Al_xGa_{1-x}N$ HEMT-type structures have been designed, in which approximately the same sheet carrier density in the 2D channel was obtained by different means. Both structures were grown on semi-insulating 4H-SiC substrate. The structure P1 was externally doped to $N_d = 2 \times 10^{18}$ cm^{-3} and had a low piezoelectrically induced charge density due to the small Al content in the barrier layer, resulting in a small strain ε_{xx} ($\varepsilon_x x \approx x$, where x is the Al mole fraction). The device structure F2 was not doped externally ($N_d = 0$) but had a higher piezoelectrically induced charge density due to the high Al content (more than two times) used in the barrier layer. It has been shown experimentally by Asbeck et al. [38] that the piezoelectric charge density is linearly proportional to the aluminum content in the barrier layer. As a result, the stronger piezo-effect in the structure F2 approximately made up the loss of the sheet density due to the absence of the external doping. The only major differences for structures P1 and F2 were the doping density and the Al content. The layer structure of both devices was the same as used in the design of high electron mobility transistors (HEMTs) and is shown in Figure 5.10.

The actual aluminum content was determined by photoluminescence (PL) and Rutherford backscattering (RBS) and was found to be 14% for the doped channel device (P1) and 33% for the undoped channel device (F2). The external doping of the barrier and channel regions, unavoidable background charges, and the piezo-effect resulted in a sheet electron concentration of about 1.1×10^{13} cm^{-2} and 1.2×10^{13} cm^{-2} for P1 and F2 device structures, respectively. Electron Hall mobility at room temperature was determined to be 616 cm^2/(V·s) for the doped HFETs, and 1339 cm^2/(V·s) for the undoped HFETs. Electron Hall mobility at temperature $T = 77$ K was determined to be 1037 cm^2/(V·s) for the doped HFETs, and 5365 cm^2/(V·s) for the undoped HFETs. This significant difference in the mobility is expected because the regular external channel doping introduces additional scattering centers in the channel, which deteriorates electron mobility. Despite this difference, both types of devices had rather similar electrical characteristics (breakdown voltage $V_{DS} > 70$V, threshold voltage $V_{th} = -5.5$ V, and transconductance $g_m = 160$–180 mS/mm). The dc current–voltage characteristics in the range of the drain and gate

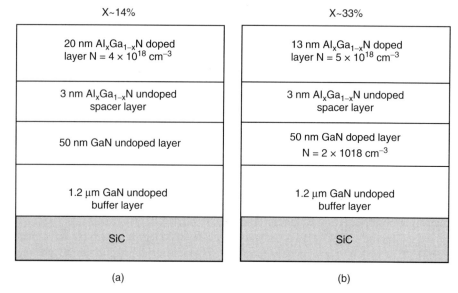

Figure 5.10. Uncapped layered structures of (a) the undoped channel and (b) the doped channel GaN/Al$_x$Ga$_{1-x}$N HFETs. Note that the undoped channel HFET has a higher aluminum content of the barrier layer and, thus, a higher piezolectric charge density to compensate for the absence of channel doping.

voltages used for the noise measurements are shown in Figure 5.11. The inset shows how dc current in these transistors changes with temperature.

One can easily verify that indeed different Al content in the barrier is responsible for modified piezoelectric field intensity and resulting different 2D channel carrier concentration. The sheet carrier concentration N_S due to the channel doping and piezoelectric effect can be written approximately as $N_S = N_P + N_D W$, where N_P is the surface density of piezoelectrically induced charges, N_D is the volume doping concentration, and W is the channel layer thickness. Knowing the resulting sheet electron concentration, we can express the difference in the piezoelectrically induced charges as $\Delta N_P = N_P^U - N_P^D = N_S^U - N_S^D + N_D W$, where superscripts U and D denote the undoped and doped channels, respectively. Substituting the appropriate numbers in the above equation, we finally obtain $\Delta N_P = 1.1 \times 10^{13}$ cm^{-2}. According to Asbeck et al. [38] N_P rises approximately linearly with the aluminum mole fraction x, following the rule $\Delta N_P / \Delta x = 5 \times 10^{13}$ cm^{-2}. Applied to the devices described above, this gives the change in the piezoelectric charge density of about 0.95×10^{13} cm^{-2}. As one can see, this number is very close to N_D, obtained from actual sheet carrier densities (within 14% error). This confirms the fact that piezoelectric charges approximately compensate the absence of the channel doping and makes the comparative noise study meaningful. Some discrepancy may be attributed to a different barrier height, which leads to different probabilities of escape for an electron.

EFFECT OF DOPING ON THE NOISE LEVEL IN GaN HFETs

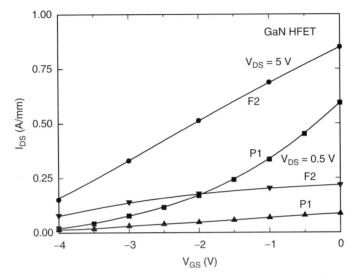

Figure 5.11. The dc current–voltage characteristics for the doped channel (P1) and undoped channel (F2) GaN HFETs shown for the same range of the drain and gate biases used in the noise measurements.

The devices selected for this study have a fixed gate length $L_G = 1.0$ μm and the gate width $W = 50$ μm. At drain–source voltage $V_{DS} = 5$ V, a doped channel device P1 has the transconductance $g_m = 160$ mS/mm at $V_{GS} = 0$ V; an undoped channel device F2 has the transconductance $g_m = 182$ mS/mm at $V_{GS} = -2.5$ V. The measurements are carried out for both the linear region of the device operation corresponding to low V_{DS}, and for the onset of the saturation region of operation (subsaturation) corresponding to $V_{DS} = 5$ V. The experimental setup used for these measurements is analogous to the one described in Balandin et al. [13,14].

The results of the noise measurements presented in Figure 5.12 revealed a significant and important difference in the noise level in the externally doped GaN HFETs and undoped HFETs. The experimental results indicate a two-orders-of-magnitude reduction in the input-referred noise spectral density of the undoped (F2) device (with a higher piezoelectric charge density) with respect to the noise density of the externally doped channel devices (P1). The threshold voltages for these particular devices are $V_T = -4.5$ V, $V_T = -5.5$ V for type P1, respectively. The results presented in Figure 5.12 were measured in the linear regime ($V_{DS} = 0.5$ V). The difference in the low-frequency noise level is particularly significant in view of the fact that these devices have comparable characteristics, for example, total sheet carrier concentration, g_m, and V_T.

At the onset of the saturation region ($V_{DS} = 5$ V), the noise spectral density for the undoped channel device is still significantly smaller than that for the externally doped channel device (see Fig. 5.13) with the same carrier density. The slope γ of the $1/f^\gamma$ dependence in all noise power density spectra is close to 1, although the noise

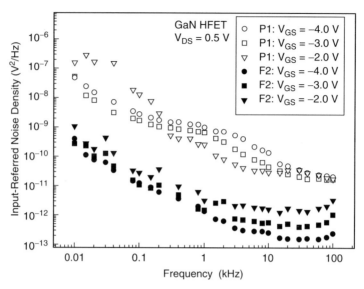

Figure 5.12. Input-referred noise spectra for the doped channel GaN HFET (P1) and the undoped channel GaN HFET (F2) in the linear regime ($V_{DS} = 0.5$ V). Note a significant difference (up to two orders of magnitude) in the input-referred spectral density between the two types of devices for all gate biases.

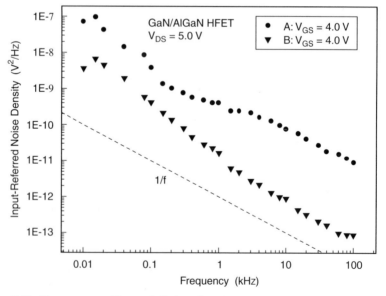

Figure 5.13. The same as Figure 5.12 but for the subsaturation regime of operation ($V_{DS} = 5.0$ V).

power varies for different devices and various gate bias values. In order to have a quantitative assessment of the overall noise in the devices, the Hooge parameter has been extracted using Eq. (5.3). The calculation procedure is similar to that reported in Balandin et al. [13,14]. The significant difference in the noise spectral densities, although offset by the mobility difference between the doped and undoped channel devices, translates into a corresponding difference in the Hooge parameters, particularly in the linear regime. At $V_{GS} = -2$ V the Hooge parameters are α_H (doped) $= 8.3 \times 10^{-3}$ and α_H(undoped) $= 7.8 \times 10^{-5}$. In the subsaturation region the difference is less pronounced, although external channel doping still degrades the noise figure: α_H(doped) $= 7.8 \times 10^{-3}$ and α_H(undoped) $= 1.3 \times 10^{-3}$ for $V_{GS} = -4$ V and $V_{DS} = 5$ V, respectively. One can also notice a trace of the G-R bulge in the noise spectra of the doped channel devices in Figures 5.12 and 5.13. The presence of the G-R peak in doped channel devices may indicate the carrier trapping–detrapping mechanism of the $1/f^\gamma$ noise in GaN/AlGaN structures, as well as the fact that dopants serve as additional trapping centers. The latter can serve as a good motivation for a low-temperature study of the flicker noise in GaN HFETs.

5.4 LOW-TEMPERATURE NOISE STUDY OF GaN TRANSISTORS

The electron (hole) states in semiconductors can be divided into two broad classes: delocalized states in the conduction (valence) band and localized states. The electrons (holes) localized on defects or impurities do not participate in the conduction of the electric current. The transition of an electron (hole) from a localized state to a delocalized state or creation of an electron–hole pair (via optical or thermal excitation) is called generation. The inverse process is called recombination. The terms trapping–detrapping are conventionally used when the localized state is associated with an impurity. The fluctuations of the number of charge carriers, for example, delocalized electron (hole) states due to random generation–recombination processes, lead to fluctuations of the resistance of the conducting channel. The latter produce fluctuations of the electric current passing through the channel and results in what is called generation–recombination (G-R) noise.

As we briefly described in the introduction, there exist some noise models, that closely relate $1/f^\gamma$ flicker noise to the G-R noise in semiconductors (see Section 5.1). For example, in McWhorter's model [5], fluctuations in the number of charge carriers in the conducting channel of a semiconductor device arise from the exchange of electrons between the channel and traps in the adjacent oxide layer. Transfer of electrons from the channel to traps and back occurs by means of tunneling with the characteristic inverse relaxation time given by $\tau^{-1} = \tau_0^{-1} \exp(-z/\lambda)$, where tunneling constant λ and τ_0^{-1} are the device and material specific constants. It can be shown [4,5] that due to this exponential dependence of the relaxation time on the tunneling parameter and provided that the traps (and 1) are distributed uniformly, the G-R processes can produce fluctuations with $1/f^\gamma$ spectrum where $\gamma = 1$. If the trap distribution is not uniform, the G-R processes produce fluctuations with $\gamma \approx 1$, which is gate bias dependent [34,35]. This type of dependence has been observed by Balandin et al. [14] for quarter-micron gate GaN HFETs, and it is described in Section 5.2 (see Fig. 5.5). Parameter γ decreases with increasing (more negative) gate bias for GaN transistors, which is similar to the observation made for n-type MOSFETs [34]. In order to have conclusive evidence that the dominant noise source is indeed related to the carrier trapping–detrapping by the localized states, one has to demonstrate pronounced G-R peaks in the noise spectra. It usually can be done at low temperature. In semiconductors with a high density of traps and impurities, the lifetime of charge carriers at low temperatures may be of the same order as the energy relaxation time. In this case the G-R processes have to be included in the carrier transport formalism together with all other scattering processes.

The low-temperature noise characteristics of the doped and undoped channel devices were examined in the temperature range from 77 to 300 K. Figure 5.14 shows these characteristics in the liner region of operation for the device F2 [40, 41]. As one can see there are no clear signs of G-R peaks for these devices. Although not shown in the figure, no G-R peaks have been observed in the linear region of operation for the doped channel devices either [42–44]. In the subsaturation region, a pronounced peak was observed at about 3–4 kHz for the doped channel devices (see Fig. 5.15). At the same time, no peak in spectra of the undoped channel device was observed. Moreover, the 77 K spectrum of the undoped channel GaN HFET flattens at the corner frequency of about 1 kHz due to the Johnson noise. The level of the Johnson noise of 6×10^{-16} V^2/Hz is in agreement with the value estimated from the Nyquist formula $S_G(f) = 4k_B TG(f)$, where k_B is the Boltzmann constant, T is the temperature, and $G(f)$ is the frequency-dependent conductance.

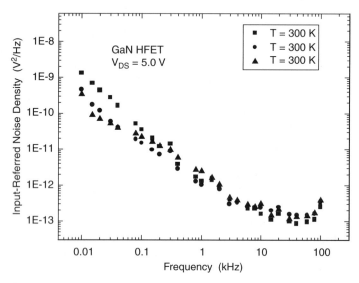

Figure 5.14. Low-temperature noise characteristics of the undoped channel GaN HFET. No generation–recombination bulges are observed.

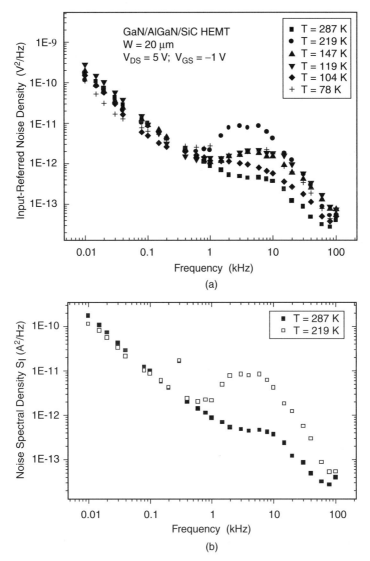

Figure 5.15. Low-temperature noise characteristics of the doped channel GaN HFET in the subsaturation regime. Generation–recombination bulges are clearly seen at frequency $f \approx 3$–4 kHz for the two different devices shown in (a) and (b).

The spectral density of the G-R fluctuations follows from the Langevin kinetic equation [4] and is given as

$$\frac{S_R(f)}{R^2} \simeq \frac{4\tau}{Vn(1+\omega^2\tau^2)}, \qquad (5.6)$$

where V is the macroscopic volume of the sample, $\omega = 2\pi f$ is the angular frequency, and R is the resistance of the sample. Fitting the Lorentzian shape, Balandin et al. [21,42] have extracted the trap activation energy $E_a \approx 0.35$ eV. The fitting procedure was based on the method described in Kirtley et al. [45]. One can also see in Figure 5.15 that $\gamma > 1.5$ for the undoped device. This unusual behavior can be attributed to several possible reasons. One possible explanation may be related to a particular trap density distribution, which in this case is strongly nonuniform. It was shown by Surya and Hsiang [35] that the γ parameter may vary over a wide range, depending on the distribution of the tunneling time constants for the carrier trapping–detrapping processes. Another possible explanation of large values of γ (close to 2) for a different material system was given by Chen [46]. The $1/f^2$ noise spectrum obtain for Al-based thin films was attributed to (1) the rate fluctuation of vacancy diffusion around the grain boundaries, and (2) the linear drift of the film resistance during the noise measurements. Similar effects, as well as temperature drift during the noise measurements reported by Balandin et al. [21], might have led to the $1/f^\gamma$ noise spectrum with a very large γ parameter value.

5.5 DEEP-LEVEL TRANSIENT SPECTROSCOPY OF GaN TRANSISTORS

Although noise spectroscopy has been used for many years to study deep levels in semiconductors, it is the deep-level transient spectroscopy (DLTS) that gives direct characteristics of deep-level defects. DLTS measures the capacitance or current change of a reversed biased junction when deep-levels emit their carriers after they were charged by a forward bias pulse. From the temperature dependence of the emission rate, the activation energy of a deep level can be deduced. Since flicker noise in the GaN/AlGaN system is most likely related to the carrier capture and emission by some traps (see Sections 5.3 and 5.4 of this chapter), we used DLTS to probe the defects.

The observation of G-R peaks in the low-temperature noise spectra of doped channel devices was an indication of the presence of some carrier traps. For this reason, we review here results of our DLTS study of the *doped* channel GaN HFETs grown on both SiC and sapphire [41–43]. A particular type of measurement that has been reported by Balandin et al. [41–43] was Fourier transform current deep-level transient spectroscopy (I-DLTS). The devices were biased in the same configuration as they would be in a circuit. Then the gate was used to pulse the drain–source current and the transient of the current pulse was used in a manner similar to the capacitance transient in DLTS. The spectra were recorded using a pulse width t_p from 100 μs to 1 ms. In Figure 5.16 we present a typical DLTS Arrhenius plot of $T^2\varepsilon$ versus $1/T$, where ε is the emission rate at a temperature T. The data are shown for the GaN doped channel HFET grown on SiC (P1). An activation energy $E_a = 0.201$ eV below the conduction band has been extracted from this Arrhenius plot [41–43].

The results of the DLTS study for other samples are summarized in Table 5.1. The bias conditions for the measurements were the following: the drain–source voltage

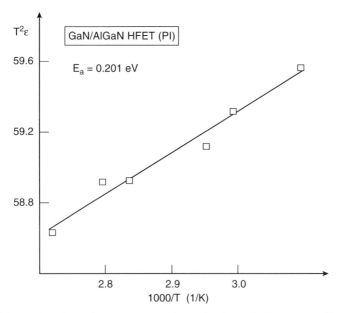

Figure 5.16. Results of DLTS measurements shown for GaN HFET grown on SiC susbtrate. The time constant for this measurement was $t_p = 100$ μs.

$V_{DS} = 1$ V for all devices; the reverse and forward biases $V_R = -7.5$ V and $V_P = 5.0$ V for L1 HFETs; and $V_R = -4.0$ V and $V_P = 3.0$ V for P1 HFETs. The values of the activation energy shown here are rather typical for the examined devices. From these values, one can conclude that the carriers are trapped and detrapped by the deep levels and not by the Si shallow states. It is also interesting to note that HFETs on sapphire substrate have an obviously higher activation energy ($E_a \approx 0.85$ eV) than devices on SiC ($E_a \approx 0.20$–0.36 eV). The origin of this difference in the deep-level activation energy is not clear since both types of devices were doped with Si and had a similar structure (see Section 5.3). Devices grown on sapphire and SiC had approximately the same flicker noise level provided that the device parameters (gate length, gate width, doping, etc.) were close. One possible explanation may be that the observed traps are not related to Si dopants but are due to some other defects introduced during growth, which are different for structures grown on SiC and sapphire substrates.

TABLE 5.1. Results of DLTS Study for GaN/AlGaN HFETs on SiC (P1) and Sapphire (L1) Substrates

Parameter	L1 (A)	L1 (B)	P1 (A)	P1 (B)	P1 (C)
E (eV)	0.854	0.846	0.201	0.288	0.365
t_p (μs)	100.0	100.0	100.0	100.0	100.0

The values of the activation energy obtained by us for the P1 device are close to the ones reported by Auret et al. [47]. They have studied proton bombardment-induced electron traps in epitaxially grown n-GaN. In their DLTS study of the control (not bombarded) sample, energies of 0.21, 0.27, and 0.45 eV have been obtained. At least two of these values are in agreement with our observations (see Table 5.1). The activation energy of about 0.42 eV was obtained by Levinshtein et al. [48,49] and Rumyantsev et al. [50]. In recent DLTS measurements, the energy of 0.412 eV was also extracted for some GaN samples [56]. Since the activation energy ($E_a \approx 0.35$ eV) extracted from the low-temperature flicker noise spectral density [21] is close to the DLTS values [41–43], it is reasonable to assume that carrier trapping–detrapping is a dominant noise mechanism in the doped GaN/AlGaN structures. At the same time, it is still not possible to exactly identify the origin of these localized states. The significant difference in the measured activation energies for the doped devices grown on SiC and sapphire substrates seems to suggest that these defects may have different origins.

5.6 THE DOMINANT NOISE SOURCES AND METHODS OF NOISE REDUCTION

From the experimental results reviewed in the previous sections of this chapter, one can see that it is still too early to present definite conclusions about the exact noise mechanisms in GaN devises, although we are able to formulate certain rules of the noise behavior and predict possible ways of its reduction. The first observation is that the noise spectral density in GaN devices is still much larger than that in their conventional counterparts, such as Si or GaAs devices. The noise level is particularly high in volume GaN samples (large resistors) with the Hooge parameters on the order of $1-10^{-2}$ [22,24,48]. The nature of this extremely high flicker noise level measured in a variety of different samples is still unknown. The lowest value of the Hooge parameter extracted so far for high-quality GaN HFETs is on the order of 10^{-4} [14]. This result was obtained independently by several groups [13,24,42,43].

In all presented spectra (see Fig. 5.3, 5.7, and 5.17), there is a clear dependence of the noise spectral density on the applied gate bias, except when the absolute value of the gate bias is so large that the channel is pinched off (see Section 5.2). This means that the flicker noise in GaN transistors indeed originates in the conducting channel, and it is not an excess noise originating in device contact regions [56]. That was particularly obvious in the GaN HFETs with improved omhic contacts. The fabrication technique and the composition of these contacts have been reported by some of us [28]. Another important observation is that the noise level is about the same in the best devices grown on sapphire and SiC. Although a great deal of effort has been spent on attempts to clarify the effect of substrate on the noise performance of GaN devices [40–44], we did not see any definite correlation. In the best device grown on both SiC and sapphire, the Hooge parameter was on the order of 10^{-4} [13, 14,21]. The Hooge parameter did not show any clear dependence on the applied gate

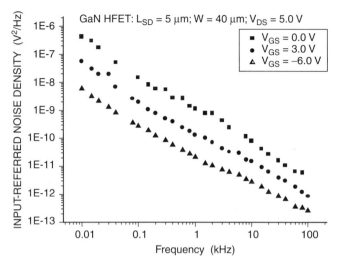

Figure 5.17. Input-referred noise spectra for the doped channel GaN HFETs. Note the strong gate bias dependence.

bias and source–drain separation distance. It was somewhat lower for devices with smaller gate length.

The results of the noise measurements by Balandin et al. [21,40–43] reviewed in Section 5.3 revealed a significant difference in the noise level in the externally doped GaN HFETs and undoped HFETs. These results indicate a two-orders-of-magnitude reduction in the input-referred noise spectral density of the undoped GaN devices (with a higher piezoelectric charge density) with respect to the noise density of the externally doped GaN devices with similar electric characteristics. This observation can lead to a potentially powerful method of noise reduction in GaN transistors, which utilizes the strong piezoelectric effect in GaN [44].

Analyzing the presented data, one can conclude that the dominant noise contribution in GaN devices is most likely related to localized states associated with defects and impurities. It is manifested by the presence of generation–recombination peaks in the low-temperature noise spectra, particularly in the subsaturation region of the externally doped GaN HFETs; and by large deviation of the γ parameter of the $1/f^\gamma$ spectrum from unity. A linear type of dependence of γ on the applied gate bias [14] may also attest to the nonuniform distribution of traps in the vicinity of the Fermi level [34,35]. This means that further improvement of material quality and reduction of the number of traps and defects may lead to significant noise suppression in GaN transistors.

So far there have been few attempts to create a theoretical description of noise in GaN devices. Levinshtein et al. [51] have attempted to extend their theory on GaN devices. According to this model, the level of $1/f$ noise should be proportional to the density of the tail states near the band edges. This density of states in the band tails depends also on the structural perfection of the material, and it is much higher in the

GaN conduction band than that in the Si or GaAs [52]. Balandin et al. [41] outline a way to modify a classical McWhorter model to apply it to GaN/AlGaN HFETs in order to obtain the observed linear dependence of γ on the applied gate bias [14].

As we briefly stated in the introduction, mobility fluctuations due to carrier scattering on lattice and crystal defects can be another important mechanism contributing to the flicker noise [8,15,18,19]. Indeed the resistance R is given as

$$R = \frac{L^2}{q\mu N}, \tag{5.7}$$

where L is the length of the resistor, q is the electron charge, and N is the number of free carriers in the sample. Thus R can fluctuate based on either μ or N or both of them. If μ and N fluctuate independently, then

$$\frac{\delta R}{R} = -\frac{\delta \mu}{\mu} - \frac{\delta N}{N}. \tag{5.8}$$

Recasting the last equation in terms of spectral density [32], we can write

$$\frac{S_R(f)}{R^2} = \frac{S_\mu(f)}{\mu^2} + \frac{S_N(f)}{N^2}. \tag{5.9}$$

If the first term on the left-hand side of this equation predominates, the low-frequency noise is due mostly to mobility fluctuations. This situation can be expected as the material quality of the GaN/AlGaN system improves. Since mobility at room temperature depends on electron scattering on phonons (acoustic and optic), one can expect that modification of electron–phonon scattering rates in heterostructures and nanostructures may bring about significant change in the noise level. It was shown by Balandin et al. [53–56] and Bandyopadhyay et al. [57] that spatial confinement of acoustic phonons in semiconductor quantum wires (narrow conducting channels) leads to strong alteration of electron–phonon scattering rates and thus can be used for noise suppression in properly designed nanostructures [53–57].

As we outlined in the introduction, flicker or $1/f$ noise in solid state devices contributes to phase noise of electronic systems, which are built on these devices. Phase noise is of great importance in phase-modulated systems, or wherever the frequency and phase stability have to meet stringent requirements, such as in oscillators and frequency synthesizers. The methods of flicker noise reduction outlined in this section are general enough and can be applied not only for GaN devices but also for other solid state devices. The latter is important for the development of a new generation of low-noise devices and electronic systems.

5.7 CONCLUSIONS

In this chapter we have reviewed recent results of the low-frequency noise measurements in GaN heterostructure field-effect transistors (HFETs). Data from the deep-level transient spectroscopy (DLTS) study of doped and undoped channel GaN/AlGaN HFETs, which may clarify the nature of the localized states in GaN material, have also been presented. We specifically concentrated our attention on the effect of doping on the flicker noise level in GaN transistors. The increase of the aluminum content in the barrier region of the undoped GaN/AlGaN HFETs, and corresponding higher piezoelectric field, compensates for the loss of the sheet carrier density due to the absence of external doping. The latter leads to a carrier density in the channel that is comparable to that induced by external doping. It was shown that the flicker noise level in such undoped devices is significantly lower than that in externally doped devices of comparable characteristics. The low-temperature noise spectra for doped GaN devices show a clear generation–recombination peak in the subsaturation region. The value of the activation energy extracted from these noise peaks is consistent with the activation energy found in our DLTS study. The latter may be an indication that the carrier trapping–detrapping mechanism is the dominant noise source in contemporary GaN devices. In the last section of this chapter we also outlined several methods that may lead to the reduction of noise in GaN transistors. The results reviewed in this chapter are important for a new generation of electronic technologies requiring a low phase noise level.

ACKNOWLEDGMENT

The authors acknowledge the financial support of MURI. We also thank Professor C. R. Viswanathan (UCLA), Professor S. Morozov (Institute of Microelectronics, Chernogolovka), G. Wijeratne (UCLA), Dr. E. N. Wang (TRW Inc.), and Dr. M. Wojtowicz (TRW Inc.) for their help with experimental measurements. We are indebted to Dr. S. J. Cai (UCLA) and R. Li (UCLA) for providing excellent GaN HFETs for this study. One of the authors (A.B.) thanks Professor M. Levinshtein (Ioffe Institute, Russian Academy of Sciences) for many illuminating discussions on $1/f$ noise in GaN devices.

REFERENCES

1. R. Pettai, *Noise in Receiving Systems*, Wiley, New York, 1984.
2. J. E. Firle and H. Winston, *Bulletin of the American Physical Society*, Vol. 30, No. 2 (1955).
3. W. R. Bennett, *Electrical Noise*, McGraw-Hill, New York, 1960, pp. 101–109.
4. Sh. Kogan, *Electric Noise and Fluctuations in Solids*, Cambridge University Press, Cambridge, U.K., 1996, p. 203.
5. A. L. McWhorter, in *Semiconductor Surface Physics*, R. H. Kingston (Ed.), University of Pennsylvania Press, Philadelphia 1957, p. 207.
6. K. S. Ralls, W. J. Skocpol, L. D. Jackel, R. E. Howard, L. A. Fetter, R. W. Epworth, and D. M. Tennant, *Physical Review Letters*, Vol. 52, 228 (1984).
7. S. Feng, P. A. Lee, and A. D. Stone, *Physical Review Letters*, Vol. 56, 1960 (1986).

8. A. van der Ziel, *Proceedings IEEE*, Vol. 76, 233 (1988).
9. M. Tacano, *IEEE Transactions on Electron Devices*, Vol. 40, 2060 (1993).
10. P. H. Handel, *Physical Review Letters*, Vol. 34, 1492 (1975).
11. P. H. Handel, *Physical Review A*, Vol. 22, 745 (1980).
12. S. J. Cai, R. Li, Y. L. Chen, L. Wong, W. G. Wu, S. G. Thomas, and K. L. Wang, *Electronics Letters*, Vol. 34, 2354 (1998).
13. A. Balandin, S. Cai, R. Li, K. L. Wang, V. Ramgopal Rao, and C. R. Viswanathan, *IEEE Electron Device Letters*, Vol. 19, 475 (1998).
14. A. Balandin, S. V. Morozov, S. Cai, R. Li, K. L. Wang, G. Wijeratne, and C. R. Viswanathan, *IEEE Transactions on Microwave Theory and Techniques*, Vol. 47, 1413 (1999).
15. J. Chang, A. A. Abidi, and C. R. Viswanathan, *IEEE Transactions on Electron Devices*, Vol. 41, 1965 (1994).
16. W. Y. Ho, C. Surya, K. Y. Tong, W. Kim, A. E. Bochkarev, and H. Markoc, *IEEE Transactions on Electron Devices*, Vol. 46, 1099 (1999).
17. F. N. Hooge, *Physics Letters*, Vol. 29A, 139 (1969).
18. L. K. J. Vandamme, *IEEE Transactions on Electron Devices*, Vol. 36, 987 (1989).
19. N. V. Dyakonova and M. E. Levinshtein, *Soviet Physics Semiconductors*, Vol. 23, 175 (1989).
20. M. E. Levinshtein and S. L. Rumyantsev, *Soviet Physics Semiconductors*, Vol. 19, 1015 (1985).
21. A. Balandin, S. Morozov, G. Wijeratne, S. J. Cai, R. Li, J. Li, K. L. Wang, C. R. Viswanathan, and Yu. Dubrovskii, *Applied Physics Letters*, Vol. 75, 2064 (1999).
22. M. E. Levinshtein, F. Pascal, S. Contreras, W. Knap, S. L. Rumyantsev, R. Gaska, J. W. Yang, and M. S. Shur, *Applied Physics Letters*, Vol. 72, 3053 (1998).
23. D. V. Kuksenkov, H. Temkin, R. Gaska, and J. W. Yang, *IEEE Electron Device Letters*, Vol. 19, 222 (1998).
24. M. E. Levinshtein, S. L. Rumyantsev, R. Gaska, J. W. Yang, and M. S. Shur, *Applied Physics Letters*, Vol. 73, 1089 (1998).
25. A. Bykhovski, B. Gelmont, and M. Shur, *Journal of Applied Physics*, Vol. 74, 6734 (1993).
26. A. Bykhovski, B. Gelmont, and M. Shur, *Journal of Applied Physics*, Vol. 81, 6332 (1997).
27. C. Skierbiszewski, T. Suski, M. Leszczynski, M. Shin, M. Skowronski, M. D. Bremser, and R. F. Davis, *Applied Physics Letters*, Vol. 74, 3833 (1999).
28. S. J. Cai, R. Li, Y. L. Chen, L. Wong, W. G. Wu, S. G. Thomas, and K. L. Wang, *Electronics Letters*, Vol. 34, 2354 (1998).
29. L. K. J. Vandamme, *Solid-State Electronics*, Vol. 23, 317 (1980).
30. L. K. J. Vandamme and H. M. M. de Werd, *Solid-State Electronics*, Vol. 23, 325 (1980).
31. F. N. Hooge, T. G. M. Kleinpenning, and L. K. J. Vandamme, *Reports on Progress in Physics*, Vol. 44, 479 (1981).
32. K. Duh, and A. van der Ziel, *IEEE Transactions on Electron Devices*, Vol. 32, 662 (1985).
33. J.-M. Peransin, P. Vignaud, D. Rigaud, and L. K. J. Vandamme, *IEEE Transactions on Electron Devices*, Vol. 37, 2250 (1990).

34. Z. Celik-Butler and T. Y. Hsiang, *Solid-State Electronics*, Vol. 30, 419 (1987).
35. C. Surya and T. Y. Hsiang, *Physical Review B*, Vol. 33, 4898 (1986).
36. S. Morozov, A. Balandin, S. Cai, R. Li, Yu. Dubrovskii, K. L. Wang, G. Wijeratne, and C. R. Wiswanathan, "Low flicker noise GaN/AlGaN HFETs with submicrometer channel," in *Nanostructures: Physics and Technology*, Zh. Alferov et al. (Eds.), Russian Academy of Sciences, St. Petersburg, Russia, 1999, p. 102.
37. K. L. Wang, R. Li, S. J. Cai, and A. Balandin, unpublished data.
38. P. M. Asbeck, E. T. Yu, S. S. Lau, G. J. Sullivan, J. Van Hove, and J. Redwing, *Electronics Letters*, Vol. 33, 1230 (1997).
39. M. S. Shur and R. Gaska, "GaN-based two-dimensional electron devices," in *Nanostructures: Physics and Technology*, Zh. Alferov et al. (Eds.), Russian Academy of Sciences, St. Petersburg, Russia, 1998, p. 524.
40. A. Balandin, R. Li, S. J. Cai, S. Morozov, and K. L. Wang, "Low-frequency noise performance of GaN/AlGaN heterostructure field effect transistors," *Bulletin of the American Physical Society*, Vol. 44, No. 1, Part I 236 (1999).
41. A. Balandin and K. L. Wang, "Piezo-doped low-noise GaN heterostructure field effect transistors," in *Proceedings of the International Meeting of the Electrochemical Society*, Honolulu, Hawaii, Oct. 1999, Vol. 99–2, p. 1864.
42. A. Balandin, R. Li, E. N. Wang, S. Cai, and K. L. Wang, "DLTS study of GaN/AlGaN heterostructures," in *Extended Abstracts of the 41st Electronic Materials Conference*, Santa Barbara, CA, June 1999.
43. A. Balandin, K. L. Wang, S. Cai, R. Li, C. R. Viswanathan, E. N. Wang, and M. Wojtowicz, "Investigation of flicker noise and deep-levels in GaN/AlGaN transistors," *Journal of Electronic Materials*, Vol. 29, No. 3, 297 (2000).
44. A. Balandin and K. L. Wang, "Piezo-doped low-noise GaN HFETs," in *Quantum Confinement VI*, M. Cahay et al. (Eds.), The Electrochemical Society, Pennington, NJ, 1999, pp. 325–332.
45. J. R. Kirtley, T. N. Theis, P. M. Mooney, and S. L. Wright, *Journal Applied Physics*, Vol. 63, 1541 (1988).
46. T. M. Chen, "Excess noise and reliability of Al-based thin films," in *Noise in Physical Systems and 1/f Fluctuations*, Vol. 285, P. Handel and A. Chung, (Eds.), AIP Conference Proceedings, St. Louis, MO, 1993.
47. F. D. Auret, S. A. Goodman, F. K. Koschnick, J.-M. Spaeth, B. Beaumont, and P. Gibart, *Applied Physics Letters*, Vol. 74, 407 (1999).
48. M. Levinshtein, private communications, 1999.
49. N. V. Dyakonova, M. E. Levinshtein, S. Contreras, W. Knap, B. Beaumont, and P. Gibart, *Semiconductors*, Vol. 32, 257 (1998).
50. S. Rumyantsev, M. E. Levinshtein, R. Gaska, M. S. Shur, A. Khan, J. W. Yang, G. Simin, A. Ping, and T. Adesida, *Physica Status Solidi A*, Vol. 176, 201 (1999).
51. M. E. Levinshtein, S. L. Rumyantsev, D. C. Look, R. J. Molnar, M. Asif Khan, G. Simin, V. Adivarahan, and M. S. Shur, *Journal of Applied Physics*, Vol. 86, 5075 (1999).
52. O. Ambacher, W. Rieger, P. Ansmann, H. Angerer, T. D. Moustakas, and M. Stutzmann, *Solid State Communincations*, Vol. 97, 365 (1996).
53. A. Balandin, K. L. Wang, A. Svizhenko, and S. Bandyopadhyay, *IEEE Transactions on Electron Devices*, Vol. 46, 1240 (1999).

54. A. Balandin, K. L. Wang, A. Svizhenko, and S. Bandyopadhyay. "The effects of low-dimensionality on the quantum $1/f$ noise," in *Proceedings of VII Van der Ziel Symposium on Quantum 1/f Noise and Other Low Frequency Fluctuations in Electronic Devices*, P. Handel (Ed.), AIP Conference Proceedings Series, American Institute of Physics, New York, 1999, p. 131.
55. A. Balandin, K. L. Wang, A. Svizhenko, and S. Bandyopadhyay, "$1/f$ noise in quantum wires," *Bulletin of the American Physical Society*, Vol. 43, No. 1, 360 (1998).
56. A. Balandin, *Electronics Letters*, Vol. 36, No. 10, 912 (2000).
57. S. Bandyopadhyay, A. Svizhenko, and M. A. Stroscio, *Physics of Low-Dimensional Structures*, Vol. X, 69 (1999).

6

POWER AMPLIFIER APPROACHES FOR HIGH EFFICIENCY AND LINEARITY

PETER M. ASBECK AND LARRY LARSON
Department of Electrical and Computer Engineering, University of California—San Diego

ZOYA POPOVIC
Department of Electrical Engineering, University of Colorado, Boulder

TATSUO ITOH
Department of Electrical Engineering, University of California—Los Angeles

6.1 INTRODUCTION

To reduce the power consumption of a wireless transceiver, greatest leverage is often provided by the output power amplifier of the transmitter. This chapter describes a series of novel concepts to increase microwave power amplifier efficiency. Within modern wireless systems, power amplifiers must meet exacting linearity specifications, which tend to be at odds with the need for high efficiency. To characterize these specifications and associated trade-offs, we first review the linearity requirements and dynamic range requirements of representative high performance communication systems. We then demonstrate concepts for power amplifier improvement that can meet the system requirements.

Output power levels and demands on efficiency and linearity are typically somewhat different for handsets (which rely on battery energy sources and have relatively modest demands on power and linearity) and base stations (which have

---

RF Technologies for Low Power Wireless Communications
Edited by Tatsuo Itoh, George Haddad, and James Harvey
ISBN 0-471-38267-1 Copyright © 2001 by John Wiley & Sons, Inc.

higher output power and more stringent linearity requirements, although their efficiency concerns are not as overriding as for handsets). The approaches outlined here can be applied to both handsets and base stations. They have particular relevance to the scenario of peer-to-peer communication, where there are no unique base stations. This scenario characterizes many communications networks for military applications.

One of the approaches discussed in this chapter is based on amplifiers operating in Class AB, used in conjunction with an efficient, rapidly modulated dc–dc converter that is capable of providing a supply voltage optimized in accordance with the instantaneous output power. It is shown that this system can provide dramatic increases in efficiency, particularly if the power amplifier is required to operate over a wide range of output power levels.

Other approaches described in this chapter are based on the operation of transistors in switching mode. By ensuring that the transistor current is zero when its output voltage is high, and conversely, that the transistor voltage is minimum when its current is high, the amplifier efficiency can in principle be increased dramatically over Class A or AB operation. Class E and F switching-mode amplifiers are also discussed, which achieve efficiency above 80% at frequencies above 5 GHz. Typically the switching-mode amplifiers have relatively poor linearity; thus they are not adequate for wireless communications applications. Here we show that a great deal of the efficiency of the Class E and F amplifier is preserved if the output is "backed off" to the level needed in representative handset applications. Also discussed are several approaches that can achieve in principle very high linearity without sacrificing the efficiency of switching-mode applications. These include the LINC (linear amplification with nonlinear components) and the bandpass Class S approaches.

An additional approach for transmitter improvement is based on a synergistic design of the power amplifier and the output antenna. By removing the typical constraint of matching each component to 50 Ω, it is possible to simplify the system, while improving its characteristics. For maximization of efficiency, the antenna design must take into account the requirements of the power amplifier impedances at various harmonic frequencies, as well as the fundamental.

6.2 LINEARITY AND EFFICIENCY REQUIREMENTS IN WIRELESS COMMUNICATIONS

Although the output signal for FM-modulated analog transmission systems (e.g., AMPS) has a constant envelope, the majority of the spectrally efficient wireless systems use modulation formats that have a variable signal envelope. For example, the output signal for filtered quaternary phase shift keying (QPSK) is shown in Figure 6.1, in both the time domain and frequency domain (for a raised-cosine filter characterized by $\alpha = 0.5$). The spectrum is well confined to a narrow frequency channel, which allows multiple users within an allocated frequency band, but it has variations in instantaneous power with a peak-to-average power ratio of 5.2 dB. If

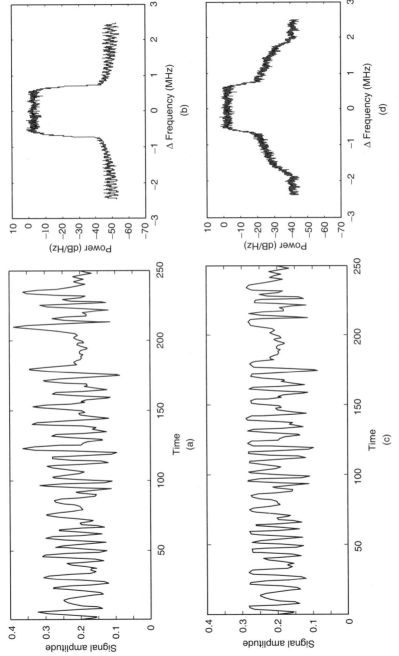

Figure 6.1. (a) Representative output envelope for a filtered QPSK signal in time domain (time units on the horizontal axis correspond to one-quarter of the bit interval); (b) same signal in frequency domain; (c) signal passed through clipping amplifier; (d) resulting output spectrum, showing adjacent channel power.

such a signal is passed through an amplifier that saturates (and thus clips the high-amplitude signals) then a distorted output waveform will be obtained (such as that shown in Fig. 6.1(c)), and correspondingly an output spectrum is obtained that is spread over a wide range of frequencies (as shown in Fig. 6.1(d)). The nonlinear action of the amplifier leads to the generation of power in adjacent channels. This power, measured with the adjacent channel power ratio (ACPR) must be kept within tight bounds (as specified by the FCC for commercial air interface standards) to prevent interference to neighboring users. The clipping also leads to a loss of modulation accuracy within the transmitted channel itself, which degrades the signal-to-noise ratio. As a result, for filtered QPSK and similar signals, the power amplifier must be operated in a mode where the clipping of large amplitude signals is kept to a minimum. This, in turn, dictates that the average output power of the amplifier must be substantially lower than its maximum (or "saturated") output power.

The signals that must be transmitted from base stations typically have even higher peak-to-average power ratios than those from handsets (and thus they require greater "output power backoff" in order to preserve linearity). A base station serving multiple users typically transmits signals at different carrier frequencies at the same time. The different carriers add with random phases, and at some instances constructively interfere in a way that considerably enhances the instantaneous peak power. This leads to high values of peak-to-average power ratio. In code division multiple access (CDMA) systems, multiple signals that have the same carrier frequency but are spread with different sequences are added together at the base station for transmission (with, in general, different power levels). This leads to a base station output signal that also has high peak power relative to the mean. To preserve the signal fidelity through the power amplifier, a relatively high value of the saturated power relative to the average power must be maintained (typically $>10\,\text{dB}$).

Table 6.1 summarizes the peak-to-average power ratio for representative wireless communications signals. Signals for the handsets have differing values of peak-to-average power, as a result of differences in modulation format. Signals from base stations tend to behave similarly for all formats and have an output power distribution that resembles that of white Gaussian noise modulation. The variation of the

TABLE 6.1. Peak-to-Average Power Ratio for Signals Used in a Variety of Wireless Communications Systems

System	Modulation	Peak-to-Average Power Ratio
AMPS (handset)	FM	0 dB
GSM (handset)	GMSK	0 dB
NADC (handset)	π/4 DQPSK	3.2 dB
CDMA IS-95 (handset)	OQPSK	5.1 dB
Multicarrier CDMA		>13 dB
Sum of eight sinusoids		9 dB
White Gaussian noise		10 dB

signal occurs on a rapid time scale, on the order of the inverse of the signal bandwidth.

Amplifiers also need to function with a wide range of average output power levels, as a result of changing characteristics of the wireless channel. The power output is varied in accordance with the changing degrees of fading and varying mobile-to-base-station distances on a time scale of milliseconds or longer [1]. Figure 6.2 shows a representative probability distribution for the transmitted power of a CDMA handset. Although the peak output power is over 600 mW, the amplifier is seldom used in this condition; the average output power is below 10 mW. Within the CDMA network, accurate control over the output power of each user is critical, since if a user employs greater power than is optimal, he/she will create interference at the base station for other users (as a result of the nonorthogonality of the signals used). Base station average power also changes significantly over time as the number of users in a given sector changes over time.

Efficiency of power amplifiers for wireless handsets is one of the most critical concerns for prolonging battery life. In base stations, power efficiency is also important, since it impacts both prime power usage and cooling requirements. While power-added efficiency (PAE) of microwave power amplifiers above 80% has been reported, in practice, the efficiencies attained in wireless networks is much lower, typically near 5–10%. A key issue limiting power efficiency in wireless communications power amplifiers is the variation in signal level that must be accommodated, as just described. Power amplifier efficiency typically drops rapidly as the amplifier output is reduced below its saturated power level, leading to low overall system

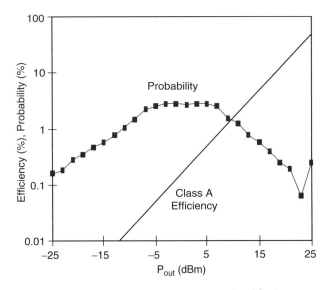

Figure 6.2. Probability distribution of the transmitted power level for a representative CDMA handset. Also shown is the efficiency of an ideal Class A amplifier at the given output power level.

efficiency. For example, Class A amplifiers have efficiency that depends linearly on the output power. Figure 6.2 shows the drain (or collector) efficiency of an ideal Class A amplifier as a function of the output power, P. The proportionality of efficiency to output power stems from the fact that the dc current and voltage used to bias the output transistor are both independent of output power—and must be selected to accommodate the maximum output power. To improve the situation, the dc current and dc voltage supplied to the amplifier (or both) must be allowed to vary as output power changes. Figure 6.3 schematically illustrates how the dc bias point of an amplifier can be varied as the power is reduced from its maximum (saturated) value, in relation to the current–voltage characteristics of the output transistor and its associated load line.

The most straightforward manner to vary the bias conditions is to alter the dc drain (or collector) current of the output transistor. In Class AB mode the current waveform is not symmetric about the quiescent bias point. As a result, the dc average current varies as the output swing is increased (as thus automatically changes with output power level). In the limit of Class B operation, the dc bias varies according to the square root of output power. As a result, power efficiency varies as $P^{0.5}$ over a narrow range (albeit at some cost in linearity). Dynamic gate biasing (changing bias conditions as a function of input power by using an input signal envelope sensitive biasing network) is an additional technique that can be used to change output current for different output powers. The number of gate (or emitter) fingers in a composite large power transistor can also be varied to optimize the current consumption (while also changing input and output impedance). The limit on dc bias current reduction is set by the trade-off of linearity and efficiency. This technique is used in many presently available power amplifiers. Figure 6.4 shows, for example, the bias current and efficiency as a function of output power for a commercial heterojunction bipolar transistor (HBT)-based power amplifier. The $P^{0.5}$ regime is clearly visible. This provides for a significantly higher efficiency when the amplifier is operated at low average power levels (which are encountered much of the time in actual handset usage).

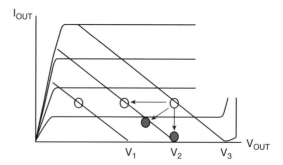

Figure 6.3. Schematic representation of output transistor characteristics and load line. The dc bias conditions at maximum power are shown, as well as desirable directions in which to vary the bias at lower output power levels.

DYNAMIC SUPPLY VOLTAGE AMPLIFIER

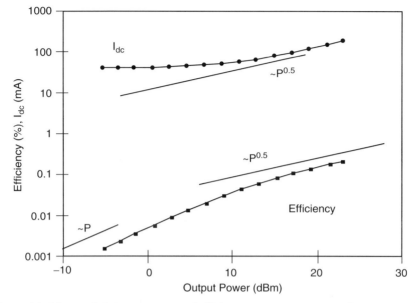

Figure 6.4. Measured dc input current and efficiency versus output power for a commercial CDMA power amplifier.

It is also possible to vary the dc supply voltage in accordance with the output power level [2–4], in order to provide additional increases in efficiency at low output power levels. The most desirable solution is to vary simultaneously dc bias current and supply voltage. In principle, efficiency that is constant over a wide range of output powers can thereby be obtained. This architecture is termed here *dynamic supply voltage (DSV) amplifier*; the term *envelope tracking amplifier* has also been applied. In the following, research results for the dynamic supply voltage amplifier are described in detail.

6.3 DYNAMIC SUPPLY VOLTAGE AMPLIFIER

The DSV amplifier structure is shown in Figure 6.5. The structure comprises a basic Class AB power amplifier and a dc–dc converter, which transforms the battery voltage to a level that is optimized for the amplifier in accordance with the instantaneous output power level. The input signal power is sensed with an envelope detector, which controls the value of V_{DD} for the power amplifier stage. The relationship between the value of V_{DD} and the output power is shown in Figure 6.6. The value of V_{DD} is chosen to be somewhat larger than the amplitude of the radio frequency (RF) signal at the drain of the device. This avoids clipping of the large voltage excursions, which would cause unacceptable signal distortion. At the same time, V_{DD} is kept as small as possible to achieve optimal efficiency.

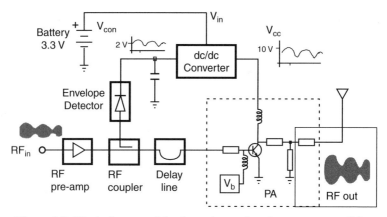

Figure 6.5. Block diagram of the dynamic supply voltage power amplifier.

It should be noted that the power amplifier remains in Class AB operation, and the overall output amplitude is governed primarily by the input signal amplitude, rather than by the value of V_{DD} (which would apply for an "envelope restoration" amplifier). Thus it is not crucial for overall system linearity that the V_{DD} value exactly replicate the input signal—only that it remain in a regime suitable to avoid clipping and increase efficiency.

Figure 6.7 shows the measured efficiency of the combined system as a function of output power. The value shown incorporates the inefficiency of the dc–dc converter as well as that of the amplifier. At the highest output power levels, the amplifier alone (without the dc–dc converter) is more efficient, since there is no loss associated with the voltage converter. At lower power levels, the system with the dc–dc converter is superior, because of its ability to tailor the power supply voltage optimally. By using the dc–dc converter with the power amplifier, a significant improvement of overall

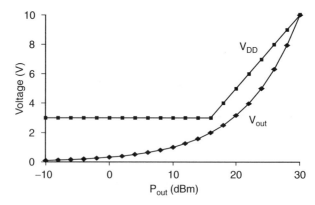

Figure 6.6. The dc–dc converter output voltage (V_{DD}) and calculated RF output voltage amplitude, as a function of output power.

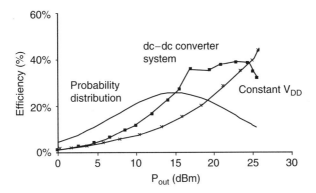

Figure 6.7. Measured efficiency of the amplifier operated at constant V_{DD} value and with dc–dc converter (to provide V_{DD} variable with input power). Also shown is a representative profile of power usage probability.

efficiency results—because the net energy consumption of the system is dominated by the low power regime, where the amplifier alone is highly inefficient. After averaging the energy consumption weighted by probability of usage, the amplifier overall efficiency was increased by 40%. Further increases should be possible, by increasing the ratio of the maximum to minimum voltage obtainable from the dc–dc converter, as discussed below. For example, by providing a power supply voltage lower than the battery voltage (as well as higher), a better average efficiency should result.

The observed improvement in efficiency is in good agreement with the result expected from simple analysis. For a Class A amplifier $\eta = \eta_{0A}(P/P_{max})$; for a Class B amplifier $\eta = \eta_{0B}(P/P_{max})^{1/2}$; and for the DSV amplifier, $\eta = \eta_{0D}$ (constant), for $P_m < P < P_{max}$, while $\eta = \eta_{0D}(P/P_{max})^{1/2}$ for $P > P_m$. Here η is the amplifier efficiency, η_0 is the maximum efficiency for the Class A, B and DSV amplifiers (achieved at the limiting power P_{max}), and P_m is the minimum power at which the supply voltage can still be effectively controlled in the DSV amplifier. The overall efficiency, averaged over the conditions of usage, is given by

$$\langle \eta \rangle = \frac{\int P_p(P)\,dP}{\int P_p(P)/\eta(P)\,dP}. \tag{6.1}$$

Here $p(P)$ is the probability that the system will have an output power P. The average efficiency depends on the characteristics of the signal. In many present applications, the average power is very much lower than P_{max}. If the average power is also much lower than P_m, then the improvement in efficiency becomes approximately $(P_{max}/P_m)^{0.5} \approx V_{max}/V_{min}$, where V_{max} and V_{min} are the maximum and minimum values of power supply voltage that can be provided by the converter. This result illustrates the importance of achieving a large V_{max}/V_{min} ratio.

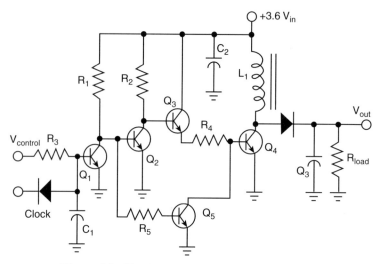

Figure 6.8. Circuit schematic of the dc–dc converter.

To enable a changing voltage supply with output power, we implemented dc–dc converters of small size, in a technology (GaAs HBT) that can be integrated monolithically with the power amplifier itself [5]. The circuit schematic for the dc–dc converter is shown in Figure 6.8. A boost topology is used, providing an output voltage in the range 3–10 V, for an input voltage of 3.3 V. The converter employs a power HBT, capable of handling up to 1 A of current. The inductor, Si Schottky rectifier, and output capacitor are external elements. The circuit incorporates a pulse-width modulator in order to allow a dc input control voltage to regulate the output voltage.

The dc–dc converter employs a high switching frequency (10–20 MHz). Typical waveforms in cellular handsets have envelope variations in the 50 KHz to 2 MHz range, according to different standards. By operating at a switching frequency of 10 MHz or above, several advantages are obtained. The output filter components may be reduced in value and size: inductors may be implemented with relatively few turns, thus reducing skin effect and dc power loss, and capacitors may be simple ceramic surface mount devices, easily located on the power circuit layout. A second benefit is that the dynamic response of the power supply has greater bandwidth. A switching frequency of 10 MHz allows for transient response time less than 1 μm. A third advantage is that the electromagnetic interference sent back on the input line (which could interfere with other circuitry powered from the same point) is much easier to remove by filtering.

The power HBT utilized in this converter exhibited a current gain of 60. Breakdown voltage exceeded 20 V. Figure 6.9 shows a microphotograph of the fabricated power switch HBT integrated circuit (IC). Emitter area for this device was over 5700 μm^2. The overall area of the converter IC was 750 μm by 900 μm. It is noteworthy that the converter was fabricated with the same process that is used to

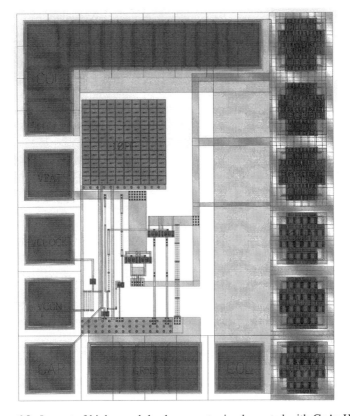

Figure 6.9. Layout of high speed dc–dc converter implemented with GaAs HBTs.

manufacture microwave (1–2 GHz) power amplifiers, which may facilitate the cointegration of these converters with microwave power circuits.

Measured efficiency of the integrated HBT-based converter was nearly flat over an output power range of 0.3–1.65 W, as shown in Figure 6.10. In the region to the left of the data point shown in the figure, the converter is off and no boost occurs.

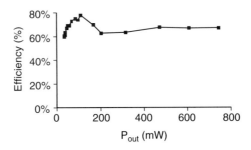

Figure 6.10. Measured efficiency of the dc–dc converter under various load conditions.

The highest efficiency measured was 80.3% at a load resistance of 20 Ω. Output ripple was less than 0.3 Vpp. Efficiency of the converter is limited principally by dissipation in the switching HBT, both on a transient and dc basis. We estimate that the switching loss, P_{0v}, due to simultaneous current and voltage overlap is 0.15 W at peak output power and 10 MHz clock frequency. Since, at full load, the converter was operated in the discontinuous mode, that is, the inductor current at switch turn-on was zero, overlap losses only occur during the switch turn-off. The dc (or resistive [4]) HBT loss is due to the large V_{CE} saturation voltage of the HBT and its series resistance. The HBT had a measured offset voltage, V_{OFFSET}, of 0.25 V, as a result of the difference in band structure between the AlGaAs emitter and the GaAs collector. The use of HBTs with wide bandgap collectors or other structural modifications could greatly reduce this value, as described elsewhere in this book. Series resistance in the power switch path due to the inductor, conductor traces, contact resistance, and the internal resistance of the HBT was estimated to be less than 0.4 Ω. The overall loss from these contributions is estimated to be 0.15 W for a duty cycle ($t_{ON} f$) of 40%. The small 0.3 V forward drop of the Schottky diode in the output path limits the loss of this device to only 50 mW. Input power for the PWM and driver was measured to be 30 mW. The overall power efficiency was thus estimated to be about 80%, which compares well with the experimentally observed value.

The DSV amplifier has significant benefit in increased efficiency. It also has a few potential problems for widespread application in wireless communications systems. One concern involves the ripple output of the 10 MHz converter. In order to increase response time, the output capacitor was reduced in size, increasing the output ripple to over 10%. However, it was found that very little conversion of V_{DD} amplitude variation (at 10 MHz) to gain variation resulted. This is evident in the spectral response of the amplifier output with a single sinusoidal tone input; there are sidebands in the output spaced at 10 MHz from the fundamental output, whose amplitude is lower by −60 dBc (an acceptably low value).

Another issue for the DSV amplifier concerns linearity. There is an inherent source of nonlinearity for this amplifier that is not present in conventional amplifiers, which must be kept within tolerable bounds. The new mechanism for nonlinearity arises because as the input power is varied and as the drain voltage of the RF output transistor correspondingly changes, in general the RF gain also varies [6]. The signal-dependent gain constitutes a mechanism for AM to AM conversion, which can cause intermodulation distortion, or exacerbate the adjacent channel power ratio (ACPR).

To avoid introducing distortion by the power supply time dependence, we explored two techniques. In one technique, in addition to varying the drain supply voltage V_{DD}, we also controlled the gate bias voltage V_{gg} in accordance with the signal level, in such a way to compensate for the variation in gain brought about by V_{DD}. This required implementing a circuit in which the gate bias voltage was generated with an operational amplifier, one of whose inputs was derived from the (signal-dependent) V_{DD}. This technique allowed improvement of the linearity to the level required for IS-95 CDMA signals; that is, the ACPR was measured to be better than −42 dBc.

An alternative, potentially very powerful, approach to overcome the inherent nonlinearity in DSV amplifiers is to use digital control of the dc–dc converter, coupled with digital predistortion of the amplifier input signal. In fact, the digital control can also be used to control V_{DD} to maximize the efficiency at the same time as it corrects the linearity problems. Application of DSP thus allows flexibility in the algorithm for power supply selection, ability to equalize the delay and the frequency response of the dc–dc converter, and ability to predistort the RF signal to overcome the associated AM–AM conversion [7, 8]. The resulting amplifier architecture, shown in Figure 6.11, is a significant step towards realization of a "smart" power amplifier, in which amplifier parameters are dynamically varied in accordance with the signal requirements to maximize performance.

The digital signal processor (DSP) computes an appropriate control voltage for the dc–dc converter according to the signal envelope, taking into account the dc and ac response of the converter. It also computes a predistorted input signal for both the I and the Q channel, to feed into the amplifier. The DSP computations are carried out at baseband. The I and Q signals must then be upconverted and amplified to provide the inputs to the power amplifier. In a cellular phone realization the computations can be performed in the same DSP that generates the conventional signal. In our experiments, the signals were precomputed using MATLAB in a conventional PC and then stored in the memory of arbitrary waveform generators (high speed digital-to-analog converters). They were subsequently read out and upconverted to feed into the power amplifier.

Figure 6.12(a) shows the power amplifier gain as a function of input power at various power supply voltages, for the uncorrected metal semiconductor field effect transistor (MESFET)-based power amplifier. No particular design optimization of the MESFET-based amplifier was done to minimize the gain variations. From these data, substantial changes in gain as a function of output power could be inferred, leading to objectionable AM–AM conversion. The DSP predistortion function was

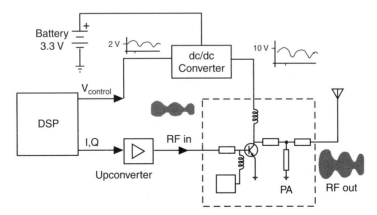

Figure 6.11. Amplifier architecture using a digital processor for dynamic supply voltage control and equalization, as well as input signal predistortion.

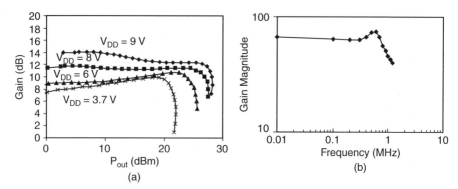

Figure 6.12. (a) Measured variation of the power amplifier gain with V_{DD}; (b) measured frequency response of the dc–dc converter.

computed to compensate for this variation (as well as for the corresponding phase variation). The frequency response of the dc–dc converter is shown in Figure 6.12(b). The equalization employed in the DSP was computed to compensate for this response (together with the nonlinear dc response and the delay of the dc–dc converter).

With the use of the DSP, the output frequency spectrum for an IS-95 CDMA input signal was considerably improved. A representative output spectrum is shown in Figure 6.13 (together with the spectrum obtained without predistortion for comparison). An improvement of 8 dB in ACPR was observed. The value of the ACPR with predistortion is −44 dBc (which exceeds the IS-95 minimum specification of −42 dBc) at the largest output power (and is better at lower output power). The linearization by DSP allows increasing the output power by about 4 dBm (and thus improving efficiency) while staying within the specifications of the IS-95 signals.

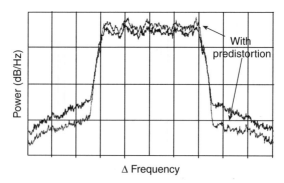

Figure 6.13. Measured output power spectra for an IS-95 CDMA signal with and without predistortion and equalization (10 dB/div, 30 kHz resolution bandwidth, 3 MHz span). The corresponding ACPR values are −44dB and −36dB.

The digitally controlled DSV architecture has considerable promise for future development. It uses DSP control over amplifier input signal as well as power supply voltage. The system can flexibly implement a variety of algorithms for system optimization. Application of DSP to amplifier control is becoming increasingly cost effective [7]. This approach requires integrated design of the overall transmitter, however, since the DSP function and the power amplifier must be closely coupled.

6.4 SWITCHING-MODE AMPLIFIERS

The power dissipation within the output transistor can be minimized by operating it as a switch, that is, by ensuring zero current when the voltage is high, and developing near-zero voltage (limited typically by on-resistance) when the current is nonzero. Nominally, with ideal, lossless passive elements, all the power delivered by the power supply reaches the load, and the efficiency is 100%. At lower RF values, transistors can be operated efficiently as switches driven by square waveforms, leading to amplifiers that have measured efficiencies over 90% when their operating load line consists of the *I* and *V* axes of the *I* to *V* curves [9]. At higher microwave frequencies, however, the output capacitance and resistance of the transistor, when driven into saturation, limit the switch performance: microwave transistors are not good switches. Switched-mode amplifiers can still be implemented at microwave frequencies, but the designer needs to keep the switch limitations in mind. If the large signal output capacitance is known, it can be used as part of the output tuning circuit. Knowing the on-resistance allows the designer to predict the maximum obtainable efficiency. An important issue for microwave PA designers is the highest frequency at which a given device can operate efficiently. A rule of thumb is that for switched-mode operation (i.e., when output power is not sacrificed at the expense of efficiency), the device can operate at high efficiency at about one-third of the cutoff (small signal unity gain) frequency. More details are given in the next section, specifically for Classes E and F, for which an interesting conclusion can be reached about the degradation of efficiency as one operates the device above the optimal efficiency frequency. The efficiency is found to decrease slowly as frequency increases, degrading first to "suboptimal" switched-mode operation [10], and then into the familiar Class AB mode.

6.4.1 Class E and F Microwave Amplifiers

In this section, we describe briefly the main properties and experimental results obtained with Class E and F microwave amplifiers, ranging in frequency from 0.5 to 10 GHz. The differentiation between E and F classes can be made in several different ways. We choose to define them with respect to output harmonic termination: in Class E, all harmonics are presented with an open circuit (or high impedance), while in Class F all even harmonics are shorted and odd harmonics open at the output.

Equations governing the design of Class E amplifiers can be derived from an approximation based on high-Q output tuned circuits. These equations include expressions for Z_{out}, the output impedance required for Class E operation; f_{opt}, the maximum frequency at which the device can operate in "optimal" class E mode; and η_d, the drain efficiency (ratio of RF output power to dc input power) [11]:

$$Z_{out} = \frac{0.28015}{\omega C} e^{j49.0524°} \quad (6.2)$$

$$f_{opt} = \frac{I_{max}}{56.5 C V_d} \quad (6.3)$$

$$\eta_d = \frac{1 + (\pi/2 + \omega CR)^2}{1 + \pi^2/4(1 + \pi\omega CR)^2} \quad (6.4)$$

In the above approximate formulas, C is the output capacitance of the device, L is the output inductance (usually due to the package), R is the on-resistance, ω is the operating (switching) frequency, I_{max} is the maximum current the device can handle, and V_d is the bias voltage. These formulas are derived from a high-Q approximation of the tuned output circuit, in which the current through the load is assumed to be purely sinusoidal. Even though this assumption is not strictly valid in many microwave amplifiers, it is a very useful starting point in the design. When designing a Class E amplifier at a microwave frequency, C and L need to be incorporated in the output circuit impedance Z_{out}, as well as in any harmonic tuning impedance. As an example, the approximate Class E circuit schematic with no on-resistance is shown in Figure 6.14(a); the hybrid implementation is shown in Figure 6.14(b). Here, the output circuit meets the Class E conditions at the fundamental and second harmonic, while it is assumed that the transistor has low or no gain at the higher harmonics. Transmission line l_2 is a quarter-wavelength long at the second harmonic so that an open is transformed to a short at plane AA'; l_1 along with L and C is designed to be also a quarter-wavelength effectively, to translate the short to an open at the transistor (switch) output. Lines l_1 to l_4 provide Z_{out} at the fundamental. However, it is commonly the case that a single-stub match is sufficient to provide Z_{out} at the fundamental while simultaneously providing a high enough impedance at the second harmonic, eliminating the need for an extra stub and reducing a portion of the losses associated with the transmission line sections.

In Class F amplifiers, since the even harmonics are shorted and the odd open-circuited, the voltage waveform at the drain is ideally a square wave and in practice contains a limited number of odd harmonics that approximate a square wave. At the fundamental, the device should be matched for maximum saturated power transfer to the load, which can be shown to be the same as the maximum linear power match, $R_{opt} = 2V_d/I_{dss}$, where I_{dss} is the maximum allowable drain current.

To date, several amplifiers have been implemented at microwave frequencies. They are described in detail in Refs. 11–17 and are summarized in this chapter. Experimental results are tabulated in Table 6.2, followed by a discussion of their salient features.

The first three Class E amplifiers were designed around a Siemens CLY5 MESFET, for which a modified Materka–Kacprzak nonlinear model was available.

SWITCHING-MODE AMPLIFIERS

TABLE 6.2. Summary of Experimental Switched Mode Microwave Amplifiers

f (GHz)	Device	P_{out} (mW)	G_{sat} (dB)	PAE (drain eff.)%	Reference	Comments
0.5	CLY5	550	15.3	80 (83)	12	
1.0	CLY5	940	14.7	73 (75)	12	
2.0	CLY5	530	9.1	56 (62)	12	$f_{opt} = 1.4$ GHz
2.5	FLK052	100		70	13	$V_d = 1$ V
4.5	FLK052	150	15	75 (80)	14	$V_d = 3$ V, gate biased on
4.5	FLK052	100	10	86 (95)	14	$V_d = 3$ V, gate pinched off
5.0	FLK052	610	9.8	72 (81)	15	
5.0	FLK202	1800	7.6	60 (73)	15	Four times larger device
8.4	FLK052	685	7.4	60 (72)	16	Class F ($f_{opt} = 1.5$ GHz)
8.35	FLK202	1700	5.3	48 (69)	16	Four times larger device
10.0	AFM042	100	10	62 (74)	17	Active nonresonant antenna

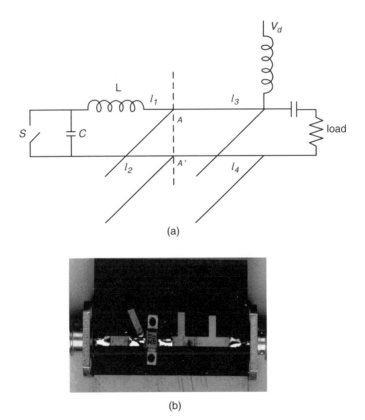

Figure 6.14. (a) Approximate Class E power amplifier circuit in which the transistor is represented as an ideal switch, and transmission line sections are used for the output match and harmonic tuning. (b) Hybrid implementation with single stub.

Harmonic balance simulations at these low frequencies are in good agreement with the measurements: the simulated PAE is within 2%, and the simulated power within about 1 dB [15]. Details on the nonlinear simulations are given in Markovic et al. [18]. At the higher microwave frequencies, we found that better nonlinear models are needed for switched-mode operation, especially for higher saturation levels. The design in these cases is done starting from a linearized device model, and then the essential large signal parameters are de-embedded from measurements, giving the final design in the next iteration.

The harmonic balance simulations also predict time-domain waveforms that correspond to the definition of Class E operation, as shown in Figure 6.15(a). At 0.5 GHz, one can measure the drain waveform using a large valued resistor probe [11], but at higher frequencies this is not possible. In order to confirm the class of operation of the X-band amplifiers, a very high impedance photoconductive sampling probe was used, which enables almost noninvasive waveform measurements at any point in the circuit that can be contacted by the probe [19]. An example of such a measurement is shown in Figure 6.15(b).

It is clear that the X-band amplifiers operate in suboptimal switched mode. The Fujitsu FLK052 MESFETs are intended to operate in the Ku band with 30% PAE, so even the 5 GHz amplifiers using these devices operate beyond the optimal theoretical frequency for Class E operation. However, they still have enough gain at the harmonics to allow for waveshaping. In the lower frequency amplifiers using the CLY5 device, the optimal frequency for Class E operation is 1.4 GHz. It can be seen from Table 6.2 that the 2 GHz amplifier has considerably lower power and efficiency, due to the fact that a higher peak drain current is required than the device can provide.

In Mader et al. [15], Class E, F, and saturated A(AB) amplifiers are experimentally compared at 0.5 GHz. The conclusions are: (1) AB and F have essentially the same saturated output power, with Class F having 15% higher efficiency; (2)

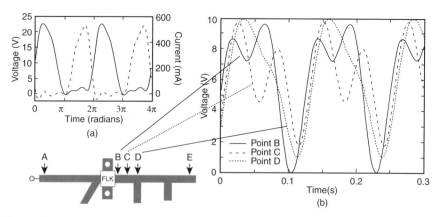

Figure 6.15. (a) Simulated Class E waveforms for the 0.5 GHz amplifier using a modified Materka–Kacprzak device model and harmonic balance show salient features of the Class E mode. (b) Measured Class F waveforms at three points at the output of an X-band amplifier using a photoconductive sampling probe.

Class E has the highest efficiency; (3) the gain of the Class E amplifier compresses at a lower input power level than the Class F gain (since the second harmonic short in the Class F amplifier flattens the switch voltage waveform, reducing the peak stress on the device); (4) for a given high efficiency, the Class F amplifier produces more power; and (5) for the same maximum output power, the third-order intermodulation products are 10 dB lower for Class F than for Class E, while the extrapolated intercept point occurs at the same power level.

If greater power is needed than available from a given device, power combining of several amplifiers is usually the solution. Some interesting conclusions can be made about types of power combining appropriate under different conditions. One can accomplish power combining at the chip level (increasing the gate periphery); by using circuit combiners; by using spatial combining; or with any combination of the three. As seen in Table 6.2, two amplifiers of different power levels were demonstrated at both 5 and 8.4 GHz, one using a device with four times larger gate periphery than the other in each case. At 5 GHz, the two amplifiers were both designed to operate in Class E mode, while at the X-band, one was Class E and the other was Class F (which corresponded to the design that gives for each the highest efficiency and power of which they were capable).

At 5 GHz, although the larger device delivers a three times higher output power, it requires five times the input power to operate in Class E mode. The on-chip power-combining efficiency is about 65% [11], while a four-element spatial combiner with the smaller device yielded a combining efficiency of 84%.

The measured frequency and power sweeps of the two X-band amplifiers are shown in Figure 6.16. In these plots, overall efficiency, defined as $P_{out}/(P_{dc} + P_{in})$, is plotted, since it was felt that this was a better indicator of low power dissipation than the conventional PAE. The chip-level combining efficiency in this case is 89%, but the bandwidth is much narrower for the larger device. The larger device yields a 2.5× higher output power, but about 3.5× higher heat power than the smaller device, with a 12% lower PAE and 2 dB lower gain. Even though on-chip combining is high, this approach is limited by propagation delays as the device size increases and by the reduced input impedance resulting from an increased number of gate fingers. From a more detailed combining efficiency comparison based on these two demonstrated amplifiers [16], the following general conclusions can be drawn: (1) in circuit combiners, a small number of larger devices is more efficient and takes up less real estate; (2) in spatial combiners, assuming the combining efficiency does not significantly drop with the number of devices combined, it is more efficient to use a large number of smaller devices; and (3) in applications that require an antenna array at the transmitter output, distributed amplification (spatial combining) is more efficient with respect to the total EIRP and thermal management as compared to circuit combiners.

As discussed below, integration of antennas with amplifiers, particularly switched-mode amplifiers, has several advantages. The antenna can be used effectively as the output tuned circuit, thereby eliminating the need for a matching network, which in turn reduces the losses and the volume taken up by the antenna and PA. This was first demonstrated by Mader et al. [15], where a second resonant slot antenna

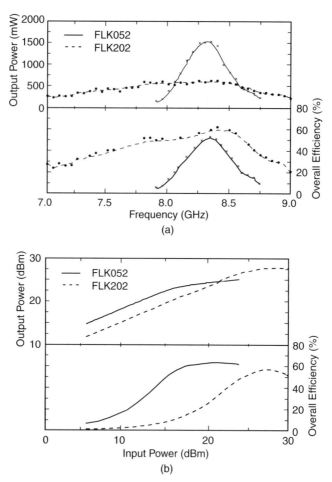

Figure 6.16. (a) Measured output power and overall efficiency versus frequency for two X-band switched-mode amplifiers. Input power levels were 20 and 26 dBm for the smaller and larger devices, respectively, and the drain and gate bias voltages in both cases were 7 and −0.9 V, respectively. (b) Measured output power and efficiency versus input power around 8.4 GHz for the two amplifiers under equal bias conditions.

provided the correct harmonic termination (high impedance at the second harmonic) to the Class E amplifier. This active antenna radiates 0.67 W with a drain efficiency of 80%, PAE of 71%, and a compressed gain of 9.3 dB at 5 GHz, which is almost identical to the microstrip circuit with no antenna. A four-element two-dimensional array of such active antennas all biased in parallel radiates 2.4 W with 74% drain efficiency estimated from the measured EIRP and radiation pattern. Based on the Class E and F X-band amplifiers described by Figure 6.16, several active patch antenna arrays were demonstrated with EIRPs up to 2.6 kW from 36 elements [13].

A 10-GHz Class E integrated antenna with an overall area of 0.4 free-space wavelengths squared is discussed in detail in Weiss and Popovic [17]. The antenna is a nonresonant microstrip-fed slot designed to have an impedance at the 10 GHz fundamental frequency equal to the Class E optimal load impedance given by Eq. (6.1), and a large capacitive impedance at the 20 GHz second harmonic. As seen in Table 6.2, this active antenna had a record drain efficiency of 74% and a PAE of 65% at 10 GHz while sacrificing only 1 dB of the manufacturer specified maximum output power of 21 dBm. A sketch of this active antenna and the passive antenna impedance at 10 and 20 GHz is shown in Figure 6.17.

6.4.2 Future Trends of Switched-Mode Power Amplifiers

The switched-mode circuit approach has also been applied at microwave frequencies to oscillators, frequency multipliers, and dc–dc power converters with switching rates in the gigahertz range. The oscillator [20] was based on the 5 GHz Class E

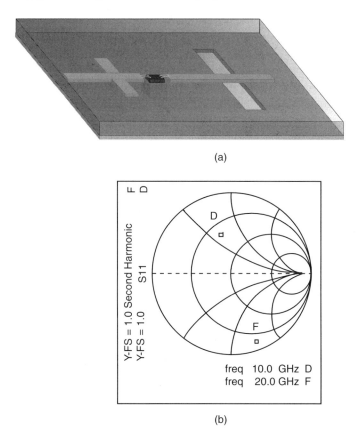

Figure 6.17. (a) Sketch of 10 GHz Class E active antenna with chip device, and (b) fundamental and second harmonic impedance of passive antenna designed to act as the optimal Class E load.

amplifier with added feedback for optimal conversion efficiency (equal in theory to the amplifier PAE) and exhibited a maximum conversion efficiency of 59% (with 300 mW power) and a maximum power of 600 mW (with 48% conversion efficiency). Several multipliers with conversion gain were demonstrated by designing the output circuit to be a Class E match at the second harmonic. For example, 5.2 dB conversion gain with 30% PAE and 330 mW of power was produced at 5 GHz with a 100 mW, 2.5 GHz input signal. It was found using photoconductive sampling [19] that a substantial harmonic power was reflected back at the input of this circuit, pointing to harmonic tuning at both input and output for improved efficiency and conversion gain.

An interesting direct application of a Class E amplifier is the dc–ac conversion stage of a dc–dc switching power converter [21], which is followed by a rectifier (ac–dc) stage. The converter switching speed was 4.5 GHz, and the overall dc–dc demonstrated conversion efficiency was 64%. This switching rate is orders of magnitude higher than other existing high speed converters. The advantages of this feature are a completely planar (low-volume) structure amenable to monolithic integration, the absence of magnetic components, and excellent dc to dc isolation provided by electromagnetic (capacitive) coupling between the dc–ac and ac–dc stages. The amplifier performance is given in Table 6.1 for two different gate biases. In this application, it is appropriate to bias the gate of the transistor in pinch-off, so that the device is off when no input switching signal is present. The diode rectifier efficiency in this circuit was 98% (defined as the ratio of the dc output power to the RF input power), and its conversion efficiency, which takes into account the reflected RF input power, is 83%. In this design, 6% is lost in total dc–dc conversion efficiency when integrating the Class E amplifier and diode rectifier.

There are a number of important and interesting issues that remain to be understood for switched-mode high efficiency microwave circuits: how to push the frequency into the millimeter-wave range (which will require devices with cutoff frequencies above 60 GHz); characterization of switched-mode linearity for different transmitter modulation schemes; efficient dynamic bias and drive control associated with signals with varying envelopes; use of devices with lower bias voltages; and combining strategies for high power transmitters.

6.5 LINC AMPLIFIER

While switching-mode amplifiers offer attractive improvements in efficiency, their output linearity is inherently reduced from that of Class A operation. There are several amplifier architectural approaches than can in principle achieve ideal linearity while providing the efficiency of switching-mode circuits. These are the out-phased (or LINC) approach and the bandpass Class S approach.

The out-phased power amplifier concept dates back to the early 1930s as an approach for the simultaneous realization of high efficiency *and* high linearity amplification [23]. It has been revived recently for wireless communication applications under the rubric of LINC (linear amplification with nonlinear components)

LINC AMPLIFIER

and many recent papers have developed the concept further [24, 25]. The LINC concept takes an envelope modulated bandpass waveform and resolves it into *two* out-phased constant envelope signals, which are applied to highly efficient—and highly nonlinear—power amplifiers, whose outputs are summed. The advantage of this approach is that each amplifier can be operated in a very efficient "switching" mode, and yet the final output can be highly linear—a key consideration for bandwidth efficient wireless communications. This is shown schematically in Figure 6.18. In theory, the efficiency of this scheme can approach 100% without any degradation of linearity.

The out-phased amplifier of Figure 6.18 takes a general baseband representation of an RF signal,

$$s(t) = a(t)e^{j\phi(t)},$$

where $a(t)$ is the instantaneous amplitude and $\phi(t)$ is the instantaneous phase of the input signal, and resolves it into two *constant envelope* vectors $s_1(t)$ and $s_2(t)$ through what is known as a signal component separator (SCS), such that

$$s_1(t) = s(t) + x_1(t),$$
$$s_2(t) = s(t) - x_1(t),$$

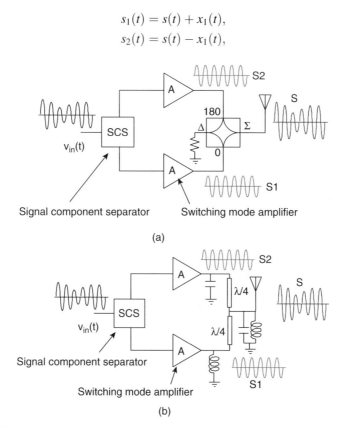

Figure 6.18. (a) Schematic diagram of LINC amplifier with hybrid output power combiner; (b) similar schematic of LINC with Chereix output power-combining technique.

and

$$x_1(t) = ja(t)\sqrt{\frac{a_{\max}^2}{|a^2(t)|-1}},$$

where $a_{\max}$ is the maximum value of the signal a. These signals $s_1(t)$ and $s_2(t)$ can then be upconverted in the traditional manner and sent to highly efficient *constant envelope* power amplifiers, whose outputs are summed in a hybrid combiner. After summing, the $x_1(t)$ components of the signals cancel, and the desired $s(t)$ is recovered.

Despite its apparent attractiveness, the LINC approach has several disadvantages that have limited its applicability. The first is that the power is typically summed with a hybrid power-combining network, as shown in Figure 6.18(a). That portion of the power delivered to the hybrid that is not delivered to the antenna is dissipated in the 50 Ω terminating resistor. As a result, the amplifier only achieves its peak operating efficiency at maximum output power, and its efficiency decreases linearly as the output power decreases. This efficiency behavior is comparable to a Class A amplifier, which is known to have a very poor overall efficiency. Of course, the *peak* efficiency of this LINC approach is much higher than the Class A amplifier, but it would be desirable to do even better.

The power-combining approach of Figure 6.18(a) allows power to be wasted in the power-combining network. It is possible, however, to derive some benefit from the power dissipated in the 50 Ω load of Figure 6.18(a). By introducing a rectifier in the circuit, we have shown that significant power can be recycled, returning the "waste" power to the battery [26]. This can improve the LINC amplifier efficiency by 1.5 times over a range of output powers.

Another approach to improved power combining relies on replacing the hybrid combiner with the so-called Chireix power combiner, illustrated in Figure 6.18(b) [23]. In this case, two quarter-wave transmission line impedance transformers and a shunt susceptance are added in order to improve the efficiency. The analysis of the Chireix system is complicated, but it has been studied extensively by Raab [24] and the benefit is substantial. The added shunt susceptance (which is inductive in one branch and capacitive in the other branch) cancels out the varying susceptance seen by each amplifier at one particular output power.

A second problem associated with the LINC approach is that gain and phase mismatch lead to severe intermodulation and distortion [27]. Typical requirements for CDMA applications are on the order of 0.3 degree phase matching and 0.5 dB gain matching, a near impossibility in most practical cases. As a result, several compensation or calibration schemes have been proposed. These techniques have not achieved wide application due to their inherent complexity and lack of flexibility. For the future, the application of digital signal processing techniques to LINC calibration is promising.

6.6 DELTA–SIGMA AMPLIFIERS

An alternative technique to achieve high linearity with a switching-mode amplifier is the Class S technique. With a Class S amplifier, output signals with excellent linearity can be produced by using an appropriate pulse-width modulation of the input switch control voltage, followed by filtering after the switching-mode amplifier. For example, Figure 6.19 illustrates the schematic structure and waveforms used in a representative Class S amplifier for audio frequencies. The output voltage is binary in nature (switched between output levels V_{DD} or 0), but the current is constrained to the desired low frequency output as a result of the output lowpass filter. Consequently, there is no power dissipated at the high switching frequency, only at the desired output frequency; thus, in principle, the Class S amplifier can have an efficiency of 100%. The application of this approach in the microwave region is complicated by the need for extremely high switching speed. We have investigated novel Class S amplifiers appropriate for microwave signals, in which the input control signals have a bandpass delta–sigma modulation, and the overall switching rate is not much greater than the frequency to be amplified [28].

The structure of a bandpass delta–sigma Class S amplifier is shown in Figure 6.19(b). The control voltage for the amplifier switch is a binary (digital) signal derived from the input signal through the use of a bandpass delta–sigma (BPDS) A/D converter, whose structure is shown schematically in Figure 6.20. Within an approximate, linearized analysis framework, the signal input to the BPDS system is found to be filtered by the passband of a resonant filter within the converter, while the quantization noise is spectrally shaped by a function that effectively eliminates it from the band of interest. The digital clock rate of the converter is typically four times the signal frequency. Practical implementations of BPDS converters operating in the microwave regime have recently been demonstrated [29].

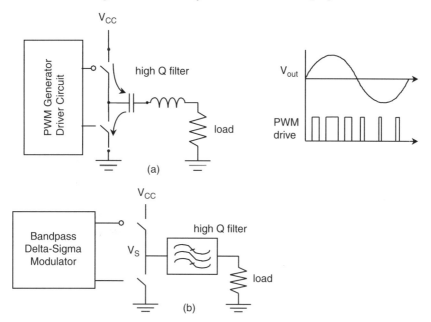

Figure 6.19. (a) Block diagram and representative waveforms in a conventional Class S amplifier for audio signals; (b) corresponding diagrams for the bandpass Class S amplifier for microwave signals.

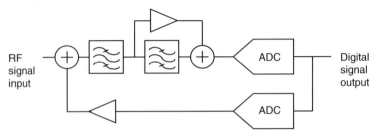

Figure 6.20. Block diagram of a bandpass delta–sigma modulator, and representative frequency response of the modulator for the signal and the quantization noise.

To demonstrate experimentally the operation of the bandpass delta–sigma Class S amplifier, we have constructed a "scaled model amplifier" operating at a low frequency (10 MHz) at which it is easy to obtain appropriate delta–sigma digital signals (from an appropriately programmed logic analyzer) and complementary switching FETs can be used. Figure 6.21 illustrates representative signals measured with this amplifier, in the frequency domain. Figure 6.22 contains the output spectrum recorded for a two tone input, illustrating the relatively low value of third order intermodulation products (-40 dB below the fundamental). The efficiency of the output amplifier under this condition was 51% (which is dramatically higher than attained for a Class AB amplifier at this level of linearity).

Amplifier efficiency is affected by voltage drop across the switches (R_{on}), dissipation during on–off transients, and discharge of input and output capacitances per cycle. To increase efficiency, it is important to use large switching devices for low R_{on}, to minimize the drain capacitance of the switching transistors, and to appropriately time the signals through the two switching transistors to avoid periods where both devices are simultaneously on. An additional factor is dissipation within the filter associated with signal components outside the frequency band of interest. To prevent current flow in the filter, the output impedance must be high at all frequencies other than the desired band. In conventional filters, the reflection coefficient is high everywhere outside the passband, but this is achieved by combinations of low impedance and high impedance. Insertion of extra series resonance elements can optimize the filter impedance profile.

Results suggest delta–sigma Class S power amplifiers are promising for high, efficiency, high linearity applications. It is envisioned that, in the future, DSP techniques can be used to generate the switch control voltages starting with baseband signals. Such an approach will allow considerable flexibility and software control over the output.

6.7 ACTIVE INTEGRATED ANTENNA APPROACH FOR POWER AMPLIFIERS

Most of this chapter has focused on increased transmitter efficiency by using various techniques to increase the performance of the power amplifier itself, which

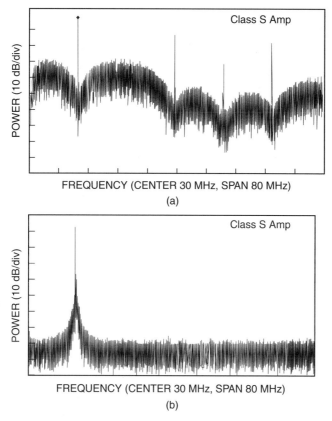

Figure 6.21. Representative measured spectra of a bandpass delta–sigma amplifier: (a) after the switching amplifier and (b) at the output.

consumes the majority of dc power in a typical handset. Another possibility for improving the efficiency of the *overall* transmitter is to focus on the insertion loss of the components between the power amplifier output and free space, including matching circuits for the antenna and power amplifier, isolator, harmonic filter, and antenna. The losses associated with each of these components will directly increase the required amplifier power output for a given radiated transmitter power. Therefore a logical technique for increasing the overall transmitter front-end efficiency is the elimination of as many components as possible between the PA output and the antenna without sacrificing system performance. The extreme of this technique, elimination of all components (except for possibly direct matching between the PA output and antenna), is known as the *active integrated antenna* (AIA) approach for power amplifiers. In order to keep transmitter performance from being sacrificed, it is necessary that the antenna take on new functionality in addition

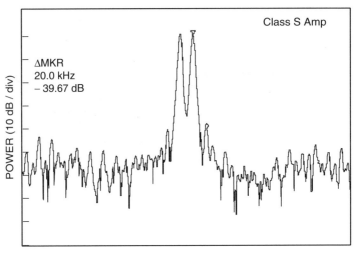

Figure 6.22. Output spectrum of bandpass delta–sigma amplifier, for input of two tones separated by 20 kHz.

to its original role as radiator. If designed properly, the resulting topology yields high transmitter efficiencies, compact size, and reduced component count. By using a planar antenna, the transmitter and antenna can be integrated onto a single printed circuit board (PCB). Therefore, the AIA approach delivers a transmitter front-end with maximum possible efficiency at minimum cost. More about the AIA approach, including more detailed design methodology and measurement techniques, is presented in Chapter 11.

Several classes of planar antennas have recently been developed for the AIA approach [30]. These antennas were designed with the goal of eliminating harmonic radiation and increasing the PAE of the transmitter through harmonic tuning. For high efficiency operation, the load impedance should provide a reactive termination at higher harmonics to reflect the power back to the MESFET with proper phase. Ideally, an open or short-circuit should be placed at the drain of the device for proper harmonic termination. However, at microwave frequencies, due to device parasitics, the higher harmonics are should be reflected back to the FET with appropriate phasing to resonate the drain-to-source capacitance. Therefore the output circuit should have a reflection coefficient of magnitude 1.0 at undesired harmonic frequencies and appropriate phase chosen to maximize efficiency. Traditionally, adding a short-circuited stub that is one-quarter wavelength at the fundamental frequency is done for this purpose, where the stub is added at the output (most often at the drain bias line) [31,32]. At low operating frequencies, this stub may be quite large. Alternatively, chip capacitors can be used for reflection of the second harmonic if they have a self-resonance near this frequency [33]. However, the AIA approach uses specifically designed antennas to achieve harmonic tuning.

In this section, the concept of AIA harmonic tuning is illustrated by two design examples. Specific emphasis is placed on the engineering of specific antennas to enhance the high efficiency performance of these front ends. Typically, when considering an antenna, the antenna designer is interested in the antenna characteristics only at the system operating frequency. However, when the antenna is used for harmonic tuning, antenna characteristics over a broad frequency span must be considered. These characteristics fall into two categories. The first characteristic, input impedance at operating frequency and higher harmonics, must be chosen to maximize amplifier performance. This means that the input impedance at the fundamental frequency should be close to the optimum output impedance of the amplifier for maximum PAE or P_{out}. Additionally, the antenna should present a reactive termination at undesired harmonics. As will be seen, this second constraint is contrary to practice in many planar antennas. The second of category characteristics covers the radiation properties of the antenna; these properties are chosen for optimal wireless performance and will depend on the application at hand.

Harmonic radiation is another important issue in wireless systems. Harmonic radiation is caused when the power amplifier generates significant harmonics, which may radiate through the antenna and consequently degrade system performance. One example of this is co-site interference, which may occur when a number of antennas operating at different frequencies are mounted in close proximity. Employing an additional filter to solve the resulting EMI problem is not only expensive but decreases the transmitter efficiency and may degrade the receiver noise figure. The harmonic tuning techniques presented in the following examples not only improve efficiency but also reduce unwanted harmonic radiation.

6.7.1 AIA with Modified Patch Antenna

One of the most commonly used planar antennas at microwave frequencies is the microstrip patch antenna. The patch antenna is a resonant-type antenna formed by a wide open-circuited microstrip conductor that is $\lambda/2$ in length at the first resonance. This causes standing waves in the cavity, which will radiate at the edges. Figure 6.23 shows the geometry of a standard rectangular patch antenna and the mode profile for the first three radiating modes: $TM_{1,0}$, $TM_{2,0}$, and $TM_{3,0}$. The operating frequency is chosen slightly away from resonance of the $TM_{1,0}$ mode to avoid overly large input impedance. When integrated with an amplifier, the second harmonic would therefore fall near the resonance of the $TM_{2,0}$ mode, where an appreciable radiation resistance exists. Therefore undesired harmonic radiation will effectively radiate from this structure and the structure in its current state is unsuitable for harmonic tuning.

Figure 6.24 shows the geometry of a rectangular patch antenna modified to eliminate the $TM_{2,0}$ mode. Referring to the mode profiles of the standard patch in Figure 6.23, the $TM_{2,0}$-mode electric field peaks at the center of the antenna. At this point, the electric field of the $TM_{1,0}$ mode is at a null. Therefore the $TM_{2,0}$ mode can be eliminated by properly modifying the boundary conditions along the centerline of the patch without affecting the $TM_{1,0}$ mode. This has been done by inserting a row of shorting pins along the centerline [34]. It was found that nine evenly placed pins

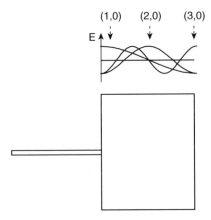

Figure 6.23. Geometry of standard microstrip patch antenna and corresponding mode profile of the first three modes.

effectively suppress the $TM_{2,0}$ mode without significantly perturbing the $TM_{1,0}$, as seen in the plot of the input impedance of the modified structure shown in Figure 6.25. From this, the resonance of the $TM_{2,0}$ mode has been eliminated, and the real part of the input impedance is almost zero where the $TM_{2,0}$ resonance should occur. Therefore this modified structure is suitable for harmonic tuning. Additionally, the characteristics of the $TM_{1,0}$ mode of the patch are virtually unchanged. This has been verified by measuring the normalized copolarization levels for both E and H planes, as shown in Figure 6.26. The cross-polarization was measured to be below -17 dB in all directions. The measured gain was 7.9 dB. These results are consistent with a conventional patch of the same dimensions operating near the $TM_{1,0}$ resonance.

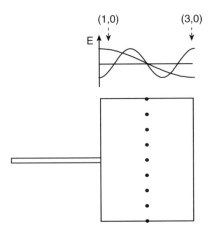

Figure 6.24. Geometry of patch antenna modified to eliminate the $TM_{2,0}$ radiation mode. Note line of vias down centerline of geometry.

ACTIVE INTEGRATED ANTENNA APPROACH FOR POWER AMPLIFIERS

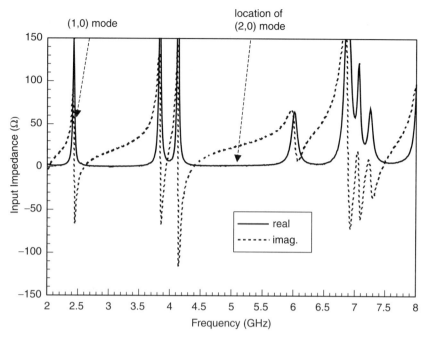

Figure 6.25. Input impedance of modified patch antenna. Elimination of $TM_{2,0}$ radiation mode is evident by negligible radiation resistance at twice the fundamental resonant frequency.

Next, two Class B AIA amplifiers were designed using a Fujitsu FLL351ME GaAs FET. The first reference amplifier used the standard patch antenna while the second used the modified patch with shorting pins for second harmonic suppression. The amplifiers are designed by incorporating the patch antenna S parameters directly into a circuit simulator (in this case, Hewlett-Packard Microwave Design System). The patch is modeled as a one-port network and data from a broad span (0.13–10 GHz) are incorporated to include effects of the first three harmonics. Then, harmonic balance is used to design the amplifier.

Both AIA amplifiers were then measured in an anechoic chamber using the Friis transmission equation to determine the radiated power. Note that the antenna gain has been calibrated out of the amplifier gain calculations for a fair comparison. It was found that the tuned AIA amplifier gave 0.5 dB higher output power when compared to the nontuned AIA with standard patch antenna, resulting in a 7% improvement in PAE [34]. This is illustrated in Figure 6.27, which shows measured PAE versus input power for both AIA modules. Maximum measured PAE is 55%. From this, we see that a very simple modification of the antenna can result in significant gains in efficiency.

6.7.2 AIA with Circular Segment Patch Antenna for Harmonic Tuning

One disadvantage to the previous structure is radiation of higher odd harmonics. The shorting pins will eliminate all even modes and therefore suppress even harmonic

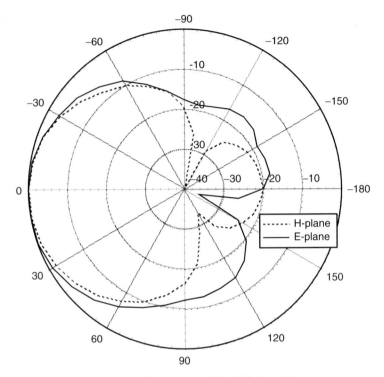

Figure 6.26. Radiation patterns of modified patch antenna.

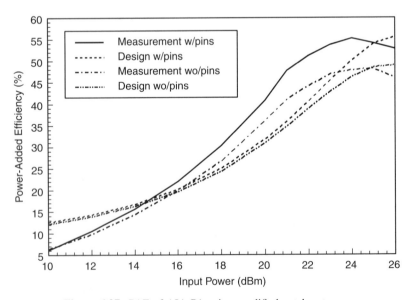

Figure 6.27. PAE of AIA PA using modified patch antenna.

radiation. However, odd modes remain unaffected, which can result in undesired radiation at these frequencies, depending on the nonlinearity of the amplifier. Designing an antenna where the higher radiation modes do not correspond to harmonic frequencies of the transmitter can eliminate this and is therefore highly desired.

Figure 6.28 shows a circular sector patch antenna used for this purpose along with its input impedance characteristics. For a circular geometry, the resonances of this antenna will be given by circular harmonics, which are not multiples of the fundamental frequency. By carefully optimizing the geometry, radiation at the second and third harmonics of the amplifier can be almost entirely eliminated. The optimized antenna consists of a circular patch of radius 740 mils. Note that the antenna is fabricated on a standard RT/Duroid of permittivity 2.33 and a thickness of 31 mils, as was the antenna of the previous example. A 120° sector of the antenna has been removed for optimal impedance. A microstrip feed is placed 30° from the edge of the voided sector. Referring to the input plot of the input impedance in Figure 6.28 resonances of the antenna structure occur where an appreciable resistance exists and the imaginary part of the input impedance crosses through zero. If the operating frequency is chosen slightly off the fundamental resonance, radiation resistance at second and third harmonics is negligible. Therefore harmonic radiation will be suppressed at these frequencies and harmonic power will be reflected back to the amplifier to add to the PAE of the transmitter [35]. It should also be noted that the circular geometry antennas feature a broad radiation pattern similar to the more

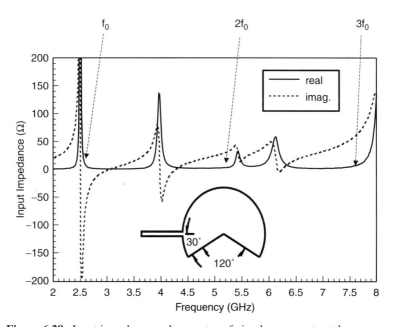

Figure 6.28. Input impedance and geometry of circular segment patch antenna.

standard rectangular patch antennas. The measured radiation characteristics of this particular antenna include 5.8 dB gain and −16 dB cross-polarization.

Using the same design procedure as with the patch antenna based AIA amplifier, one-port S parameters of the circular segment patch antenna are incorporated directly into a harmonic balance simulation. This time, a Class F AIA is designed for. A simple matching circuit between the device drain and antenna brings the antenna impedance to an optimal value at the fundamental and first two harmonics required for the Class F design. A microwave technology MT-8HP power GaAs FET biased at 10% I_{DSS} is used as the active device. Maximum measured PAE at 2.55 GHz is 63% at an output power of 24 dBm, as shown in Figure 6.29. We can also see that this hybrid-simulation method shows excellent agreement with measurements at both small-signal and large-signal operation. Additionally, this amplifier satisfies the IS95/98 ACPR linearity requirement for OQPSK (ACPR < -42 dBc at 1.25 MHz offset) even when operated at a high PAE of 43%.

Harmonic suppression has also been examined for this particular AIA power amplifier. This is observed by measuring the second and third harmonics frequency radiation. Since the shape of the pattern is typically different at harmonic frequencies, it is necessary to measure the entire pattern to get an idea of the harmonic suppression level. Measured E and H plane cuts of the circular segment patch antenna AIA amplifier are shown in Figure 6.30. The input power is set for maximum PAE. From the figure it is apparent that the second and third harmonics

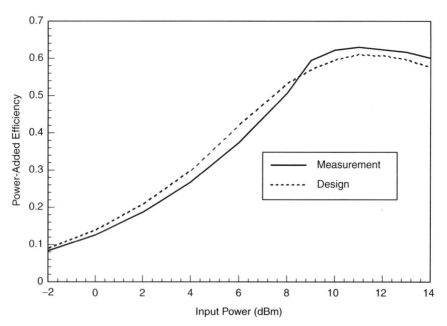

Figure 6.29. PAE of AIA PA using circular segment patch antenna at an operating frequency of 2.55 GHz.

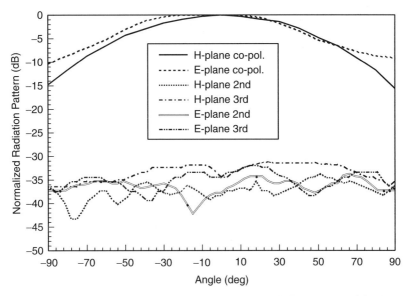

Figure 6.30. Normalized E and H plane copolarization radiation pattern of fundamental, second, and third harmonics for the Class F amplifier integrated with circular segment antenna.

are 33.8 dB and 31.4 dB below peak measured fundamental power, respectively, for all angles. Note that it would be necessary to integrate the power in the entire radiation pattern for each of the harmonics to determine the absolute harmonic suppression value. The output power has been calibrated using the receiving antenna gain and the Friis transmission formula at corresponding frequencies. Therefore the fundamental and harmonic power levels are referenced at the output of the antenna integrated with the amplifier.

6.7.3 Trends for Active Integrated Antennas

In both of the preceding examples, an antenna was developed that was nonradiating at one or more higher harmonics, the benefit being harmonic tuning via the reflected harmonic power and increased harmonic suppression. Therefore, in the AIA approach for power amplifiers, the job of the antenna designer is to develop an antenna with an input impedance characteristic that can be used as a tuned load at the operating and harmonic frequencies. Additionally, the antenna must maintain sufficient radiation characteristics for the application at hand. The result is a compact transmitter front end with extremely high realized efficiency.

In many cases, the AIA technique can be combined with other techniques presented in this book for even greater gains in efficiency or linearity. One example is an AIA integrated with a novel harmonic tuning structure. The novel tuning structure is based on a simple form of the periodic structure referred to as photonic bandgap (PBG) in Chapter 8 [36,37]. The structure consists of periodically etched holes in the microstrip ground planes. At the resonant frequency of the period (spacing equal to one-half a guided wavelength), a deep stopband effect is observed in the insertion loss characteristic of the structure. In particular, this structure has been used to implement third harmonic tuning in a new version of the patch with pins AIA power amplifier presented in the first example of this section. Since the PBG structure is designed for third harmonic tuning, it is relatively compact at the fundamental frequency where the spacing would correspond to one-sixth of a guided wavelength. A total of four periods were used. It was found that the peak PAE for this AIA reached 61% at an input power of 14 dBm [33]. The operating frequency was 2.45 GHz. Additionally, the measured second and third harmonics are 33.8 dB and 31.4 dB below peak measured fundamental power, respectively, for all angles, indicating that third harmonic tuning has successfully been implemented. Therefore this technique is a simple and effective method of tuning additional harmonics that results in increased efficiency and low harmonic radiation levels.

6.8 SUMMARY AND OUTLOOK

The power amplifier is the most critical individual circuit for overall communications system efficiency. However, to obtain significant performance improvements, it is beneficial to optimize the power amplifier together with other parts of the wireless system. This chapter has focused on improvement approaches that have resulted from an interdisciplinary study carried out as a multiuniversity research program. The interdisciplinary themes critical for amplifier improvement include the following:

1. The amplifier efficiency is strongly related to system requirements for transmitted power variability and linearity (which in turn stem from the modulation format).
2. To optimize the system in an environment of large power variations (resulting from variations of transmission distance, fading, or envelope variations caused by the modulation), it is highly advantageous to implement a variable power supply voltage.
3. For even higher efficiency, switching-mode power amplifiers offer many benefits. The success of these circuits relies, however, on improving the high frequency characteristics of the transistors used, since the switching transitions must be very fast relative to the period of the transmitted signal.
4. Switching-mode amplifiers with high linearity require novel signal processing concepts. To achieve these, and for a variety of other benefits, it is worthwhile to employ digital signal processing.
5. Finally, to optimize the overall transmitter system, it is useful to integrate the design of the power amplifier and the transmitting antenna.

We expect that, in the future, the themes outlined here will be pursued further with continued success. There are other, related, themes that have not adequately been described in this chapter. These themes include the improvement of amplifier linearity by specific architectures such as adaptive predistortion and Cartesian feedback and feed-forward. Such techniques can allow trading off linearity for efficiency in a bare amplifier, and subsequently recovering the linearity through the auxiliary circuitry. Another approach that is promising for the future is based on using switching-mode amplifiers together with low frequency amplifiers that control the envelope of the output signal (the envelope elimination and restoration technique, or Kahn amplifier approach). These advanced architectures, as well as the ones described in this chapter, rely on the continuing advance of transistor technology that will enable cost-effective manufacturing of increasingly complex and powerful signal processing circuits to accompany the basic output power transistors.

ACKNOWLEDGMENTS

The work reported here has been carried out under the sponsorship of the U.S. Office of the Secretary of Defense and the Army Research Office under the "Low Power, Low Noise Electronics Technologies for Mobile Wireless Communication" MURI, monitored by Dr. J. Harvey, Dr. R.Trew, Dr. J. Mink, and Dr. D. Woolard. The authors are grateful to them, and to their MURI team colleagues, for numerous discussions of the issues presented. This chapter describes the work of many graduate students and researchers. The authors are additionally grateful to G. Hanington, M. Iwamoto, C. Fallesen, A. Metzger, W. Deal, V. Radisic, J. Sevic, K. Gard, M. F. Chang, W. Hooper, and H. Finlay for numerous helpful conversations during the course of this work. HBTs that were used in the dc–dc converter were provided by HRL Laboratories and Conexant Systems Corporation.

REFERENCES

1. J. Sevic, "Statistical characterization of RF power amplifier efficiency for IS-95 CDMA digital wireless communication systems," in *Proceedings 2nd Annual Wireless Communications Conference*, 1997.
2. T. Nojima, S. Nishiki, and K. Chiba, "High efficiency transmitting power amplifiers for portable radio units," *IEICE Transactions*, Vol. E74, 1563 (1991).
3. C. Buoli, A. Abbiati, and D. Riccardi, "Microwave power amplifier with envelope controlled drain power supply," in *25th European Microwave Conference*, 1995.
4. G. Hanington, P. F. Chen, V. Radisic, T. Itoh, and P. M. Asbeck, "Microwave power amplifier efficiency improvement with a 10 MHz HBT dc–dc converter," *1998 IEEE MTT-S International Microwave Symposium Digest*, Vol. 2, 589–592 (1998).
5. G. Hanington, A. Metzger, P. Asbeck, and H. Finlay, "Integrated dc–dc converter using GaAs HBT technology," to be published.
6. S. Bouthillette and A. Platzker, "High efficiency L-band variable output power amplifiers for use in communication systems," *1996 IEEE MTT-S International Microwave Symposium Digest*, Vol. 2, 563–566 (1996).

7. F. Zavosh, M. Thomas, C. Thron, T. Hall, D. Artusi, D. Anderson, D. Ngo, and D. Runton, "Digital predistortion techniques for RF power amplifiers with CDMA applications," *Microwave Journal*, Oct. 1999.
8. C. Fallesen, G. Hanington, and P. Asbeck, "Improved linearity of a dynamic supply voltage power amplifier using digital predistortion," in *1999 IEEE Topical Workshop on Power Amplifiers for Wireless Communications*.
9. N. A. Sokal and A. D. Sokal, "Class E—a new class of high-efficiency single-ended switched power amplifiers," *IEEE Journal of Solid State Circuits*, Vol. SC-10, 168–176 (June 1975).
10. F. H. Raab, "Suboptimum operation of class-E power amplifiers," in *Proceedings RF Technology Exposition*, Santa Clara, CA, pp. 85–98, Feb. 1989.
11. T. B. Mader," Quasi-optical class E power amplifiers," Ph.D. dissertation, Department of Electrical and Computer Engineering, University of Colorado, Boulder, Aug. 1995.
12. T. B. Mader and Z. Popovic, "The transmission-line high-efficiency class-E amplifier," *IEEE Microwave and Guided Wave Letters*, Vol. 5, No. 9, 290–292 (Sept. 1995).
13. E. Bryerton, "High-efficiency switched-mode microwave circuits," Ph.D. dissertation, Department of Electrical and Computer Engineering, University of Colorado, Boulder, June 1999.
14. S. Djukic, D. Maksimovic, and Z. Popovic, "A planar 4.5-GHz dc–dc power converter," *IEEE Transactions on Microwave Theory and Techniques*, Vol. 47, No. 8, 1457–1460 (Aug. 1999).
15. T. B. Mader, E. W. Bryerton, M. Markovic, M. Forman, and Z. Popovic, "Switched-mode high-efficiency microwave power amplifiers in a free-space power-combiner array," *IEEE Transactions on Microwave Theory and Techniques*, Vol. 46, No. 10, 1391–1398 (Oct. 1998).
16. E. Bryerton, M. Weiss, and Z. Popovic, "Efficiency of chip-level versus external power combining," *Special Issue on Low-Power/Low-Noise Circuits of the IEEE Transactions on Microwave Theory and Techniques*, 1482–1485 (July 1999).
17. M. Weiss and Z. Popovic, "A 10-GHz high-efficiency active antenna," in *1999 IEEE IMS Symposium Digest*, Anaheim, CA, pp. 663–666, June 1999.
18. M. Markovic, A. Kain, and Z. Popovic, "Nonlinear modeling of Class-E microwave power amplifiers," *Journal of the RF and Microwave Computer-Aided Engineering*, Vol. 9, Issue 2 (Mar./Apr. 1999).
19. M. Weiss, M. Crites, E. Bryerton, J. Whitacker, and Z. Popovic, "Time domain optical sampling of nonlinear microwave amplifiers and multipliers," *IEEE Transactions on Microwave Theory and Techniques*, Vol. 47, No. 12, 2599–2604 (Dec. 1999).
20. E. Bryerton, W. Shiroma, and Z. Popovic, "A 5-GHz high-efficiency class-E oscillator," *IEEE Microwave and Guided Wave Letters*, Vol. 8, No. 12, 441–443 (Dec. 1996).
21. S. Djukic, D. Maksimovic, and Z. Popovic, "A planar 4.5-GHz dc to dc power converter," *Special Issue on Low-Power/Low-Noise Circuits of the IEEE Transactions on Microwave Theory and Techniques*, 1457–1460 (July 1999).
22. M. Weiss, Z. Popovic, and F. H. Raab, "Linearity characteristics of X-band power amplifiers in high-efficiency transmitters" (unpublished work).
23. H. Chireix, "High power outphasing modulation," *Proceedings of the IRE*, Vol. 23, No. 11, 1370–1392 (Nov. 1935).

24. F. H. Raab, "Efficiency of outphasing RF power-amplifier systems," *IEEE Transactions on Communications*, Vol. COM-33, No. 10, 1094–1099 (Oct. 1985).
25. F. Casadevall and J. Olmos, "On the behavior of the LINC transmitter," in *Proceedings IEEE Vehicular Technology Conference*, pp. 29–34, 1990.
26. R. Langridge, T. Thornton, P. M. Asbeck, and L. E. Larson, "A power re-use technique for improved efficiency of outphasing microwave power amplifiers," *IEEE Transactions on Microwave Theory and Techniques*, Vol. 47, 1467–1470 (Aug. 1999).
27. L. Sundstrom, "The effect of quantization in a digital signal-component separator for LINC tranmsitters," *IEEE Transactions on Vehicular Technology*, Vol. 45, No. 2, 346–352 (May, 1996).
28. A. Jayaraman, P. F. Chen, G. Hanington, L. Larson, and P. Asbeck, "Linear high-efficiency microwave power amplifiers using bandpass delta-sigma modulators," *IEEE Microwave and Guided Letters*, Vol. 8, No. 3, 121–123 (1998).
29. A. Jayaraman, P. Asbeck, K. Nary, S. Beccue, and K.-C. Wang, "Bandpass delta–sigma modulator with 800 MHz center frequency," in *IEEE Gallium Arsenide Integrated Circuit Symposium, 19th Annual Technical Digest*, pp. 95–98, 1997.
30. V. Radisic, Y. Qian, and T. Itoh, "Novel architectures for high-efficiency amplifiers for wireless applications," *IEEE Transactions on Microwave Theory and Techiques*, Vol. 46, 1901–1909 (Nov. 1998).
31. J. R. Lane, R. G. Freitag, H.-K. Hahn, J. E. Degenford, and M. Cohn, "High-efficency 1-,2- and 4-W class-B FET power amplifiers," *IEEE Transactions on Microwave Theory and Techniques*, Vol. 34, 1318–1325 (Dec. 1986).
32. C. Duvanaud, S. Dietsche, G. Patanut, and J. Obregon, "High-efficiency class F GaAs FET amplifier operating with very low bias voltage for use in mobile telephones at 1.75 GHz," *IEEE Microwave Guided Wave Letters*, Vol. 3, 268–270 (Aug. 1993).
33. E. Camargo and R. M. Steinberg, "A compact high power amplifier for handy phones," *IEEE MTT-S International Microwave Digest*, 565–568 (June 1994).
34. V. Radisic, S. T. Chew, Y. Qian, and T. Itoh, "High efficiency power amplifier integrated with antenna," *IEEE Microwave Guided Wave Letters*, Vol. 7, 39–41 (Feb. 1997).
35. V. Radisic, Y. Qian, and T. Itoh, "Class F power amplifier integrated with circular sector microstrip antenna," *IEEE MTT-S International Microwave Digest*, Vol. 2, 687–690 (June 1997).
36. V. Radisic and T. Itoh, "Active antenna power amplifier with PBG," in *28th European Microwave Conference Digest*, Vol. 1, Amsterdam, The Netherlands, pp. 156–160, Oct. 1998.
37. V. Radisic, Y. Qian, and T. Itoh, "Broad-band power amplifier integrated with slot antenna and novel harmonic tuning structure," *IEEE MTT-S International Microwave Digest*, Vol. 3, 1895–1898 (June 1998).

7

CHARACTERIZATION OF AMPLIFIER NONLINEARITIES AND THEIR EFFECTS IN COMMUNICATIONS SYSTEMS

JACK EAST, WAYNE STARK, AND GEORGE I. HADDAD

Department of Electrical Engineering and Computer Science
The University of Michigan, Ann Arbor

7.1 INTRODUCTION

7.1.1 Chapter Overview

Amplifier are a critical component of all communications systems. They are usually the components that require the most prime power, thus setting the efficiency performance of the communications system, and they are also the most nonlinear component, thus setting the interference and distortion performance of the system. Many of the requirements of a communications system are determined by the design of the output power amplifier and its interaction with the signal waveform. Although there is a long history of amplifier design and characterization in the field of microwave engineering, recent implementations of more complex digital waveforms for personal communications applications require a new set of tools and parameters to characterize amplifier operation. The goal of this chapter is to describe some of these new system performance requirements within the context of microwave amplifier operation and characterization and to briefly discuss some of the tools

---

RF Technologies for Low Power Wireless Communications
Edited by Tatsuo Itoh, George Haddad, and James Harvey
ISBN 0-471-38267-1 Copyright © 2001 by John Wiley & Sons, Inc.

being developed to better understand amplifier operation in modern communications systems.

This chapter is arranged into five sections. This first section will continue with a brief summary of power and linearity trade-offs in communications systems. Interference, distortion, and other problems associated with nonlinear operation and the trade-offs between nonlinearity and efficiency will be discussed. A new figure of merit will be developed to quantify these trade-offs for low energy communications. Section 7.2 will continue the discussion of analytic descriptions of amplifier operation. These formulations are fast and easy to use so that a wide variety of operating conditions can be investigated. The trade-off is simplicity and quick simulation time versus limited accuracy compared with more detailed computer simulations and measurements. This approach uses simplified amplifier descriptions to investigate a wide variety of operating conditions. As an example of these analytic tools the performance of a simple saturating amplifier in a communications system will be discussed. Finally, the use of a novel bias scheme to further improve the efficiency of the system without introducing additional distortion will be described. Section 7.3 will discuss more detailed numerical tools to investigate nonlinear communications systems. Earlier tools were extensions of existing microwave simulations. However, these tools are not able to handle the number of frequencies needed to correctly represent typical modern digital information and new implementations are needed. Two simulation tools will be discussed: an envelope approximation that is useful to describe low level nonlinear operation and a complete large signal harmonic balance simulation tool. Section 7.4 will describe the experimental characterization needed to accurately implement device models for the simulation tools. A new large signal characterization system will be described and the system will be used to compare the predictions of a variety of existing large signal device simulations with experimental large signal data. Finally, Section 7.5 will look at future requirements, open issues, and unanswered questions in the field and propose some areas for future work.

7.1.2 Efficiency, Distortion, and Bandwidth Trade-offs in Amplifiers

The trade-off between power efficiency and linearity or distortion sets the performance of an amplifier in a communications system. This trade-off can be explained with the help of Figures 7.1 through 7.3. Figure 7.1 shows the voltage gain of a typical field-effect transistor (FET) amplifier as a function of the input voltage drive level. For modest drive levels the gain is approximately constant. Since the output power is proportional to V_{out}^2 and the gain is approximately constant, the drain efficiency is proportional to V_{in}^2. This is the case for drive voltages less the 0.25 V in Figure 7.2. Conditions change with higher drive levels. The gain compresses, the operating point changes, and the dc power increases. The resulting efficiency versus drive level first increases linearly, reaches a maximum, and then begins to decrease with increasing radio frequency (RF) drive level. One of our goals in communications system design is to operate near the peak amplifier efficiency point. This will optimize the power consumption and battery life. However, the peak efficency point

INTRODUCTION

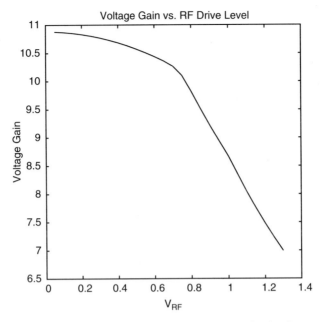

Figure 7.1. Voltage gain versus RF voltage drive level.

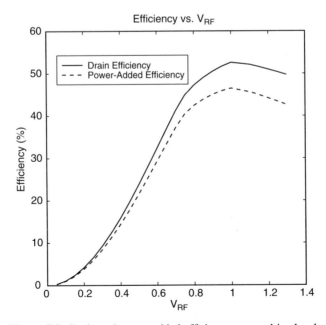

Figure 7.2. Drain and power-added efficiency versus drive level.

in this example is in the gain saturation operating region of the amplifier. If the input signal waveform has a nonconstant envelope then the amplified output waveform will be distorted. The amount of distortion will depend on the choice of the waveform and the operating point of the amplifier. An example of amplifier distortion is spectral regrowth or adjacent channel power generation for digitally modulated signals. The adjacent channel power is proportional to the drive level, as shown in Figure 7.3. The allowed adjacent channel power ratio (ACPR) is usually limited by the communication protocol being used. The required ACPR is usually reached by reducing the input drive level or "backing off" the amplifier until an acceptable ACPR is obtained. Clearly, reducing the drive level to reduce the ACPR contradicts our goal of optimum efficiency.

The distortion associated with the nonlinear amplifier is caused by the variable gain with signal level, so a possible solution is to use a constant envelope or amplitude data waveform. A possible modulation is mimimum shift keying (MSK). This modulation shifts the phase but not the amplitude of the time domain waveform. The resulting time domain amplitude is constant. This signal could pass through a nonlinear amplifier without suffering spectral regrowth. A typical nonconstant envelope modulated signal is quadrature phase shift keying (QPSK). The time domain response of these two modulations is shown in Figure 7.4. The large peak-to-average

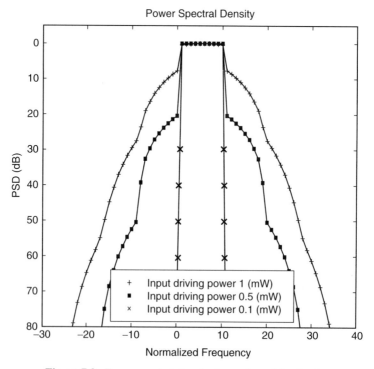

Figure 7.3. Power spectral density for various drive levels.

INTRODUCTION

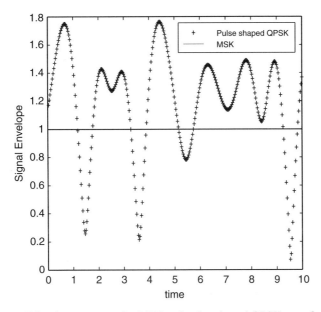

Figure 7.4. Time response for MSK and pulse shaped QPSK waveform.

ratio of the QPSK signal, when passed through a nonlinear amplifier, will create the spectral regrowth and adjacent channel power that needs to be avoided. However, there are other trade-offs in the choice of the waveform. The frequency domain power spectral density (PSD) of the constant envelope MSK signal and the nonconstant envelope QPSK signal are shown in Figures 7.5 and 7.6. The frequency scale in these two figures is normalized to T_b, the time duration of the data bits. The constant envelope signal occupies a much wider spectrum than the nonconstant envelope signal with the same data rate. However, the available bandwidth is usually specified and limited. The amplifier performance and the modulation waveform result in spectral regrowth and spectrum usage and need to be optimized for best performance. In the next section we will discuss a new figure of merit to optimize these trade-offs for low energy communications systems.

7.1.3 Energy Based Figure of Merit for Communications Systems

The last section described the trade-offs between amplifier linearity, distortion, and spectral efficiency in communications systems. In this section a figure of merit that optimizes the energy performance will be described. Additional details of this analysis are given by Liang et al. [1]. One conventional figure of merit for a communications system is the bit error rate (BER). This is usually a decreasing function of E_b/N_0 where E_b is the received energy per bit of information and N_0 is the two-sided power spectral density of the noise in an additive white Gaussian noise (AWGN) channel. The relationship between E_b, the average received power $\bar{P}_r$, and

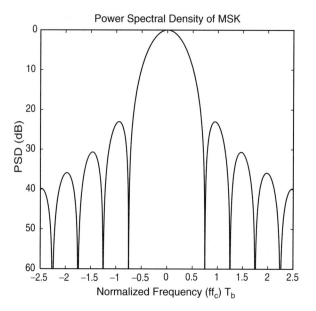

Figure 7.5. Power spectral density for an MSK waveform.

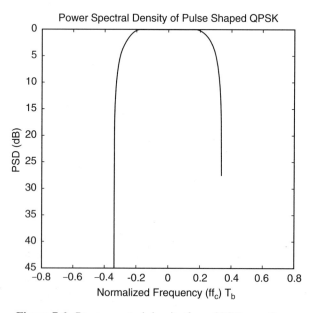

Figure 7.6. Power spectral density for a QPSK waveform.

INTRODUCTION

the data rate R_b is

$$\frac{\bar{P}_r(\text{J/s})}{R_b(\text{bits/s})} = E_b(\text{J/bit}). \tag{7.1}$$

One possible goal is to minimize E_b/N_0 for a given BER. This will depend on the choice of the modulation scheme. The power spectral density of the signal will also depend on the choice of the modulation, with some modulations occupying more spectrum than others. A larger data rate R_b usually requires a wider frequency band. An alternative is to send as many bits as possible for a given bandwidth W. This will define a bandwidth efficiency

$$\frac{R_b}{W} \; (\text{bits/H}_z). \tag{7.2}$$

These figures of merit are widely used in communications. However, our goal is to minimize the total energy used by the communications system in order to conserve battery life, so we need to use a modified figure of merit to optimize the power and spectral efficiency of the system.

The conventional measure R_b/N_0 only accounts for the energy received at the receiver. We are interested in the energy required to transmit the information as well. The instantaneous power-added efficiency of the amplifier is defined as

$$P_{ae}(t) = \frac{P_{RF}(t) - P_{in}(t)}{P_{dc}(t)}, \tag{7.3}$$

where $P_{in}(t)$ and $P_{RF}(t)$ are the input and output RF powers, respectively, and $P_{dc}(t)$ is the dc power. The total average power into the device, $\bar{P}_t$, is

$$\bar{P}_t = \bar{P}_{dc} + \bar{P}_{in} \tag{7.4}$$
$$= \bar{P}_{RF}\left(1 + (1 - \bar{P}_{ae})P_{dc}\right)/\bar{P}_{RF} \tag{7.5}$$
$$= \bar{P}_{RF}(1 + w), \tag{7.6}$$

where an overbar denotes an average value and w is the fraction of the average power that is not converted into RF power. This represents loss associated with the nonideal efficiency of the amplifier. The energy needed to transmit a bit of information becomes

$$E_t = \bar{P}_t/R_b \tag{7.7}$$
$$= \bar{P}_r(1 + w)/R_b \tag{7.8}$$
$$= E_b + E_w, \tag{7.9}$$

where we are assuming no propagation loss so that $\bar{P}_{RF} = \bar{P}_r$, E_b is the required energy per bit for a given BER and N_0 and E_w is the additional energy per bit

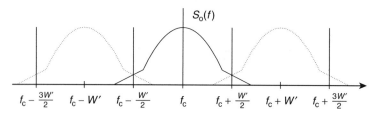

Figure 7.7. Power spectral density for spectrum efficiency.

associated with the inefficiency of the amplifier. The fraction of the energy that is not transmitted will depend on the operating conditions of the amplifier and the resulting amplifier efficiency versus input drive level. We will see that improving the efficiency with a large input drive level will also increase the noise associated with N_0 due to distortion in the amplifier.

The bandwidth or spectral efficiency is R_b/W', where W' is the frequency separation between two adjacent channels as shown in Figure 7.7. We will also assume that the neighboring channels are using the same amplifier and modulation schemes. Interference due to either overlap of the PSD of two ideal signals or interference due to spectral spreading of a signal due to nonlinear amplifier effects will both cause adjacent channels to interfere with each other and increase the noise N_0. Equation (7.9) and the expression for the spectral efficiency can be combined to describe both energy and spectral efficiency. The resulting new figure of merit is

$$D = \frac{E_t/N_0}{R_b/W'}. \tag{7.10}$$

We can use this new figure of merit to investigate the trade-offs between amplifier linearity, distortion, and noise for a variety of modulation waveforms.

7.1.4 Some Results

The figure of merit in Eq. (7.10) was used to investigate the energy performance of a simple communications system. A Class AB FET amplifier was used. LIBRA was used to predict the AM–AM performance and the amplifier efficiency. The amplifier is nearly linear at small input power levels and saturates at higher drives. The AM–PM phase distortion is neglected in this simulation. A signal is modulated with various modulation schemes, passed through a pulse shaping filter, and then passed through the amplifier. The modulations include two constant envelope signals, minimum shift keying (MSK) [2], and modified MSK with quasi-band limited (QBL) pulses [3] and three nonconstant envelop signals, quaternary phase shift keying (QPSK) [2], π/4 QPSK [4], and orthogonal frequency division multiplexing (OFDM) [5]. The details of the calculations are given in Liang et al. [1].

The figure of merit for a variety of operating conditions is shown starting with Figure 7.8. Here the figure of merit on a decibel scale is plotted versus the amplifier

INTRODUCTION

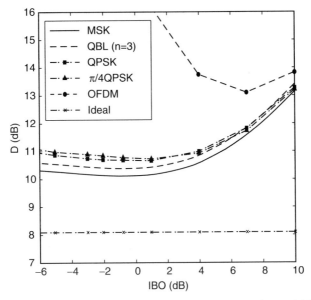

Figure 7.8. Figure of merit for various modulation schemes ($W'/R_b = 1.35$ and BER = 10^{-3}).

input back off (IBO). The input backoff is defined to be the ratio of the input power at the beginning of saturation to the average input power,

$$\text{IBO} = 10 \log (P_{\text{in}}^{\text{sat}} / \bar{P}_{\text{in}}). \tag{7.11}$$

The IBO is a measure of the degree of saturation in the amplifier. The W'/R_b in the figure is 1.35. This is equivalent to having a large center-to-center channel spacing. There are several forms of interference for the signals in Figure 7.8. When the nonconstant envelope signal is large enough to be in the nonlinear operating region of the amplifier, spectral regrowth occurs. This additional signal can spill over into the adjacent channel, causing interference and increasing the noise term in the figure of merit. The nonlinear amplifier can also create in-band noise interference. Finally, there is "noise" associated with the spectral width of the signal. The constant envelope signals occupy a wider bandwidth than the nonconstant envelope signals. The nonlinear amplifier doesn't distort these signals, but the wider spectrum can still interfere with the adjacent channel signal. All the figure of merit curves in Figure 7.8 start at a level, decrease with increasing input drive, reach a minimum value, and then begin to increase. At low signal levels the figure of merit is dominated by the amplifier efficiency. The amplifier efficiency is low at low drive levels so more dc power is needed to produce a received signal with a given BER. Increasing the drive level increases the efficiency, reduces the power needed, and reduces the figure of merit. Finally, the efficiency saturates and begins to decrease and the noise associated with the nonlinear amplifier distortion begins to increase. Both these effects degrade the figure of merit. In Figure 7.8 the constant envelope MSK modulation has the best figure of merit over the entire input operating range. The large channel spacing reduces the effect of the wider MSK spectrum and no additional interference is generated by the amplifier. The OFDM modulation has the worst performance for this case and for most of the conditions to be described. Since this modulation requires larger bandwidth for its orthogonal signals, its spectral efficiency and figure of merit are degraded. However, its wider bandwidth and frequency diversity are an advantage in a fading or multipath propagation environment. Figure 7.9 shows the figure of merit plot for a channel spacing that is approximately half the one in Figure 7.8. There are important differences compared with the results in Figure 7.8. The two nonconstant envelope modulations now have the best figure of merit. The figure of merit for these two modulations has improved by approximately 2 dB. The improvement is due to the narrower channel spacing and resulting improved spectral efficiency. Since the "noise" associated with the nonconstant envelope signals is due to nonlinear amplifier operation, there is little noise at lower drive levels. However, the performance of the constant envelope MSK modulation has degraded. The smaller bandwidth increases the spectral overlap of the wider MSK signals and greatly increases the "noise" associated with the adjacent channel interference. Even with the improved spectral efficiency, the MSK modulation is ~ 1.5 dB worse in this case.

These results show some of the trade-offs in the optimization of communications systems and also the importance of investigating the hardware and the signal processing issues together. Even the simple models used here show some of the

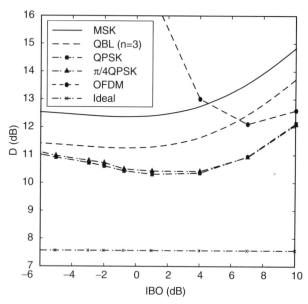

Figure 7.9. Figure of merit for various modulation schemes ($W'/R_b = 0.675$ and BER = 10^{-3}).

important trade-offs. However, we need more information to better understand the optimization process. One approach is to continue to use analytic models and to further investigate modulation and interference in more detail. This is the subject of the next section of this chapter. Another approach is to better understand the effects of the nonlinear device on complex signals. This is the subject of Section 7.3. Finally, all of the results in this chapter depend on the accuracy of the nonlinear device simulation being used. Section 7.4 describes measurements to better understand the limitations of existing device models.

7.2 ANALYTIC TECHNIQUES FOR COMMUNICATIONS SYSTEMS SIMULATIONS

A variety of analytic tools are available for communications systems simulations. In this section we will use this analytic approach to further investigate energy trade-offs. As mentioned in Chapter 1 and in the introduction of this chapter, there is a trade-off between power efficiency and bandwidth efficiency even without considering nonlinearities from the power amplifier. The nonlinear amplifier introduces another trade-off between power efficiency and bandwidth efficiency. A typical power amplifier is most efficient in converting dc power into RF power when operating close to saturation. However, when the amplifier is close to saturation it operates in a nonlinear mode. As such, the amplifier generates intermodulation products that cause in-band distortion and out-of-band signals. The in-band distortion degrades performance while the out-of-band signal generation causes interference ith adjacent channels. If we decompose the output signal of the amplifier into a desired signal (related linearly to the input of the amplifier)and an undesired signal (related nonlinearly to the input), then the efficiency of the amplifier as measured by the amount of dc power translated to desired output power (as opposed to undesired output) is larger the closer the amplifier is to saturation. The goal then is to operate the amplifier to minimize the total power required to obtain a given bit error rate.

Typically, an amplifier output must be operated at many different power levels. For example, in a cellular system mobile users close to a base station must turn down their power in order not to cause excessive interference with other users. Mobile users at the edge of a cell must operate at maximum power in order to maintain a given signal-to-noise ratio. Because of this it is not always possible to operate the amplifier at the most power efficient operating point. However, by controlling the bias on an amplifier in relation to the input signal the efficiency when the input is low can be improved significantly.

In order to quantify the performance with a variable bias controlled amplifier we need to introduce several new metrics. Consider an amplifier with a maximum dc power supply level of $P_{dc,m}$ with power gain G. Assume the output saturation power is related to the maximum dc power as $P_{sat} = \alpha P_{dc,m}$, where α is a constant less than one. Let $P_{dc}(t)$ be the instantaneous power supplied to the amplifier and P_{dc} be the average power supplied. Also let P_o be the average output power. The output backoff is defined as $OBO = P_{sat}/P_o$. The smaller the OBO of an amplifier the closer the amplifier is to operating near saturation and thus the amplifier is power efficient but nonlinear. Define $S = P_{dc,m}/P_{dc}$. With these definitions we can relate the average dc power to the average output power as

$$P_o = \frac{P_{sat}}{OBO} = \alpha \frac{P_{dc,m}}{OBO} = \alpha \frac{SP_{dc}}{OBO}, \qquad (7.12)$$

$$P_o(dB) = \alpha(dB) + S(dB) - OBO(dB) + P_{dc}(dB). \qquad (7.13)$$

The amount of received power is some fraction of the amount of transmitted power, which for our purposes we set to one. If the amplifier is perfectly linear, then the probability of a bit being in error (the bit error rate) with only additive white Gaussian noise is a function of the energy received per bit E_b and the noise power spectral density $N_0/2$. However, because of nonlinear amplification there is some additional interference that degrades the performance. The amount of degradation depends on the amplifier operating point. For large output backoff there is little interference generated because of nonlinearities and thus the error probability is the same as the error probability with a linear amplifier, whereas for small backoff the amplifier is more nonlinear. Let $\Delta E_b/N_0$ denote the increase in average received signal energy-to-noise ratio necessary to maintain a given bit error probability when the effects of the nonlinearity are considered. Here the energy is the product of the received power and the data bit duration. This $\Delta E_b/N_0$ is clearly a function of the operating point of the amplifier. We can combine the two performance measures by defining the total dc power degradation (TDD) as

$$TDD = OBO(dB) - S(OBO)(dB) + \Delta E_b/N_0(dB). \qquad (7.14)$$

The metric TDD (which depends on the desired bit error rate) takes into account the power inefficiency (via $OBO - S(OBO)$) of the amplifier as well as the nonlinear distortion of the amplifier via $\Delta E_b/N_0$. The goal of the system design is to minimize TDD. The metric TDD does not take into account the input power of the amplifier, which, for reasonable gain systems, is quite small.

As mentioned earlier, conventionally designed amplifier do not operate efficiently (from a power conversion point of view) when the output backoff is large. This can be compensated for by varying either the gate or drain voltage as a function of the input RF signal envelope. In a single bias controlled amplifier the gate voltage is varied with the input signal, whereas in a dual bias controlled amplifier both the gate and drain voltages are varied. The dc power supplied thus varies as the input signal varies in a single and dual bias controlled amplifier. The relation of dc power to input power is shown in Figure 7.10 for fixed bias schemes, single bias controlled schemes, and dual bias controlled schemes. In this figure we have normalized the parameters so that the gain (in the linear region) is 500 (27 dB). Clearly, when the input power is small, the single and dual bias schemes consume significantly less power than the fixed bias schemes but yield the same output power. The savings in average dc power is small when the backoff is small. We can quantify the power

ANALYTIC TECHNIQUES FOR COMMUNICATIONS SYSTEMS SIMULATIONS 241

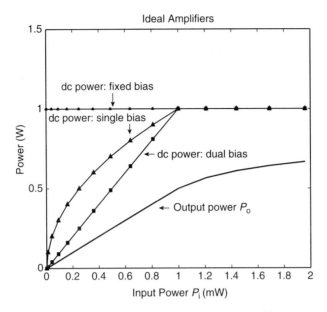

Figure 7.10. Characteristics of a power amplifer.

savings when the backoff is large by computing the difference between OBO and S (in dB). Consider a model for an amplifier consisting of an ideal soft limiter followed by a bandpass filter, which passes only signals around the fundamental frequency. By doing a Fourier analysis the overall input-output relation can be written as

$$V_o(t) = \begin{cases} V_{o,m} \dfrac{V_i(t)}{V_{i,m}}, & V_i(t) \leq V_{i,m} \\ \dfrac{2V_{o,m}}{\pi} \left[\dfrac{V_{i,m}}{V_i(t)} \arcsin\left(\dfrac{V_{i,m}}{V_{i(t)}}\right) + \sqrt{1 - \left(\dfrac{V_{i,m}}{V_i(t)}\right)^2} \right], & V_i(t) \geq V_{i,m} \end{cases}$$

(7.15)

The limiting output voltage for this model is $4V_{o,m}/\pi$. Then the ratio of saturation power to average output power can be written (for large backoff or small $V_i(t)$) as

$$\frac{P_{\text{sat}}}{P_o(t)} = \frac{V_{\text{sat}}^2}{V_0^2} \qquad (7.16)$$

$$\approx \frac{16 V_{o,m}^2/(\pi^2)}{V_{o,m}^2 \int_0^1 e^2\, de} \qquad (7.17)$$

$$= \frac{16^2}{\pi^2 e^2}, \qquad (7.18)$$

where $e(t) = V_i(t)/V_{i,m}$ represents a normalized input. The denominator in the above expression is an approximation to the mean squared output power, where we have assumed the input envelope is less than one. A similar analysis shows that the ratio of maximum dc power to average dc power can be written as

$$S = \frac{P_{dc,m}}{P_{dc}} \approx \begin{cases} 1, & \text{fixed bias} \\ 1/e, & \text{single bias control} \\ 1/e^2, & \text{dual bias control.} \end{cases} \quad (7.19)$$

Combining the above two terms the part of total dc power degradation due to amplifier inefficiency can be written as

$$\text{OBO(db)} - S(\text{db}) = \begin{cases} \text{OBO(db)}, & \text{fixed bias} \\ 20\log(4/\pi) + 10\log(\bar{e}/e^2), & \text{single bias control} \\ 20\log(4/\pi) = 2.1\,\text{db}, & \text{dual bias control.} \end{cases} \quad (7.20)$$

Thus for very large output backoff the loss of dc power due to amplifier inefficiency with dual bias controlled amplifier reaches a maximum of 2.1 dB, whereas with fixed bias design the loss increases linearly as the output backoff increases. Figure 7.11 shows the total dc power degradation for an orthogonal frequency division multiplexing (OFDM) scheme with 64 carriers for different bias controlled schemes. For each scheme there are two curves. The lower curve is the degradation due just to

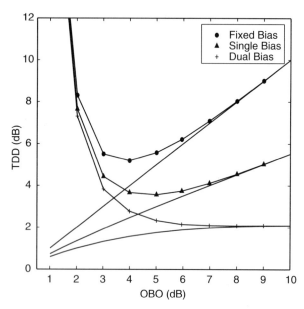

Figure 7.11. Total dc power degradation as a function of output backoff for orthogonal frequency division multiplexing ($M = 64$).

amplifier inefficiency while the upper curve accounts for both the amplifier inefficiency and the increased power necessary to maintain a given bit error probability (10^{-4}) due to the effects of the nonlinearity (intermodulation distortion).

At small output backoff the amplifier is operating near saturation so the amplifier inefficiency term is small while the nonlinear distortion term is large. At large backoffs the amplifier is nearly linear so the main component of the power degradation is due to the amplifier inefficiency. The asymptotic limiting value of 2.1 dB is clearly shown for the dual bias controlled scheme. In addition, for the dual bias controlled scheme the optimum operation (output backoff) is much larger than the other schemes and makes the power efficiency of OFDM as good as QPSK. Figure 7.12 shows the total dc power degradation for square-root raised cosine filtered QPSK modulation for different bias controlled schemes. For QPSK the single or dual bias scheme saves minimal power consumption. However, the effects of adjacent channel interference have not been considered. The adjacent channel interference is the ratio of power in the adjacent bands to the power in the desired frequency band. In Figure 7.13 we plot the adjacent channel interference as a function of output backoff for QPSK and OFDM. If we constrain the adjacent channel interference to be lower than some desired value, then a lower bound is set on the possible output backoff. Because of this for QPSK it is likely that we will not be able to operate the amplifier as close to saturation as desired to minimize TDD. For example, if the acceptable adjacent channel interference is -35 dB, then the minimum output backoff is 3.5 dB. This results in a dc power degradation of about 3.5 dB with a fixed bias amplifier, whereas it results in a TDD of only 2 dB with a dual bias amplifier, a savings of 1.5 dB.

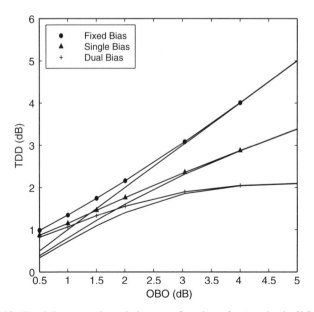

Figure 7.12. Total dc power degradation as a function of output backoff for QPSK.

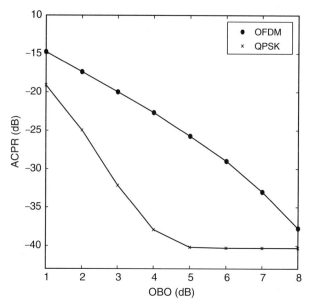

Figure 7.13. Adjacent channel power ratio as a function of output backoff for OFDM and QPSK.

7.3 NUMERICAL MODELING TECHNIQUES

7.3.1 Introduction

The analytic tools described in the first portion of this chapter provide very useful insight into the trade-offs in communications systems design. However, more detailed models are needed to accurately predict performance. Wireless designers face a number of challenging and opposing issues: system designs have become more complex, linearity and efficiency specifications have become more stringent, active devices have been pushed closer to their performance limits, and time-to-market requirements have shrunk. The engineering challenges have prompted a demand for accurate, computationally efficient, robust, and versatile simulation tools. Exciting advances have been made toward this goal, but a number of open problems remain. The purpose of this section is to cover existing methodologies, introduce recent developments, and discuss some open problems.

7.3.2 Nonlinear Simulation and Digital Modulation

Trade-offs between linearity and power consumption are among the most important design considerations for modern communications systems. The power amplifier is critical in this regard, having a profound effect on system linearity and efficiency. In principle, the amplifier can always be made more linear, but the improvement comes at the expense of higher power consumption or a more complex design; effective nonlinear simulators are the key to proper exploration of such engineering trade-offs.

Modern wireless systems process digital modulation signals that pose serious problems in nonlinear circuit analysis. These signals are of the general form

$$x(t) = \text{Re}\{\tilde{x}(t)e^{j\omega_c t}\}, \qquad (7.21)$$

where $\tilde{x}(t)$ is the complex envelope and ω_c is the carrier frequency. For the purposes of simulation and spectral analysis, the envelopes are approximated by periodic waveforms but, due to the random nature of input data, the period is chosen to extend over several hundred data bits. Short sequences result in significant run-to-run variations of simulation results and 512 data bits is usually accepted as the shortest period for satisfactory results. The length of the envelope period, coupled with a large carrier-to-envelope bandwidth ratio, creates numerical difficulties that prevent the application of classical simulation tools. Instead, special methods have been devised to cope with the problems more efficiently.

The available methods for nonlinear simulation can be grouped into two broad classes: behavioral (also known as system-level or black-box) and circuit-level. The aim of behavioral methodology is to devise an explicit input–output relationship for the nonlinear circuit. In doing so, the intricate details of circuit structure are usually disregarded. On the other hand, circuit-level methods are based on the numerical solution of equations that arise from fundamental circuit laws. These methods have their merits and drawbacks, which will be discussed in the rest of this section.

7.3.3 Behavioral Simulation

The primary purpose of behavioral methods is to allow fast analysis at the system level. For example, studies of BER performance involve numerical processing of data sequences that may exceed a million bits; fast and accurate calculation of the response of each system component, including the amplifier, is critical to achieving acceptable simulation times.

The most widely used behavioral approach is based on the bandpass memoryless model of the amplifier nonlinearity. *Bandpass* indicates that the output component centered at the carrier frequency is the only one considered; and *memoryless* describes the nature of the input–output envelope relationship. In accordance with the memoryless assumption, the response to a slowly modulated carrier is approximated by interpolation between swept-amplitude sinusoidal responses. The amplitude and phase of the fundamental component of an amplifier's sinusoidal steady-state response are nonlinear functions of input amplitude, and these functions can be represented by numerical fitting of a discrete set of amplitude and phase data to a suitable functional series.

Formally, let α denote the amplitude of the input sinusoid, and suppose that the amplitude and phase functions are expanded as

$$f(\alpha) = \sum_{n=1}^{N} b_n \alpha^n, \qquad g(\alpha) = \sum_{n=1}^{N} c_n \alpha^n. \qquad (7.22)$$

Functions $f(\alpha)$ and $g(\alpha)$ are frequently referred to as AM–AM and AM–PM characteristics. Now, let the input be a general modulated signal of the form

$$x(t) = \text{Re}\{x(t)e^{j\omega_c t}\} = \text{Re}\{|\alpha(t)|e^{j[\omega_c t + \phi(t)]}\}. \tag{7.23}$$

By assuming that the response to a swept-amplitude sinusoid closely approximates the response to a slowly varying envelope, we obtain the following input-output model:

$$y(t) = \text{Re}\{f(|\alpha(t)|)e^{j[\omega_c t + \phi(t) + g(|\alpha((t)|)]}\}. \tag{7.24}$$

Equation (7.24) is the most popular behavioral simulation approach in use, although several equivalent forms appear in the literature; variations include the rectangular (in-phase and quadrature-phase) representation of the model and the use of different basis functions for the expansion of AM–AM and AM–PM characteristics.

The appeal of the aforementioned procedure is apparent: the input–output relationship is explicit and its computation simple and efficient; there are relatively few model parameters that are easily determined by common linear least-squares techniques; the swept-sinusoidal responses are easily available from standard circuit simulators or by measurement.

Despite its appeal, this method has drawbacks. Dynamic behavior of the complex envelope (envelope memory) is exhibited by some amplifier, but it is not modeled by Eq. (7.24), nor can its presence be detected by single-frequency sinusoidal testing. The model does not account for possible changes in the source or load conditions due to, for example, antenna loading; in general, any change in the amplifier structure must be accompanied by a separate behavioral model.

7.3.4 Circuit-Level Methods

Circuit-level methods overcome the disadvantages of behavioral simulation at the expense of computational speed. They rely on the numerical solution of a system of equations obtained from Kirchhoff's laws and are rigorous in the sense that the equations arise from fundamental physical laws. It should be noted, however, that the accuracy is limited by the quality of active device models. In fact, circuit simulation is meaningful only if it is accompanied by robust and reliable transistor models.

7.3.5 Time Domain Simulation

Time domain simulators, such as SPICE, are popular tools for analog circuit analysis and are capable of efficient simulation of transient and sinusoidal steady-state response. Although it is usually impractical to use SPICE for modulated signal analysis, a brief introduction to this method provides a useful foundation for future discussion.

NUMERICAL MODELING TECHNIQUES   247

Assume that the circuit equations can be written as a system of differential equations:

$$\mathbf{f}(\mathbf{x}(t), d\mathbf{x}(t)/dt, \mathbf{u}(t)) = \mathbf{0}, \tag{7.25}$$

where $\mathbf{x}(t)$ is a real vector of K circuit variables (usually node voltages and some branch currents), $\mathbf{u}(t)$ is the vector of sources, and $\mathbf{f}$ is a vector of K nonlinear algebraic functions. The specific form of the circuit equations is chosen only for simplicity in presentation, and all of the following results apply to other forms as well.

Equation (7.25) is solved numerically at a discrete set of time samples $\{t_i\}$ over the time interval of interest. The first step in the solution is the approximation of derivative operations by one of the available formulas such as backward-differencing. Backward-differencing is based on an approximation of the form

$$\frac{d\mathbf{x}_i}{dt} \approx \alpha_i\, \mathbf{x}_i + \beta_i, \tag{7.26}$$

where $\mathbf{x}_i$ is short for $\mathbf{x}(t_i)$ and α_i and β_i are known constants. Substitution of Eq. (7.26) in the circuit equation yields a set of algebraic nonlinear equations, one for each sample t_i.

In the case of modulated signals, the high frequency carrier is followed over a comparatively very long envelope period. In typical cases, the circuit must be solved over an interval of millions of carrier cycles. Since the integration algorithm takes many small steps per carrier cycle in order to maintain accuracy, the time domain approach leads to impractically long simulation times.

7.3.6 Frequency Domain Simulation

Frequency domain simulation, and in particular the method of *harmonic balance*, has long been the workhorse tool for nonlinear design of microwave and RF circuits. Engineers are accustomed to using harmonic balance simulators mainly for the purpose of sinusoidal and two-tone steady-state nonlinear analysis. We briefly discuss the principles of harmonic balance and its application to modulated signal analysis.

Following Kundert and Sangiovanni-Vincentelli [6], suppose that the input signal and the response are periodic. By periodicity, each waveform in the circuit (hence, each element $x_k(t) \in \mathbf{x}(t)$ and $u_k(t) \in \mathbf{u}(t)$) can be written as a Fourier series:

$$x_k(t) = \sum_{n=-N}^{N} X_{k,n} e^{jn\omega_c t}, \qquad u_k(t) = \sum_{n=-N}^{N} U_{k,n} e^{jn\omega_c t}, \tag{7.27}$$

where it is assumed that frequency components beyond order N are negligible. If we define the vector of nodal Fourier coefficients as

$$\mathbf{X} = [X_{1,-N}, \ldots, X_{1,N}, \ldots, X_{K,-N}, \ldots, X_{K,N}]^T,$$

$$\mathbf{U} = [U_{1,-N}, \ldots, U_{1,N}, \ldots, U_{K,-N}, \ldots, U_{K,N}]^T, \tag{7.28}$$

the substitution of the assumed form of the response yields an equation for the elements of $\mathbf{f}(t)$ of the form

$$f_k(\mathbf{x}, d\mathbf{x}/dt, \mathbf{u}) = \sum_{n=-N}^{N} F_k(\mathbf{X}, \mathbf{U}, n) e^{jn\omega_c t}. \tag{7.29}$$

Making use of the well-known orthogonality relationships for the Fourier basis, Eqs. (7.29) and (7.25) yield the harmonic balance system

$$\mathbf{F}(\mathbf{X}, \mathbf{U}) = \mathbf{0}. \tag{7.30}$$

$\mathbf{F}$ is defined in analogy to $\mathbf{X}$ and $\mathbf{U}$. We point out that a similar equation can be derived if the input consists of a finite number of sinusoidal signals that are not harmonically related.

The harmonic balance system is a nonlinear algebraic system of $(2N+1)K$ equations in the unknown Fourier coefficients. The equations are generally complex, but note that for all positive n, $F_{k,n}$ is the complex conjugate of $F_{k,-n}$. It is thus possible (and desirable) to formulate the harmonic balance system as $(2N+1)K$ real equations in order to reduce the computational effort, and the interested reader is referred to Ref. 6 for details and a rigorous treatment. In the interest of notational simplicity, we continue our presentation based on Eq. (7.30).

Classical methods solving the harmonic balance system are based on the Newton–Raphson iteration:

$$\mathbf{X}^{\text{new}} = \mathbf{X}^{\text{old}} - \mathbf{J}_\mathbf{F}^{-1}(\mathbf{X}^{\text{old}}) \mathbf{F}(\mathbf{X}^{\text{old}}), \tag{7.31}$$

where $\mathbf{J}_\mathbf{F}$ is the Jacobian matrix of $\mathbf{F}$, the matrix of derivatives of the elements of $\mathbf{F}$ with respect to the unknown nodal Fourier coefficients; the Jacobian is square of order $(2N+1)K$.

It is, in principle, possible to apply the method of harmonic balance to modulated signal excitations. To arrive at a proper signal representation, recall that the envelope is assumed periodic, and assume that it is sampled at $2L+1$ instants, at a rate of $\omega_0/2\pi$ over a period of at least 512 data pulses. With the carrier products included, we have a series representation of the form

$$x_k(t) = \sum_{m=-M}^{M} \sum_{l=-L}^{L} X_{k,m,l} e^{j(m\omega_c + l\omega_0)t} \equiv \sum_{n=-N}^{N} X_{k,n} e^{j\omega_n t}. \tag{7.32}$$

Knowing that the number of frequencies in Eq. (7.32) can exceed 10,000 for typical applications, the problem of applying harmonic balance to modulated signals is apparent. The total number of unknowns in the harmonic balance system is formidable, exceeding 100,000 even for small circuits. The storage and factorization of the Jacobian matrix, required by direct linear solvers, is usually impossible. It is possible, however, to apply iterative linear solvers to the Newton–Raphson iteration, and this procedure has been applied successfully to modulated carrier excitations [7]; nevertheless, the simulations remain slow and memory consuming.

Another harmonic balance technique, termed the *modified fixed-point method* [8], has recently been proposed as an alternative to iterative Newton methods in the solution of large problems such as modulated analysis of nonlinear amplifiers. It is based on the notion that the harmonic balance equation can be written as

$$\mathbf{YX} = \mathbf{F}_{\mathrm{NL}}(\mathbf{X}, \mathbf{U}), \tag{7.33}$$

where $\mathbf{Y}$ is a $(2N+1)K$ matrix and $\mathbf{F}_{\mathrm{NL}}$ is a nonlinear function. As it turns out, $\mathbf{Y}$ is block diagonal, consisting of $K \times K$ blocks of $(2N+1) \times (2N+1)$ diagonal matrices. Because of the structure of $\mathbf{Y}$, it is tempting to define the iteration

$$\mathbf{YX}^{\mathrm{new}} = \mathbf{F}_{\mathrm{NL}}(\mathbf{X}^{\mathrm{old}}, \mathbf{U}), \tag{7.34}$$

because Eq. (7.34) reduces to $2N+1$ *independent* linear systems of $K \times K$ equations, a much more tractable problem than Eq. (7.31) in cases when N or K are large.

Iterations of this type are known as fixed-point iterations. Although fixed-point iterations have inferior convergence properties to Newton–Raphson, they offer the advantage of having smaller memory requirements, thus being applicable to very large problems. The fact that $\mathbf{Y}$ does not depend on the iteration step makes it possible to factorize and store the $2N+1$ equivalent coefficient matrices at the beginning of the iteration process; subsequent iterations only require triangular substitutions, which are simple and efficient numerical steps.

7.3.7 Envelope Simulation

Envelope simulation is a promising new technology for simulation of nonlinear circuits with modulated carrier excitations [9–11]. To arrive at its formulation, suppose that the circuit is driven by a modulated source so that any circuit waveform can be written as

$$x_k(t) = \sum_{n=-N}^{N} X_{kn}(t) e^{jn\omega_c t}, \tag{7.35}$$

where $X_{kn}(t)$ is a complex envelope. If the assumed form of the response is substituted in Eq. (7.25), each element of $\mathbf{f}(t)$ has the form

$$f_k(\mathbf{x}, d\mathbf{x}/dt, \mathbf{u}) = \sum_{n=-N}^{N} F_k(\mathbf{X}, d\mathbf{X}/dt, \mathbf{U}, n) e^{jn\omega_c t}, \tag{7.36}$$

where $\mathbf{X}(t) = [X_{1,-N}(t), \ldots, X_{1,N}(t), \ldots, X_{K,-N}(t), \ldots, X_{K,N}(t)]^T$ is the vector of envelope components of the node voltages and a similar definition applies to $\mathbf{U}(t)$.

The key assumption in this method is that the signals are narrowband; that is, the envelopes are sufficiently constant over the carrier period $2\pi/\omega_c$ that for any envelope waveform in the circuit, say, $\tilde{y}(t)$, we have the approximation

$$\int_0^{2\pi/\omega_c} y(t) e^{jn\omega_c t} \, dt \approx y(t) \int_0^{2\pi/\omega_c} e^{jn\omega_c t} \, dt. \tag{7.37}$$

Thus the envelope components may be considered nearly orthogonal in the Fourier basis, and a procedure similar to what was used to arrive at the harmonic balance equation yields the envelope equation

$$\mathbf{F}(\mathbf{X}(t), d\mathbf{X}(t)/dt, \mathbf{U}(t)) = \mathbf{0}. \tag{7.38}$$

The envelope equation can now be solved by approximating derivatives as in the case of time domain methods. This yields, at each time step t_i, a set of nonlinear algebraic equations that are structurally similar to the harmonic balance system Eq. (7.30). The power of the envelope method lies in the fact that the size of each system is only $(2N+1)K$, where N is the order of carrier distortion products (a number between 5 and 11 in typical applications). It is also important to observe that the samples are spaced on the scale of envelope variation, not carrier variation as in the case of time domain methods. The envelope is nominally sampled at the Nyquist rate, though smaller steps, if necessary, may be taken by adaptive differential solvers. In either case, the method is approximately equivalent to the solution of several thousand harmonic balance systems of the kind that is typically encountered in simulation of nonlinear amplifier with single-tone excitations. This is a tractable problem, though the large number of systems that need to be solved may cause long simulation times, on the order of several minutes for single-stage amplifier on modern workstations. When the excitation waveform is such that the amplifier may be considered weakly nonlinear, it is possible to reduce that time to seconds using the *weakly nonlinear envelope method* described next.

7.3.8 Weakly Nonlinear Envelope Simulation

Under the assumption of weakly nonlinear operation, the response of a nonlinear circuit is accurately approximated by [12,13]

$$\mathbf{x}(t) = \mathbf{x}_{dc} + \sum_{l=1}^{L} {}^l\mathbf{x}(t), \tag{7.39}$$

where $\mathbf{x}_{dc}$ is the dc solution and ${}^l\mathbf{x}(t)$ solves the linear differential equation:

$$\mathbf{L}({}^l\mathbf{x}(t), d\,{}^l\mathbf{x}(t)/dt) = {}^l\mathbf{b}(t). \tag{7.40}$$

NUMERICAL MODELING TECHNIQUES

L is a linear algebraic operator and $^l\mathbf{b}(t)$ is the vector on "nonlinear current" sources. Under the assumption of modulated signal excitation, each component $^l\mathbf{b}(t)$ and $^l\mathbf{x}(t)$ can be shown to be of the form

$$^l\mathbf{x}_k(t) = \sum_{n=-l}^{l} {}^lX_{k,n}(t)e^{jn\omega_c t}, \tag{7.41}$$

Upon substitution of Eq. (7.41) in Eq. (7.40), and using near-orthogonality arguments similar to Eq. (7.37), we arrive at the following system of linear differential equations:

$$\mathbf{L}_n({}^l\mathbf{X}_n(t), d\,{}^l\mathbf{X}_n(t)/dt) = {}^l\mathbf{b}_n(t), \quad l=1,\ldots,L, \quad n=-l,\ldots,l, \tag{7.42}$$

where ${}^l\mathbf{X}_n(t) = [{}^lX_{1,n}(t),\ldots,{}^lX_{K,n}(t)]^T$.

Note that the problem has been reduced to solving a sequence of systems of linear differential equations in K unknowns. The number of systems that need to be solved appears large, but recall that envelope components with negative frequency indices may be found by complex conjugation. In addition, some envelope components are known to be zero, and some happen not to be of interest. In all, the third order ($L=3$) weakly nonlinear envelope method requires only four linear differential system solutions [14].

Each differential system given by Eq. (7.42) can be solved by a differencing procedure similar to Eq. (7.26), resulting in an algebraic system of K linear equations in K unknowns at each time sample t_i. The numerical complexity of this problem is smaller than any of the previous circuit-level methods and results in dramatic simulation speed improvements.

7.3.9 Example

This section provides an example of typical simulation results. The circuit to be analyzed is a single-stage, 5 GHz MESFET amplifier whose simplified structure is shown in Figure 7.14. Transistor nonlinearities are described by the Statz model [25].

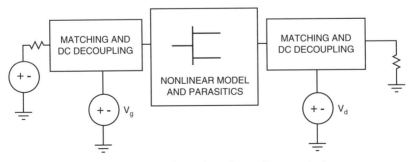

Figure 7.14. Circuit topology for nonlinear analysis.

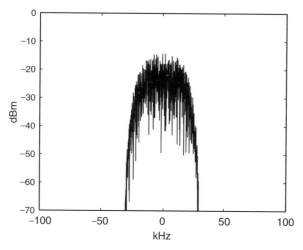

Figure 7.15. Input power spectral density.

The amplifier is driven by a 512 bit narrowband QPSK sequence whose spectrum is shown in Figure 7.15. The spectrum of the load power is shown in Figure 7.16.

It is of interest to compare the results and simulation speeds using the methods described in previous sections. Figure (7.17) shows the results of simulations obtained by the modified fixed-point (MFP) method, the nonlinear envelope (NE) method, and the weakly nonlinear envelope (WNE) method, for a range of input

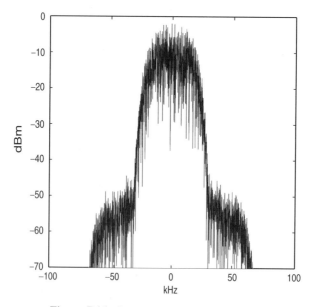

Figure 7.16. Output power spectral density.

NUMERICAL MODELING TECHNIQUES

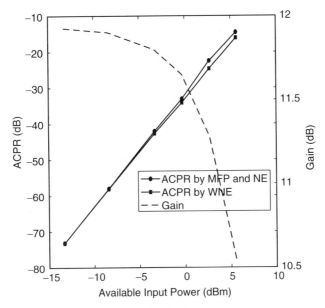

Figure 7.17. ACPR calculation comparison.

power. The MFP and NE methods yield, as expected, identical results. The WNE method tracks very closely the results of the two more general methods up to approximately 1 dB of gain compression. Averaged over the input power range, the simulation speed ratio of the NE, MFP, and WNE methods is approximately 60:30:1, though the numbers may vary depending on details of implementation.

7.3.10 Conclusions

We have discussed the application of behavioral and circuit-level methods to modulated signal simulation of nonlinear circuits. Behavioral methods are well-suited for system-level applications due to their computational efficiency, but the accuracy may be poor and there is a lack of a priori estimates of model quality. Circuit-level techniques offer a high degree of accuracy, limited mainly by the validity of active device models. Their main drawback is poor numerical efficiency and large memory requirements relative to behavioral methods.

We have presented several methods, recently developed to improve the numerical efficiency of simulation with modulated signals. The envelope method offers an excellent combination of robustness, speed, and memory requirements, but it is limited to narrowband excitations. The modified fixed-point method is more general in the sense that it is not limited to narrowband excitations and it frequently outperforms the envelope method when applied to single-stage amplifier operated several decibels in gain compression. The weakly nonlinear envelope method offers dramatic speed improvements over either of the above, but it is limited to narrowband excitations and weakly nonlinear operation.

7.4 NONLINEAR TRANSISTOR CHARACTERIZATION

7.4.1 Introduction

Transistors are one of the nonlinear components in most communications systems. Sections 7.2 and 7.3 have discussed the modeling of transistor amplifiers for a variety of modulation conditions. However, the nonlinear nature of the problem is determined by the nonlinear description of the transistor. An accurate transistor description is needed in order to obtain accurate system performance predictions. The "accuracy" of a model depends on the application. The simplest models are small signal equivalent circuits. These models are usually based on the physical construction of the device and include resistors, capacitors, and current sources, along with parasitic elements such as series resistances, bond pad capacitances, and lead inductances. These models are usually obtained by fitting small signal frequency domain S parameter measurements over a range of frequencies to the equivalent circuit. This is usually a two-step process. The first characterization is a set of "cold FET" measurements. These measurements involve S parameter measurements with zero V_{ds} or V_{ce} and characterize the capacitances in the transistor and the parasitics. Combinations of low and high frequency measurements can be used to separate parasitic inductance, resistance, and capacitance effects. Base–emitter or gate–source bias can also be used in the characterization. This initial set of data, along with a cold device equivalent circuit, can be used to fit the parasitic element values. A least squares fit of the data versus frequency is usually used to confirm the results.

The next level of complexity is frequency- and bias-dependent small signal S parameter fitting. A network analyzer is used to obtain multibias S parameters. The data are then converted into intrinsic transistor information by removing the parasitic element characterized earlier. The data from different bias points can then be used to fit a bias-dependent small signal model. The bias-dependent elements of an FET model, for example, include the transconductance g_m, the gate to source capacitance C_{gs}, the output resistance R_d, and the gate to drain capacitance C_{gd}. This type of model is a good starting point for small signal amplifier design. However, there is no signal-level-dependent information or nonlinear element involved.

A third level of complexity involves large signal RF characterization. This involves driving the transistor with a varying RF signal. This requires the measurement of both reflection coefficients and power levels and involves a more complex error correction. Two types of measurements are possible: a constant impedance system and a load-pull system. A constant impedance system uses the impedance presented by the measurement system, usually $50\,\Omega$ modified by the chip carrier, bias tee, and other components between the device and the load. This approach requires only a small extension of existing equipment and software. However, the resulting data are also limited. Many large signal measurements are used to characterize power devices. The information that is needed is the large signal performance under realistic conditions. This is usually at source and load conditions different from $50\,\Omega$. Power devices usually operate at impedance levels of only a few ohms, so the $50\,\Omega$ load of a typical measurement system doesn't give information at a realistic operating point. This limitation can be overcome by "load pulling" the transistor.

7.4.2 Large Signal Characterization of Transistors

Load-pull measurement of transistors gives the most information available on the large signal nonlinear operation. A detailed investigation of load pull for transistor characterization has been carried out. Load pull involves changing the load impedance seen by a transistor during a measurement and measuring the resulting transistor response. Both passive and active load-pull systems have been developed. A passive load-pull system uses a mechanical or electrical system to change the load. A typical mechanical system uses a movable probe in a coaxial line. The location and insertion of the probe can be carefully moved to reproduce most of the Smith chart. The impedance versus probe position can be measured as part of a calibration and then later used as part of the transistor measurement. An electrical load can also be used. PIN diode attenuators and phase shifters can be used to represent loads. Again, the bias settings for various load reflection coefficients can be calibrated and used in the measurements. A variety of commercial systems are available using passive load pull. They are available with the characterization software needed for turnkey operation. However, there is a problem with these passive systems. Since there is always loss between the transistor terminals and the load, the resulting reflection coefficient will always be less than 1. Since many power transistors operate near the low impedance edge of the Smith chart, this can be a problem. A possible solution is an active load-pull system.

Active load pull uses an external signal to represent the load [16–18]. The basic idea behind an active load-pull system is shown in Figure 7.18. The equation in this figure defines the reflection coefficient for a microwave measurement. In a passive measurement system Γ_{load} is determined by the physical load in the system. In an active pull system the b term in the numerator is controlled electronically. Conventional active load-pull systems use the configuration shown in the Figure 7.18(a). This system uses a signal derived from the input signal and injected into the output

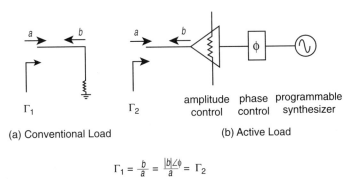

Figure 7.18. Active load realization.

port. The magnitude and phase of the injected signal can be controlled by the measurement system. Since the "reflected" b term in the figure is determined by the amplitude and phase control, any value of b and thus any Γ_{load} can be created. Realistic limitations are imposed by the amount of power available to the active load. Since the system is active, we can use the external power from the measurement system to overcome measurement system loss and present reflection coefficients covering the entire Smith chart. This is particularly useful for measuring low impedance power devices. This flexibility comes with some limitations. The active load usually requires a more complex microwave setup, and additional microwave power is needed to create the load. Since the device under test usually has a low impedance, much of the load power is reflected. The impedance presented by the active load depends on the output impedance of the transistor, so presenting a given load to the transistor sometimes involves measurement iteration. Finally, commercial active load systems are not available. Most of these systems are still in research laboratories.

An alternative active load configuration is shown in Figure 7.18(b). The signal from the source is replaced with a phase locked synthesizer. The phase reference is obtained from the input synthesizer of the measurement system. This has several advantages. Microwave cables between the input and output ports are replaced with cables at the 10 MHz reference frequency. The expensive microwave phase control is replaced with a simple 10 MHz phase shifter. Finally, this configuration can be used to create an active load at the second or higher harmonic frequencies. This configuration has been used to realize a large signal harmonic active load-pull system [15].

7.4.3 A Time Domain Active Load-Pull System

Most existing large signal characterization systems are based on network analyzers in the frequency domain. However, several time domain systems have been proposed [19–21]. There are several advantages with time domain measurements. First, we would like to characterize transistors using both fundamental and harmonic tuning. Time domain measurements can measure several frequencies at the same time. This approach also allows phase information between the fundamental and the harmonics to be measured. Many of the physics-based device simulation codes run in the time domain, so corresponding time domain measurements allow a more direct comparison with these results. Finally, the combination of the time domain measurements with the active loads described in the last paragraph allow very broad-band measurements without filters or diplexers.

A block diagram of the new system is shown in Figure 7.19. The system is based on a microwave transition analyzer instead of a network analyzer. The system includes the transition analyzer, an input and output phase bridge, an input synthesizer and amplifier, an output fundamental and harmonic active load, and switches to control the signal path. The transistor under test is measured in a 2.4 mm connector based fixture. The hardware has a dc to 40 GHz bandwidth. The active loads are generated with power amplifier having 2–8 GHz bandwidths. The figure does not

NONLINEAR TRANSISTOR CHARACTERIZATION

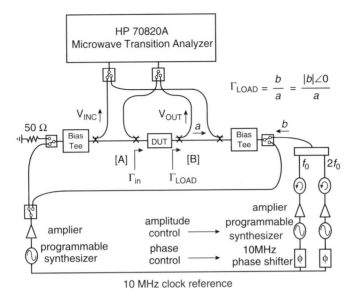

Figure 7.19. Time domain measurement system.

include couplers on the input and output sides for power level calibration. The switches in the signal paths are used to convert between large signal measurements and conventional small signal S parameter measurements.

Calibration and small signal and large signal characterization use a combination of frequency and time domain information. The raw measurements are taken in the time domain. The information required to error correct the system is then converted to the frequency domain using a Fourier transform. The system is calibrated using conventional standards, and a set of error coefficients are calculated and stored for the frequencies under consideration. A typical measurement consists of a time domain measurement of a voltage waveform, a conversion to the frequency domain using a Fourier transform, error correction of the fundamental and harmonic frequencies, and a conversion back to the time domain for display. Since vector impedance information is also available, the time domain current waveforms can also be constructed. The entire error correction and measurement sequence is under software control. The measurement system accuracy can be tested by measuring the properties of known passive structures. A typical measurement involves the transmission characteristics of a 50 Ω through line. These results confirm the performance of the system over most of the Smith chart, with modest errors only near active loads corresponding to near-short conditions. With this new measurement system available, we can begin the investigation of large signal transistor operation.

7.4.4 Transistor Modeling and Characterization

The goal of transistor modeling and characterization is to develop accurate circuit level transistor models to use in circuit simulators. Since we will be using these

258   AMPLIFIER NONLINEARITIES AND THEIR EFFECTS IN COMMUNICATIONS

models to investigate nonlinear effects of amplifier in communication systems, we need an accurate nonlinear device model. A variety of FET device models have been described. These include publications by Curtice [22], Curtice and Ettenberg [23], Materka and Kacprzak [24], and Statz et al. [25]. A recent book by Golio [26] gives an excellent overview of large signal device characterization. These models differ in the choice of functions used to model the drain current, in the choice of the form of the gate source and gate drain capacitance, and on the number of parameters that need to be fitted in order to realize the model. The model parameters are found by fitting a combination of dc measured drain currents and small signal measured S parameter data. The small signal data are used to obtain bias-dependent capacitances, resistances, and transconductance. Cold FET modeling is used to remove the effects of parasitic elements, allowing direct modeling of the intrinsic FET.

These models were the starting point for an experimental evaluation of large signal transistor operation. The experiment involved experimental dc and small signal measurements, followed by careful parameter extraction for the different FET models. The transistor used was a pHEMT [27]. The small signal and measured dc drain current versus voltage characteristics were then least-squares fit to the various large signal transistor models. These models were then used to predict the large signal transistor current and voltage waveforms for a variety of operating loads. Typical interesting loads include 50 Ω terminations at the fundamental and the harmonics, a load equal to the maximum small signal gain load at the fundamental and a 50 Ω load at the harmonics, a 50 Ω load at the fundamental, and a range of second harmonic loads to investigate second harmonic tuning effects. These loads were used with input power sweeps to investigate the large signal operation of the transistor. A typical dc drain current match is shown in Figure 7.20. The figure shows

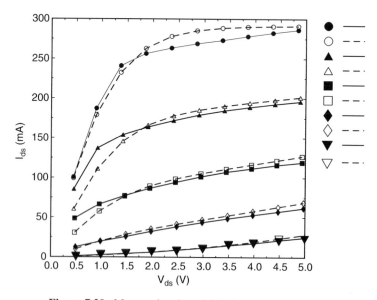

Figure 7.20. Measured and modeled drain characteristics.

the experimental data and a best fit to a Curtice-Ettenberg model. This fit can be used to predict the output conductance and transconductance during large signal operation. Similar plots are available for the other published large signal models. All the large signal models produce reasonable fits to the drain characteristics. Fits of modeled and measured C_{gs} and C_{gd} produce poorer matches. Simple models that predict the gate source capacitance from the gate bias alone do not produce the proper variation with drain source voltage. Modified models such as the EEHEMT model available in LIBRA do a better job. A comparison of the measured and modeled C_{gs} for the EEHEMT model is shown in Figure 7.21. The information from the various models can be used to predict the large signal performance of the transistor. Since amplitude and phase information is available at the fundamental and second harmonic, the time domain device current and voltage waveforms can also be found. A typical measured waveform on the output plane of the transistor is shown in Figure 7.22 The current versus voltage trajectories are plotted on the output plane of the transistor. The three curves correspond to a low initial transducer gain point (7.5 dB), the point of maximum transducer gain (8.7 dB), and a point 3 dB into compression from the maximum gain point. This type of curve would be very difficult to obtain with a conventional frequency domain load-pull system. The time domain system can also measure current and voltage amplitudes at the device terminals. Figure 7.23 shows the measured RF current amplitude at the fundamental versus the input drive power for a Class AB bias point. The open circles correspond to the measured data and the remaining curves correspond to predictions from several of the large signal models. All of the models do a reasonable job in this case.

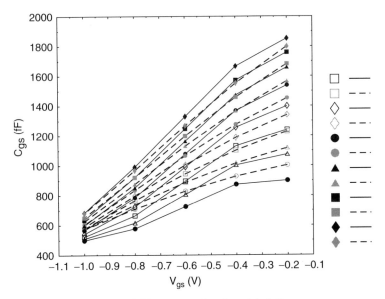

Figure 7.21. Measured and modeled C_{gs}.

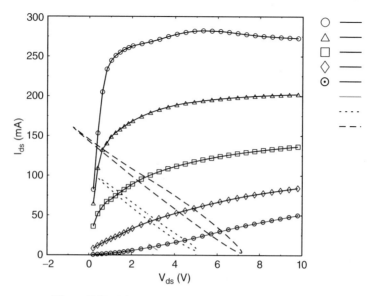

Figure 7.22. Measured transistor output characteristics.

Figure 7.24 shows a similar plot for the second harmonic current. Notice the much wider range of predicted currents for this measurement. These measurements are the starting point for a more detailed nonlinear description of transistors. The next step is to make similar comparisons for ACPR predictions to further quantify the performance of nonlinear transistor models.

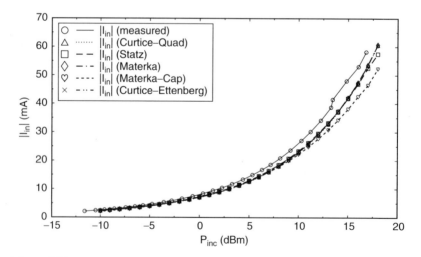

Figure 7.23. Measured and modeled fundamental drain current versus input power.

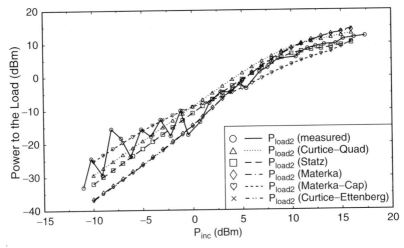

Figure 7.24. Measured and modeled second harmonic drain current versus input power.

7.5 OPEN ISSUES AND CONCLUSIONS

This chapter has described some of the issues, trade-offs, and design tools being developed to better realize energy efficient digital communications systems. Although much progress has been made, several important areas of research remain. One problem is the changing nature of communications systems. New modulation schemes and more restrictive limitations on signal interference are constantly occurring. Designs and systems that worked well in the past need to be revised and redesigned. These changes all require better and more accurate tools. These better tools will come from an even closer interaction between the fields of communications, digital signal processing, and microwave design. With this in mind, several interesting areas of further research include:

1. Better behavioral models.
2. Better measurement techniques that correlate time and frequency domain measurements.
3. Improved numerical tools for a variety of distortion conditions.
4. Improved large signal models.

Better behavioral models are needed to allow accurate simulations in reasonable run times. Memoryless models need to be replaced with models that better describe actual circuit performance. Better measurement techniques that correlate time and frequency domain measurements are also needed. In fact, these two points are part of the same problem. Our communication amplifier must operate over a wide frequency range, with a carrier at a microwave frequency and a modulation envelope that has components near dc. In order to accurately understand the circuit operation, we need to have a correspondingly wide range of measurement information. One approach would be to measure the response with the actual digital waveform as an input. Although this would provide information about the response, it would also require additional measurements for each new operating condition. A better approach would be to develop simpler combinations of time and frequency domain waveforms that could be used to accurately predict performance over a wide range of operating conditions.

This leads to the third and fourth points, better numerical tools and better descriptions of transistor operation. Better tools imply both faster and better "matched" to the different problems at hand. Faster means the ability to obtain useful information in a reasonable amount of time. "Matched" means the ability of the tool to solve the correct problem. A calculation of the ACPR at levels of -40 dB is different than predicting large signal gain saturation effects. We need to carefully match the tools, the experimental inputs to the simulation and the range of conditions to the problem. Finally, we must develop a new set of device models, parameter extractions, and simulations to match our tools with experimental results.

ACKNOWLEDGMENT

The work described in this chapter as supported by the MURI program in Low Power/Low Noise Electronics, contract DAAH04-96-1-0001 and the MURI program in Low Energy Electronics for Mobile Platforms, contract DAAH04-96-1-0377. We would like to thank the following graduate students for their significant contributions to this work: Vuk Borich, Je-hong Jong, James Kempf, and Paul Liang.

REFERENCES

1. C. Liang, J. Jong, W. Stark, and J. East, "Nonlinear amplifier effects in communications systems," *IEEE Transactions on Microwave Theory and Techniques*, Vol. 47, 1461–1466 (1999).
2. J. G. Proakis, *Digital Communications*, McGraw-Hill, New York, 1995.
3. F. Amoroso, "The use of quasi-band limited pulses in MSK transmission," *IEEE Transactions on Communications*, Vol. COM-27, 1616–1624 (Oct. 1979).
4. K. Pahlavan and A. H. Levesque, *Wireless Information Networks*, Wiley, New York, 1995.
5. W. Zou and Y. Wu, "COFDM: an overview," *IEEE Transactions on Broadcasting*, Vol. 41, 1–8 (Mar. 1995).
6. K. Kundert and A. Sangiovanni-Vincentelli, "Simulation of nonlinear circuits in the frequency domain," *IEEE Transactions CAD*, 521–535 (1986).
7. V. Rizzoli, C. Cecchetti, D. Masoti, and F. Mastri, "Nonlinear processing of digitally modulated carriers by the inexact-Newton harmonic-balance technique," *Electronics Letters*, Vol. 33, 1760–1761 (Oct. 1997).
8. V. Borich, J. East, and G. Haddad, "A fixed-point harmonic balance approach for circuit simulation under modulated carrier excitation," *Proceedings ISCAS 99*, Vol. 6, 346–349 (1999).

REFERENCES

9. P. Feldmann and J. Roychowdhury, "Computation of circuit waveform envelopes using an efficient, matrix-decomposed harmonic balance algorithm," *Proceedings ICCAD 96*, 295–300 (1996).

10. E. Ngoya and R. Larcheveque, "Envelope transient analysis: a new method for the transient and steady state analysis of microwave communication circuits and systems," *Proceedings MTT-S 96*, 1365–1368 (1996).

11. V. Rizzoli, A. Neri, F. Mastri, and A. Lipparini, "A modulation-oriented piece-wise harmonic-balance technique suitable for transient analysis and digitally modulated signals," *Proceedings EuMC* 546–549 (1996).

12. J. Bussgang, L. Ehrman, and J. Graham, "Analysis of nonlinear systems with multiple inputs," *Proceeding IEEE*, Vol. 62, 1088–1119 (Aug. 1974).

13. S. A. Maas, *Nonlinear Microwave Circuits*, Artech House, Norwood, MA, 1988.

14. V. Borich, J. East, and G. Haddad, "The method of envelope currents for rapid simulation of weakly nonlinear communications circuits," *Proceedings MTT-S 99* 981–984 (1999).

15. J. Kempf, "Time domain characterization of transistors," Ph.D. thesis, The University of Michigan, 1999.

16. D. Teeter, J. East, and G. Haddad, "Large Signal HBT characterization and modeling at millimeter wave frequencies," *IEEE Transactions on Microwave Theory and Techniques*, Vol. 41, No. 6, 1087–1093 (June/July 1993).

17. D. Yang and D. Peterson, "Large signal characterization of two-port nonlinear active networks," *International Microwave Symposium Digest*, 345–347 (1982).

18. R. Actis and R. McMorran, "Millimeter wave load pull measurements," *Applied Microwaves*, 91–102 (Nov./Dec. 1989).

19. M. Siplia, K. Lehtinen, and V. Porra, "High-frequency periodic time domain waveform measurements," *IEEE Transactions on Microwave Theory and Techniques*, Vol. 36, No. 10, 1397–1405 (Oct. 1988).

20. G. Kompa and F. Van Raay, "Error-corrected large signal waveform measurement system combining network analyzer and sampling oscilloscope," *IEEE Transactions on Microwave Theory and Techniques*, Vol. 38, No. 4, 358–365 (Apr. 1990).

21. C. J. Wei, Y. E. Lan, C. M. Huang, W. J. Ho, and J. Higgins, "Waveform-based modeling and characterization of microwave heterojunction transistors," *IEEE Transactions on Microwave Theory and Techniques*, Vol. 43, No. 12, 2899–2903 (Dec. 1995).

22. W. Curtice, "A MESFET model for use in the design of GaAs integrated circuits," *IEEE Transcations on Microwave Theory and Techniques*, Vol. 28, No. 5, 448–455 (May 1980).

23. W. Curtice and M. Ettenberg, "A nonlinear GaAs FET model for use in the design of output circuits for power amplifier," *IEEE Transactions on Microwave Theory and Techniques*," Vol. 33, No. 12, 1383–1393 (Dec. 1985).

24. A. Materka and T. Kacprzak, "Computer calculation of large-signal GaAs FET amplifier Characteristics," *IEEE Transcations on Microwave Theory and Techniques*, Vol. 33, No. 2, 129–134 (Feb. 1985).

25. H. Statz, P. Newman, I. Smith, R. Pucel, and H. Haus, "GaAs FET device and circuit simulation in SPICE," *IEEE Transcations on Electron Devices*, Vol. 34, No. 2, 160–168 (Feb. 1987).

26. J. M. Golio, *Microwave MESFET's and HEMT's*, Artech House, Norwood, MA, 1991.

27. Private communication with Dr. D. Teeter, Raytheon Microelectronics, Research Division.

8

PLANAR-ORIENTED PASSIVE COMPONENTS

YONGXI QIAN AND TATSUO ITOH
Department of Electrical Engineering, University of California—Los Angeles

8.1 INTRODUCTION

Passive components, including transmission lines, interconnects, couplers, filters, resonators, and various types of transistors and impedance transformers, are indispensable parts in any microwave and millimeter-wave front-end circuitry. Although metallic waveguides were used predominantly in earlier microwave systems, planar microwave integrated circuits, particularly those based on microstrip lines and coplanar waveguides (CPWs), have become the most preferred technologies in designing modern wireless systems for both military and commercial applications. These planar circuits, either in the form of hybrid or monolithic integration (HMIC/MMIC), offer numerous advantages over traditional waveguide-based systems, such as compact size, reduced weight, lower profile, easy mounting of active devices, and greatly reduced fabrication cost since they can simply be printed on a circuit board or semiconductor wafer in large quantities instead of being machined piece by piece.

One of the fundamental disadvantages of planar structures when used at high frequencies, however, is their relatively high loss, or low quality factor (Q), in comparison to metallic waveguides. As wireless applications move toward higher microwave and millimeter-wave frequencies, it becomes increasingly difficult to realize high-Q components using conventional microstrips or CPWs, due to a number of factors such as increased metallic and dielectric losses, as well as other electromagnetic effects including surface wave and radiation losses.

RF Technologies for Low Power Wireless Communications
Edited by Tatsuo Itoh, George Haddad, and James Harvey
ISBN 0-471-38267-1 Copyright © 2001 by John Wiley & Sons, Inc.

Following a brief overview of the basic constructing elements for modern planar microwave/millimeter-wave passive circuits, this chapter will present two innovative approaches to meet the above-mentioned technical challenges of high-Q passive components for high-frequency applications. The first approach, based on a unique process named SIMPOL (silicon/metal/polyimide), aims at realizing low-loss, high-isolation interconnects and passive components on low-resistivity, CMOS-grade silicon substrates. This SIMPOL structure is believed to be significant in paving the way toward the ultimate goal of a millimeter-wave system-on-chip (SOC), since it allows us to integrate high-performance millimeter-wave circuits on the same silicon substrate for digital ICs, while satisfying the extremely stringent dynamic range ($\sim$100 dB) requirement in these advanced mixed signal MMICs.

The second approach for high-Q passive components is based on the photonic bandgap (PBG) concept. These are basically periodic structures that can be used to control the propagation, reflection, or radiation of electromagnetic waves in a slightly different way compared to conventional waveguiding or filtering techniques. A class of planar-oriented PBG structures has been developed, including a novel uniplanar compact PBG (UC-PBG), which demonstrated several unique properties such as low insertion loss, slow-wave effect, wide stopband, and easy realization of a perfect magnetic impedance surface. A number of specific applications have been demonstrated, including low-loss slow-wave structure, lowpass filter with improved rejection, bandpass filter with intrinsic spurious suppression, harmonic tuning in microwave power amplifier, and leakage suppression in conductor-backed coplanar waveguides and striplines.

8.2 BASIC STRUCTURES FOR PLANAR PASSIVE CIRCUITS

Up until the 1970s, most microwave and millimeter-wave systems, developed predominantly for military purposes, were based on metallic waveguides or coaxial cables. However, since the mid-1960s, the advent of the integrated circuit (IC) technology started having a major impact on the microwave community as well. A family of printed circuit board (PCB)-based transmission lines, including microstrip lines, coplanar waveguides (CPWs), coplanar striplines (CPSs), and slotlines were invented, and experienced intensive research and development [1,2]. In fact, accurate analysis and design of these inhomogeneous structures proved to be such a great challenge to the electromagnetics community that it ignited the most active research period in numerical techniques during the next several decades [3]. The true driving force under such intensive research efforts, however, is the enormous advantages made possible by adopting the planar integrated circuit technology, such as compact size, reduced weight, lower profile, easy integration with semiconductor devices, and huge cost reduction due to mass production [4].

The use of planar structures such as microstrip lines and CPWs also allows the construction of lumped circuit elements such as interdigitated capacitors and spiral inductors, an advantageous feature not easily available in conventional metallic waveguide technology. This makes it possible to design a monolithic microwave

BASIC STRUCTURES FOR PLANAR PASSIVE CIRCUITS   267

integrated circuit (MMIC), which is arguably the only option to enable massive deployment of many microwave and millimeter-wave application systems. Figure 8.1 shows some examples of planar microwave passive components based on the microstrip line technology. Similar circuit components can be constructed with other types of transmission lines such as CPW and CPSs. These passive circuits are key components in modern microwave HMICs and MMICs.

One of the great challenges to modern radio frequency integrated circuit (RFIC) designers is to achieve the highest possible quality factor (Q) while keeping the planar or planar-oriented circuit topology compatible to modern semiconductor fabrication processes. Since the performance of the passive circuit components as shown in Figure 8.1 will be determined mainly by the quality of the transmission lines themselves (microstrip line in this case), we will focus our discussions on some of basic planar transmission line structures, which are the main building blocks of most passive components in modern HMICs and MMICs.

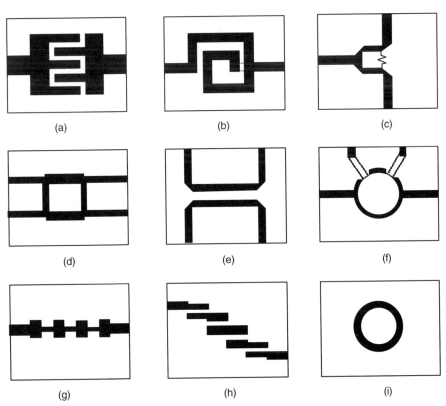

Figure 8.1. Examples of planar passive components based on microstrip line technology: (a) interdigitated capacitor, (b) spiral inductor, (c) Wilkinson power divider/combiner, (d) branch line coupler, (e) directional coupler, (f) rat-race ring coupler, (g) lowpass filter, (h) bandpass filter, and (i) ring resonator.

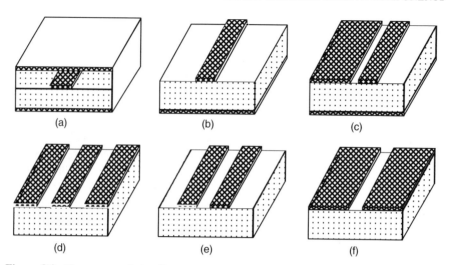

Figure 8.2. Planar transmission line structures frequently used in microwave and millimeter-wave integrated circuits: (a) stripline, (b) microstrip lines, (c) micro-coplanar stripline, (d) coplanar waveguide, (e) coplanar striplines, and (f) slotline.

Figure 8.2 shows some of the most popularly used planar transmission line structures in today's microwave and millimeter-wave integrated circuits. Except for the slotline in Figure. 8.2(f), all these planar transmission lines support the so-called quasi-TEM mode as their fundamental propagation mode, with reasonably low insertion loss and weak frequency dispersion in both their characteristic impedance values and propagation constant.

The stripline shown in Figure 8.2(a) can be considered as a planar version of the coaxial line and is perhaps the earliest type of transmission line compatible with printed circuit fabrication technology. Due to its multilayer nature and the fact that the strip conductor is sandwiched between two dielectric substrates, the stripline is not as popular for use in MMICs as the microstrip lines and CPWs. However, with rapid progress in multilayer technology, partly driven by the industrial motivation of high-density, low-cost MMICs, there has been renewed interest in stripline-like multilayer structures in recent years. One attractive feature of the stripline is that it can easily be shielded in all four lateral directions, with the help of via technology, thus providing complete signal isolation between adjacent transmission lines fabricated on the same substrate. As will be discussed in more detail in the next section, this unique feature makes stripline-like structures a very attractive technology that enables the realization of extremely high isolation (>100 dB) between the transmitting and receiving MMICs as required in a future RF system-on-chip.

The microstrip line, as shown in Figure 8.2(b), together with a large number of its variations, is probably the default technology for today's hybrid and monolithic integrated circuits up to the W-band. For HMIC designs, including both wire/ribbon bonding and flip-chip assembly approaches, the microstrip transmission line and related passive components can be fabricated on low-loss substrates such as alumina, Duroid, and quartz with appropriate thickness. However, in MMIC designs, it becomes imperative to construct the microstrip-based circuits on the same semiconductor wafer (GaAs, InP, Si, etc.) with the active devices. As the operation frequencies of the MMICs increase, the wafer has to be thinned accordingly to facilitate the fabrication of low inductance vias as well as heat sinking. As a result, the width of a typical 50 Ω microstrip shrinks and becomes more lossy at higher frequencies, making it less attractive for MMIC designs above 100 GHz.

Meanwhile, coplanar waveguides (CPWs), as shown in Figure 8.2(d), are playing an increasingly important role in planar circuit designs in the millimeter and submillimeter-wave regions. The characteristic impedance of the CPW line is relatively insensitive with respect to the substrate thickness. Another major advantage of the CPW is that the line loss can be controlled with more flexibility by optimizing the cross-sectional dimensions (center conductor and slot widths), a feature not available in the microstrip line structure. This makes CPWs widely used in low-noise amplifiers where minimizing the transmission line loss is of the utmost importance.

While the original CPW structure assumes an infinitely thick substrate, in reality the dielectric substrate or semiconductor wafer always has a finite thickness. Also, for easy assembly and packaging it is desirable to add a metal layer at the bottom of the substrate, forming the so-called conductor-backed CPW (CB-CPW). An immediate problem with the CB-CPW, however, is that it will easily excite a parallel-plate mode, which will carry a significant part of the energy away from the CPW line. A common practice adopted by MMIC designers today is to use numerous vias to suppress this parallel-plate mode. While this approach works reasonably well at microwave and millimeter-wave frequencies up to the W-band, radiation losses due to the vias become prominent at higher frequencies, putting a significant limitation on the realizable gains of amplifiers and other circuits at such frequencies.

Another major development in the CPW technology is the finite ground CPW (FG-CPW) [5]. By truncating and optimizing the lateral width of the ground planes on both sides of the CPW center conductor, it is possible to realize a nonleaky transmission line even with a metallization on the backside. Various circuits have been designed and demonstrated. Numerical simulations based on the FDTD method indicated that the FG-CPW line can be useful up to about 1 THz [6].

Furthermore, recent advances in microwave multichip assemblies require cost-effective combination of MMIC chips and other RF components using low-cost batch fabrication and assembly processes. To lower the cost with higher yield, novel design concepts other than conventional microstrip or CPW-based MMICs with chip-and-wire assembly techniques have been pursued intensively, driven partially by explosive demand for handheld communication equipment and phased-array antennas for mobile satellite communications. For example, a novel embedded transmission line (ETL) reported recently allows the mixing of different transmission-line types (i.e., coplanar and striplines) for maximum MMIC design flexibility and, in one configuration, permits the elimination of backside processing, which is indispensable in typical microstrip-based MMICs [7].

8.3 SIMPOL APPROACH FOR MIXED SIGNAL SILICON MMICs

8.3.1 Background and Motivation

The motivation underlying our research efforts on the SIMPOL structure is quite straightforward: to realize low-loss RF transmission lines and interconnects with low insertion loss and extremely high level of signal isolation (>100 dB) on CMOS-grade conductive silicon substrate, which will subsequently enable us to build entire microwave or millimeter-wave communications systems on a single silicon chip using modern SiGe and mixed signal MMIC technology.

While silicon has been the semiconductor of choice for modern digital ICs, the microwave community has focused mainly on GaAs and InP as the major semiconductor materials ever since the advent of MMIC technology in the late 1970s, in spite of the fact that these exotic compound materials are much more expensive than silicon. Historically, silicon bipolar junction transistors (BJTs) and metal oxide semiconductor field effect transistors (MOSFETs) could not deliver a high enough maximum frequency of oscillation (f_{max}) for microwave circuit operation above a few gigahertz, due to the much slower carrier mobility in silicon. However, along with fundamental advances in Si-based integrated circuits during the past few years, the landscape of microwave integrated circuits has changed significantly [8]. With the availability of advanced growth techniques such as molecular beam epitaxy (MBE) and ultra-high-vacuum chemical vapor deposition, Si/SiGe/Si heterojunction bipolar transistors (HBTs) with excellent intrinsic speed ($f_{max} > 160$ GHz) [9], power handling capability [10] [11], and low-noise characteristics [12] have been demonstrated successfully. The SiGe technology, pioneered by IBM, is starting to enter the gigahertz market formerly occupied by GaAs-based devices (FETs and HEMTs). The good thermal conductivity of silicon makes it particularly attractive for power amplifier applications.

Meanwhile, high-resistivity silicon (HRS) wafers with resistivity of over $1000\,\Omega\cdot$cm are now readily available, although at high prices, which permit Si circuits to be designed using the expertise previously developed for GaAs ICs. For planar circuits on this type of high-resistivity silicon substrate, the conductor losses due to the skin effect dominate the loss contribution, with the substrate losses in silicon accounting for a minor part [13,14]. Technical issues such as the dependence of microwave losses on dc bias have been addressed and effective solutions such inserting a SiO_2 layer with various lateral profiles between the strip metallization and HRS substrate have been proposed and investigated [15]. Extremely low-loss CPW lines with 1.25 μm aluminum metallization on passivated HRS substrates have been demonstrated recently, with a measured attenuation loss of 0.11 dB/mm at 30 GHz [16].

In order to take full advantage of the enormous capabilities of modern digital IC technology, for example, the ability to integrate a logic function including power management in RFICs, it is very desirable that RF and microwave circuits be designed and implemented on standard CMOS-grade silicon substrates, which are usually highly conductive (resistivity ranging from 0.1 to $40\,\Omega\cdot$cm). The coexistence of analog and digital signals on the same silicon substrate, also known as mixed signal ICs, will eventually lead to the ultimate system-on-chip solution, the "Holy Grail" of RFIC still dreamt of by modern RF and microwave designers [17]. To reach this noble goal of a SOC, however, tremendous technical breakthroughs must still be achieved. The extremely high dynamic range experienced in a modern wireless communications device (e.g., handheld phone), for instance, makes it almost impossible to integrate the transmitter and receiver circuits on the same semiconductor substrate, be it Si or GaAs, unless a revolutionary technology is developed to guarantee such high signal isolation level (~ 100 dB). To implement an entire mixed signal MMIC on conductive silicon, the significant microwave transmission line losses, dominated by the substrate loss, should also be reduced to an acceptable level, that is, comparable to those on semi-insulating GaAs or InP substrates.

As will be shown in detail below, the SIMPOL structure we developed recently is a very promising approach to realize extremely high isolation, low-loss RF transmission lines and interconnects for such mixed signal silicon MMIC implementation. SIMPOL takes full advantage of the latest development in multilayer microwave circuit technology, as well as the most advanced VLSI processing techniques such as inductively coupled plasma (ICP) etching, reactive ion etching (RIE), and chemical mechanical planarization (CMP) [18]. This allows us to realize excellent insertion loss (<0.25 dB/mm up to 45 GHz) and extremely high noise isolation levels (>80 dB up to 20 GHz, >40 dB up to 45 GHz) on commercial low-cost conductive silicon. In addition to simple transmission lines, a branch line (90°) coupler with 0.7 dB insertion loss at 37 GHz has also been demonstrated recently [19]. This SIMPOL architecture should play an instrumental role in realizing highly integrated mixed signal Si MMICs, which will eventually lead us toward the most challenging goal of microwave and millimeter-wave SOC.

8.3.2 SIMPOL Interconnect Concept

As described in Section 8.2, the microstrip line is currently the most widely used passive element in modern MMIC designs. When implementing microstrip lines on CMOS-grade conductive silicon substrate, however, the line losses will become excessively large, dominated by the substrate loss, as shown in Figure 8.3. In the paper by Milanovic et al. [20], the attenuation of CPWs on CMOS-grade silicon substrates with conductivities ranging from 0.4 to 12.5 $\Omega \cdot$ cm were investigated both theoretically and experimentally. At 20 GHz, the line losses were between 2 and 4 dB/mm, depending on the conductivity and dimensions of the CPWs. Such high losses make it impractical to realize high-performance transmission lines and passive components at higher microwave and millimeter-wave frequencies. It is imperative to remove the lossy silicon underneath the microstrip line by using modern micromachining techniques such as anisotropic etching [21]. Also known as microshield lines, such micromachined transmission lines can achieve excellent insertion loss, dominated by the ohmic loss of the strip conductor.

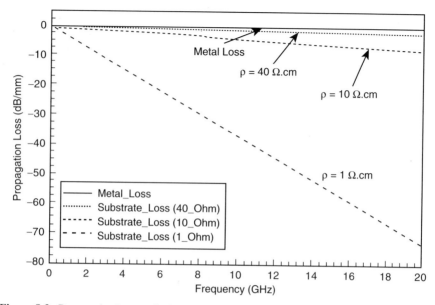

Figure 8.3. Propagation losses of microstrip lines on conductive silicon substrates ($\varepsilon_r = 11.7$, thickness = 100 µm) with various resistivities. The gold microstrip is 3 µm thick and 70 µm wide.

To achieve the required high dynamic range in mixed signal MMICs, however, it is equally important (if not more so) to realize a high degree of signal isolation, or low crosstalk, between adjacent transmission lines within a densely integrated circuit environment. Unfortunately, the microstrip line is far from the ideal solution for this purpose. Figure 8.4 plots both simulation and measurement results of the coupling between two 50 Ω ($W = 90$ mil) microstrip lines on 31 mil thick Duroid substrate with $\varepsilon_r = 2.33$, where the line spacing is $S = 90$ mil and the coupling length is 742 mils. For these types of conventional microstrip interconnects, the signal isolation is relatively poor due to coupling through both the substrate and air regions. It is found that the coupling level remains about -30 dB even when the line spacing is increased to $S = 3W$, which might be insufficient in many microwave/digital transceiver designs where a very high signal-to-noise ratio (>80 dB) is required.

As an effort to address the above problems, we have developed a novel silicon/metal/polyimide process, abbreviated as SIMPOL, as shown in Figure 8.5 [22–25]. Compared to existing interconnect/packaging technologies, the SIMPOL structure has several unique features and major advantages: (1) it is low in cost and can be built on standard, low-resistivity ($\rho = 0.1$–$40 \, \Omega \cdot$ cm) substrate instead of expensive high-ρ silicon substrate; (2) it has low loss, low dispersion, and extremely high isolation characteristics (>100 dB); (3) its performance does not depend on the conductivity of the underlying silicon substrate since the transmission lines are enclosed and shielded by metals; and (4) it can easily be fabricated using standard

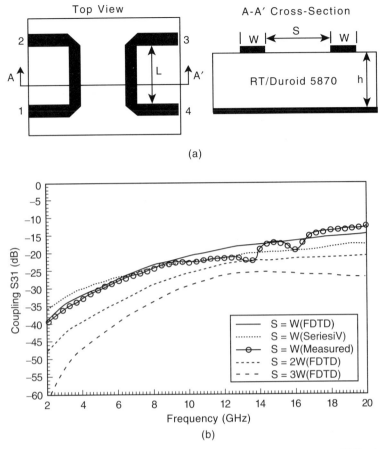

Figure 8.4. Simulation and measurement results of coupling between two 50 Ω microstrip lines with various line spacings.

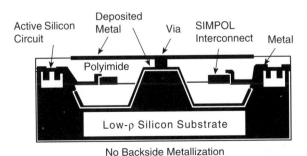

Figure 8.5. Cross-sectional view of the SIMPOL (silicon/metal/polyimide) interconnect structure.

CMOS processing techniques and requires no wafer thinning and backside processing.

The fabrication of the SIMPOL interconnect is compatible with modern VLSI processes. After forming active devices on silicon, the wafer is etched by using a downstream ICP with tailored sidewall slopes. The metals (Ti/Al/Ti) are then sputtered to cover the bottom and sidewalls of the recessed areas. Polyimide will be applied, planarized using CMP, and then etched to half-fill the recessed areas by using RIE. After that, the wafer will be patterned to lift-off transmission lines and form interconnects for integrated circuits. A second polyimide layer will be applied to bury the striplines and replanarize the wafer to support the subsequent deposition of the top interconnect metal. The top metal can be connected to the bottom metal through etched vias. When grounded, both top and bottom metals form an enclosed structure to shield the transmission lines completely from interference and noise.

The SIMPOL interconnect supports a TEM-like mode similar to that in a stripline structure, but with additional advantages that radiation and leakage losses are totally eliminated; thus it has inherently low insertion loss (~ 0.1 dB/mm at 20 GHz). In a typical stripline circuit, densely populated via hole fences have to be used to minimize both energy leakage into the parallel-plate mode and crosstalk between neighboring lines [26]. In the SIMPOL structure, however, the striplines are buried deep inside the recessed areas, thus providing much better shielding between neighboring channels without having to use closely spaced via fences. Also, in MMIC fabrication it is usually difficult to make small plated vias through a thick substrate. This problem is avoided easily in SIMPOL, since the grounding vias are relatively short, as can be seen in Figure 8.5, and do no need to go through the whole substrate as required in conventional stripline designs. Another advantage of SIMPOL is that the RF transmission line is totally isolated from the substrate, thus avoiding the problem of dc bias dependence of the propagation characteristics as seen in planar lines fabricated directly on the silicon substrate [27].

It should also be pointed out that although the SIMPOL structure resembles the three-dimensional (3D) MMIC structures reported in Nishikawa et al. [28], which mainly use thin-film microstrip (TFMS) lines for RF transmission and interconnection, the two approaches differ significantly in terms of the maximum level of signal isolation that can be achieved. While reasonably low line loss has been demonstrated for the TFMS line, it will have similar crosstalk characteristics to that of regular microstrip lines, as indicated in Figure 8.4, which is insufficient to achieve the extremely high level of signal isolation as required in the projected SOC applications. Since in the SIMPOL structure adjacent transmission lines are completely shielded and isolated, it might be the only approach to achieve >100 dB dynamic range requirement, unless more sophisticated and expensive technologies, such as conformal packaging based on micromachining and wafer bonding, are used [29].

8.3.3 FDTD Modeling and Simulation

To demonstrate the extremely low-crosstalk capabilities of the proposed SIMPOL structure, we first developed a simplified model and carried out extensive full-wave

EM simulations [23]. Due to the unique cross-sectional profile of the SIMPOL structure, 3D full-wave EM simulators are required for rigorous characterization of its performances. An in-house 3D finite-difference time-domain (FDTD) code is utilized for this purpose, because it can predict accurately and efficiently the characteristics of arbitrary 3D structures over a wide range of frequencies with one single simulation [30].

Figure 8.6 shows the simplified FDTD model where a solid metal block is used to approximate the shielding walls in SIMPOL. The FDTD simulations assume the following parameters: $h_1 = 50\,\mu m$, $h_2 = h_3 = 20\,\mu m$, $W = 40\,\mu m$, $D = 40\,\mu m$, $S = 120\,\mu m$, and the coupling length is $1000\,\mu m$. The vias are $20\,\mu m$ by $20\,\mu m$ squares that are $20\,\mu m$ thick. As shown by the simulation results in Figure 8.7, SIMPOL is essentially a low-loss transmission line that is very similar to the conventional stripline, which corresponds to the case without the shielding metal block in Figure 8.6. What is most significant, however, is that the crosstalk levels between two SIMPOL interconnects (both S_{31} and S_{41}) drop dramatically as the spacing between vias, d, decreases. An extremely high level of isolation of over $100\,dB$ can be achieved by adjusting the density of the vias, while keeping a reasonably close line spacing ($S = 3W$). As a reference, in conventional striplines assuming similar dielectric substrate and feature sizes, the coupling level is around $-40\,dB$ at the high-frequency end, as indicated by the simulation results for the case

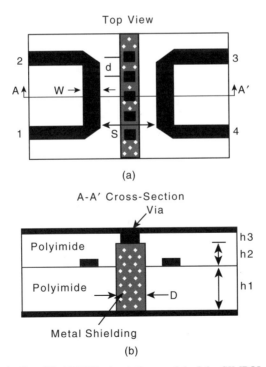

Figure 8.6. Simplified FDTD simulation model of the SIMPOL structure.

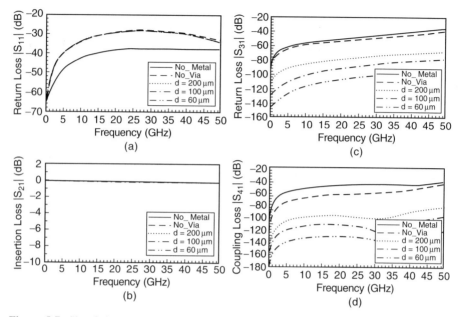

Figure 8.7. Simulation results for the simplified SIMPOL model shown in Figure 8.6: (a) return loss (S_{11}), (b) insertion loss (S_{21}), (c) far-end coupling (S_{31}), and (d) near-end coupling (S_{41}).

of "No Metal" in Figure 8.7. The line spacing between two conventional striplines has to be increased to more than $8W$ in order to realize a 100 dB isolation. This indicates that the SIMPOL structure is not only superior in noise shielding, but also very promising in increasing the integration density of MMICs, which will eventually lead to compact and low-cost circuit design. Finally, for the extreme case of $d = 0$, the vias are connected to form a solid metal bar. In this case the two channels are completely isolated. In fact, the coupling level is so low that it exceeds the dynamic range of the FDTD code, which uses single precision (4 bytes) variables.

8.3.4 SIMPOL Design and Fabrication

As a first step to evaluate the insertion loss and crosstalk characteristics of the SIMPOL interconnect, we have fabricated a simplified test structure that follows the same fabrication procedures described in Section 8.3.2, except for the first step of anisotropic etching of the silicon substrate. Since no active components are involved at this stage, the modified model should offer corresponding performances to the original SIMPOL structure with respect to line insertion loss and mutual coupling. Figure 8.8 shows the cross-sectional view of the simplified test structure we designed for fabrication. A 375 μm thick silicon wafer is first deposited with a thin layer of SiO_2, and then metallized with Ti/Al. Liquid polyimide (Du Pont 2808, $\varepsilon_r = 3.5$) is

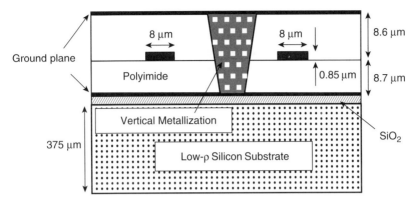

Figure 8.8. Cross-sectional view of the fabricated SIMPOL testing structure.

subsequently applied to form a polyimide layer (8.7 μm thick) by spinning and solidification. Signal lines (8 μm wide and 0.85 μm thick) are then formed on top of the first solidified polyimide layer, followed by a via process to build vertical solid metal blocks. A second polyimide layer with the identical thickness is spinned and solidified in the same way along with a similar via process, and finally the top metal is deposited to form the top ground plane for shielding and self-packaging purposes. Figure 8.9 shows the layout of a uniform line for insertion loss measurement, as well as a pair of coupled lines for evaluation of the signal isolation performance. The length of the uniform line is 1 mm, and the coupled lines have a coupling length of 0.89 mm.

The characteristic impedance and the line loss of the above stripline-like structure are 50 Ω and 0.5 dB/mm at 30 GHz, respectively. It should be pointed out that this structure is not optimal in terms of insertion loss. Figure 8.10 shows the calculated line loss of 50 Ω striplines at 30 GHz for various thicknesses of the polyimide layer by using Agilent EEsof LineCalc [31]. As can be seen, the line loss can be reduced to about 0.1 dB/mm at 30 GHz or even lower if we increase the dielectric substrate to 40 μm or thicker. In a fully implemented SIMPOL structure with anisotropically etched trenches as shown in Figure 8.5, the polyimide, or other types of polymer such as benzocyclobutene (BCB), can easily be made much thicker than that in Figure 8.8. There is a trade-off between line loss and chip integration density, however, because the 50 Ω line will become wider with increased dielectric thickness. Meanwhile, using gold instead of aluminum as the stripline conductor will also help reduce the line loss, especially at millimeter-wave frequencies.

8.3.5 Measurement Results

The *S*-parameters of the fabricated test structure were measured on an HP 8510C network analyzer, using a pair of GGB Picoprobes with 150 μm pitches (40A-GSG-150-LP). A TRL (thru-reflect-line) calibration was carried out for the frequency

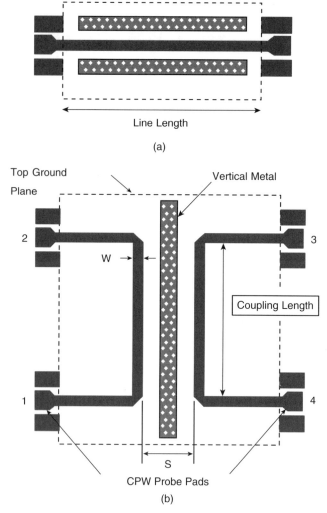

Figure 8.9. Layout configuration of (a) a uniform transmission line for propagation line loss measurement and (b) a pair of U-shaped lines for coupling measurement.

range from 20 to 40 GHz with a dedicated on-wafer calibration standard that was fabricated on the same wafer. Figure 8.11 shows the photographs of the fabricated test structures.

After calibration, we first measured the insertion loss of the 1 mm straight line as shown in Figure 8.11(a). The measured line loss and the calculated result using HP EEsof LineCalc are plotted in Figure 8.12. At 30 GHz, the measured line loss is 0.62 dB/mm, which is close to the calculated loss of 0.5 dB/mm. The relatively large discrepancy at the higher frequency end (40 GHz) might have been caused by both

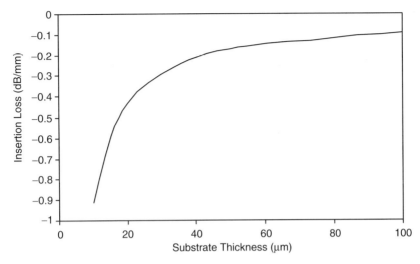

Figure 8.10. Calculated insertion loss of 50 Ω striplines for various polyimide substrate thicknesses ($\varepsilon_r = 3.5$, frequency = 30 GHz).

imperfect calibration and increased parasitic effects of the solid via blocks. It should be pointed out here that the insertion loss of the SIMPOL interconnect can be reduced significantly by increasing the thickness of the dielectric material, as indicated by the LineCalc predictions in Figure 8.10.

Meanwhile, Figure 8.13 shows the measured results for the signal crosstalk of the coupling structure in Figure 8.11(b), as well as the background noise floor. The background noise corresponds to the transmission coefficient when the two probes are suspended in air by 1 mm while maintaining identical distance to that of the SIMPOL test structure. The measured coupling level (S_{31}) is very close to the background noise up to 30 GHz (< -60 dB), and it increases monotonically to approximately -40 dB at 40 GHz.

The degradation in crosstalk at higher frequencies is believed to be caused mainly by the imperfect termination of the idling ports, which were open-circuited in the test structure. The coupling level dropped noticeably when two pieces of absorbers were placed at the two idling ports (port 2 and 4) during S-parameter measurement. In a more recent design, thick-film resistors (NiCr) were deposited at the idling ports of the coupled lines for matched-load termination [19]. Figure 8.14 plots the measured crosstalk of the improved test structure of the coupled lines with the idling ports terminated by the matched loads. Extremely low-noise crosstalk (< -80 dB up to 18 GHz, < -40 dB up to 50 GHz) has been achieved. Meanwhile, due to the use of a thicker polyimide layer (27 μm total thickness), the line insertion loss was also reduced significantly (<0.25 dB/mm up to 45 GHz).

In order to explore the feasibility of millimeter-wave MMICs using the SIMPOL architecture, a branch line (90°) coupler has also been designed and fabricated [19].

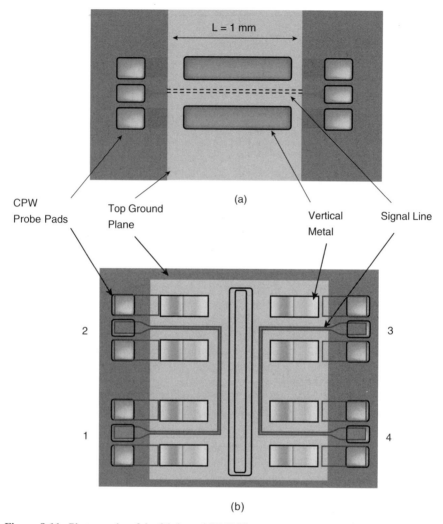

Figure 8.11. Photographs of the fabricated SIMPOL test structures: (a) top view of a uniform 1 mm straight line and (b) top view of a pair U-shaped coupled lines.

At the center frequency of 36 GHz, the coupler delivered -3.77 dB and -3.75 dB, respectively, to the two output ports. The measured bandwidth was 8.58 GHz or 24%. These measurement results are very encouraging since they indicate convincingly that it is possible to realize low-loss, low-crosstalk passive components up to millimeter-wave frequencies on top of CMOS-grade silicon, thus paving the way to the ultimate goal of building an entire wireless system, including both the RF front-end and digital circuitry, into a single chip of silicon.

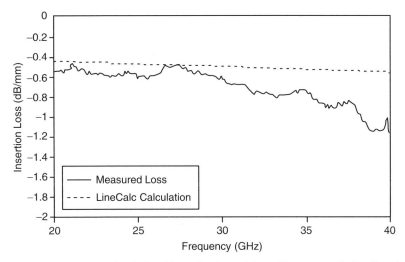

Figure 8.12. Measured and calculated insertion loss of the uniform transmission line shown Figure 8.11(a).

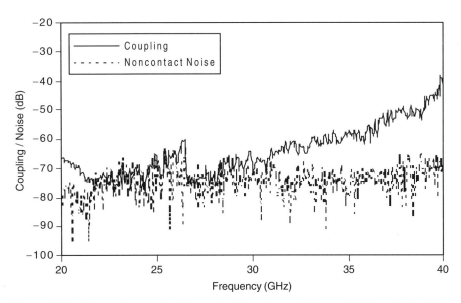

Figure 8.13. Measured crosstalk of the coupled lines shown in Figure 8.11(b), with the idling ports open-circuited.

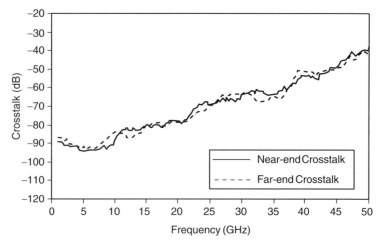

Figure 8.14. Measured crosstalk of an improved test structure of the coupled lines as shown in Figure 8.11(b), with the idling ports terminated by matched loads.

8.4 HIGH Q COMPONENTS BASED ON PHOTONIC BANDGAP STRUCTURES

8.4.1 Background

Photonic bandgap materials are periodic structures capable of prohibiting the propagation of electromagnetic waves along one or more directions within a certain band of frequencies [32]. Although research on the PBG originated mainly in the optical regime [33], the PBG structures are readily scaleable and applicable to a much wider range of frequencies including microwaves and millimeter waves. There have been intensive research efforts in recent years in the application of novel PBG structures for high-Q microwave components. In fact, microwave applications of periodic structures can be dated back to the early days of radar, with such examples as one-dimensional (1D) slotted-waveguide array antennas, as well as two-dimensional (2D) frequency-selective surfaces (FSSs) and polarization diplexers. The emphasis of recent research efforts, however, is on 1D or 2D PBG structures that are compatible with modern planar circuit fabrication technology.

For 1D periodic structures, the term PBG can only be used in a relatively loose fashion. In fact, the ubiquitous high–low impedance microstrip lowpass filter is by itself a 1D periodic or quasiperiodic structure. It should be pointed out, however, that applying the PBG concept allows one to greatly extend the horizon of imagination when conceiving novel structures to control the behavior of electromagnetic waves, whether it is a guided wave, surface wave, or radiation wave. For example, microstrip-based lowpass (band-reject) filters can be achieved by modifying the dielectric substrate [34] and/or the ground plane [35], rather than following the traditional high–low impedance line approach. For 2D periodic structures, conventional circuit theory-based analysis become almost completely invalid, and full-wave-based EM simulations play a crucial role in characterizing and designing such PBG lattices. These novel PBG structures have shown great potential applications such as slow-wave phase shifters, surface wave and leaky wave suppressors, and high impedance ground plane as well as TEM waveguides.

Figure 8.15 shows some of the planar or planar-oriented PBG structures that have been proposed and investigated at the authors' research group during the past few years. In this section we will focus our discussions on PBG structures for microstrip or CPW-based planar circuit applications. The structures to be described include (1) dielectric-based PBG microstrip lines, (2) a PBG ground plane for microstrip lines, and (3) a uniplanar compact PBG (UC-PBG) structure. The compact PBG with vias [36,37], shown in Figure 8.15(c), is described in Chapter 11 together with applications for surface wave suppression in microstrip patch antennas. The high-Q PBG image guide resonator, shown in Figure 8.15(d), can be found in Yang et al. [38].

8.4.2 Dielectric-Based PBG Microstrip Structures

One of the most straightforward approaches to create a propagation stopband along a microstrip line is to modify the dielectric substrate in a periodic manner. Several such synthesized dielectric lattices that possess distinctive stopbands for the quasi-TEM wave propagation in microstrip lines have been proposed and investigated [34]. They include (a) square-lattice, square-hole, (b) triangular-lattice, square-hole, (c) honeycomb-lattice, square-hole, and (d) honeycomb-lattice, circular hole. The inhomogeneous and hybrid-mode nature of these structures mandates the use of 3D full-wave solutions of Maxwell's equations. It was found that FDTD is a very efficient numerical approach to characterize such novel 3D structures, since it can provide their S-parameters over an extremely broad frequency range with one single simulation [30].

Figure 8.16 shows a $50\,\Omega$ microstrip line fabricated on a 1×10 honeycomb lattice of circular holes drilled through the dielectric substrate. The dielectric substrate is RT/Duroid 6010 with $\varepsilon_r = 10.2$ and a thickness of 50 mils. The honeycomb lattice size is 250 mils and the radius of the drilled holes is 50 mils. The total length of the microstrip line is 2850 mils. The PBG holes are drilled through the substrate, and a conductive tape is applied in the ground plane of the microstrip line. Figure 8.17 plots both FDTD simulation and measurement results of the S-parameters. The first stopband occurs at 8.90 GHz, in close agreement with the FDTD prediction (<2% discrepancy). Despite an increased insertion loss at higher frequencies due partly to the SMA connectors, the overall characteristics of the PBG structure agree well with FDTD simulation results.

This PBG structure has been employed subsequently for harmonic tuning in microwave power amplifiers [39]. Two Class AB GaAs FET amplifiers were designed and fabricated in the 4.4–4.8 GHz frequency range. In the first case, a dielectric PBG line was incorporated in the design to tune the second harmonic. The second design uses a $50\,\Omega$ microstrip line with no harmonic tuning. Figure 8.18

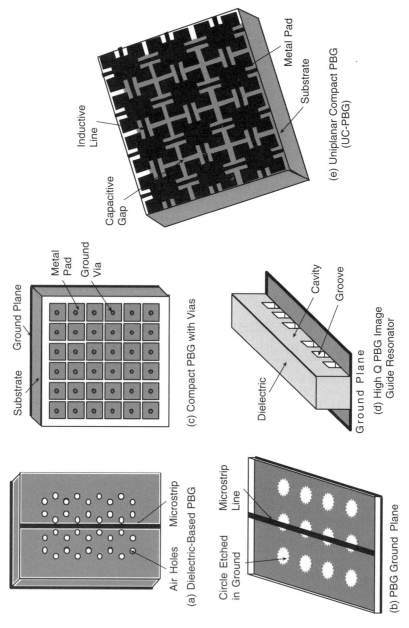

Figure 8.15. Examples of planar PBG structures for microwave and millimeter-wave circuits and antenna applications.

284

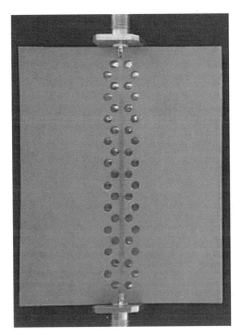

Figure 8.16. A microstrip line fabricated on honeycomb-lattice, drilled-hole PBG dielectric substrate.

shows the photograph of the power amplifier with the PBG structure. For the harmonic tuned amplifier, a 5% improvement in the power-added efficiency (PAE) was measured at the center frequency of 4.5 GHz. An increase of over 0.3 dB in output power was also measured over a 9% frequency bandwidth.

8.4.3 PBG Microstrips on Etched Ground Plane

An alternative PBG structure suitable for microstrip and other planar circuit applications is based on etching a 2D lattice of holes in the ground plane, as shown in more detail in Figure 8.19. Compared to the PBG microstrip in Figure 8.16, no drilling through the dielectric substrate is required; thus the fabrication process is greatly simplified. Moreover, a much wider and deeper stopband can be realized compared to the dielectric-based PBG microstrip structure. In this case, the stopband frequency is mainly determined by the lattice size, a. The shape of the stopband, including its width and depth, is affected by both the number of periods, and the radius, r, relative to the lattice size. Initial experiments indicate the values of r within $0.15 < r/a < 0.25$ range are optimal. Smaller r causes smaller perturbation and diminishes the stopband dip. On the other hand, an overly large r/a ratio may cause too strong a perturbation so that the passband will contain significant ripples [35].

Again, the characteristics of wave propagation along microstrip lines on this type of modified ground plane can be predicted accurately by using FDTD simulations.

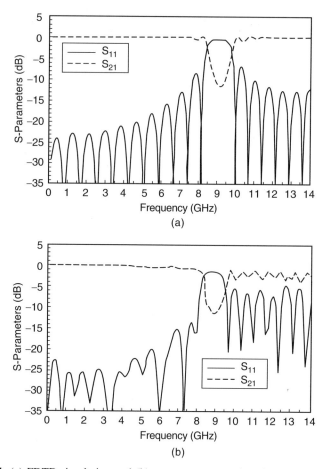

Figure 8.17. (a) FDTD simulation and (b) measurement results of the S-parameters of the PBG microstrip line shown in Figure 8.16.

Figure 8.20 shows both simulation and measurement results of the S-parameters of a 50 Ω microstrip line on a PBG structure with the following parameters: substrate thickness = 31 mils, dielectric constant = 2.33, lattice $a = 480$ mils, and radius $r = 110$ mils. The circles in the ground plane were modeled using a staircasing approximation. The maximum stopband depth is -22.5 dB at 9.08 GHz for FDTD, and -22.8 dB at 8.93 GHz for measurement. Less than 2% error is obtained between measured and FDTD data for the maximum stopband depth, and good agreement is observed over the entire frequency band.

One advantage of this PBG ground plane is that it offers a very broad and deep stopband for the microstrip line in comparison with the drilled-dielectric approach described in the previous section. This property can be used for broadband harmonic tuning of power amplifiers, which might be difficult to achieve using conventional double-stub tuners or filters [40,41]. Figure 8.21 is a photograph of a Class AB active

HIGH Q COMPONENTS BASED ON PHOTONIC BANDGAP STRUCTURES

Figure 8.18. Class AB power amplifier using dielectric-based PBG for harmonic tuning.

antenna amplifier integrated with such a PBG structure and a slot antenna. The measured PAE is better than 50% over an 8% bandwidth (3.7–4.0 GHz), with second harmonic tuning only.

8.4.4 Uniplanar Compact PBG Structure

While the planar-type PBG structures discussed in the previous sections have demonstrated interesting stopband characteristics, a key concern in their practical

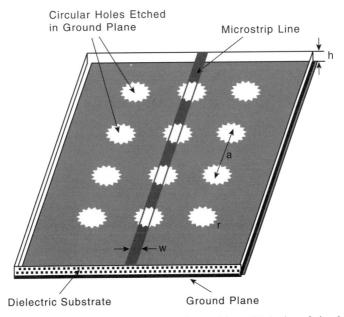

Figure 8.19. Microstrip line PBG structure based on etching a 2D lattice of circular holes in the ground plane.

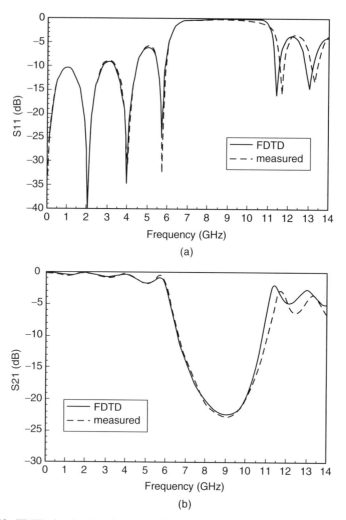

Figure 8.20. FDTD simulated and measured (a) S_{11} and (b) S_{21} for the PBG structure shown in Figure 8.19.

application is the relatively large dimensions required to achieve a decent stopband at the desired operation frequency. Great efforts have been made to reduce the sizes of PBG structures while retaining their unique stopband properties. For instance, a novel metallodielectric photonic crystal consisting of multiple circuit board/polyimide layers with vertical vias was investigated by Sievenpiper and Yablonovitch [42]. The lower end of the stopband was found to be determined by the *LC* resonant frequency, thus resulting in a much more compact PBG structure compared to those achievable previously. A slightly modified version, which is constructed by a square lattice of metal pads with grounding vias, was employed to suppress undesirable surface waves excited in the dielectric substrate [36,37]. A

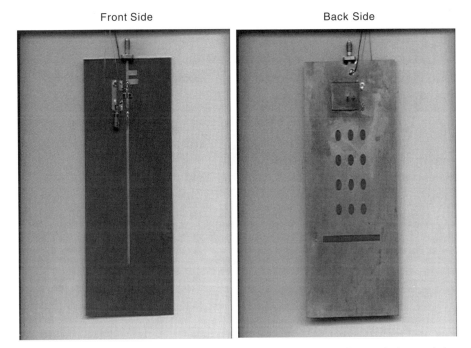

Figure 8.21. Class AB active antenna amplifier integrated with a PBG ground plane and slot antenna.

microstrip patch antenna surrounded by such a compact PBG lattice was found to have improved radiation efficiency and beam patterns. See Chapter 11 for a more detailed discussion on this topic.

Our intensive searching for novel planar, compact PBG structures has culminated in the most recent uniplanar compact PBG (UC-PBG) structure, as shown in Figure 8.22. Each unit element consists of a square metal pad with four connecting branches. The inductive lines and capacitive gaps form a 2D *LC* network that greatly reduces the lower edge of the bandgap. A microstrip line that employs this UC-PBG as its ground plane reveals several unique properties, such as low insertion loss, slow-wave effect, and extremely wide stopband. Due to the significant slow-wave effect, the period of the PBG lattice can be reduced to as small as $0.1\lambda_0$ at the cutoff frequency, resulting in the most compact PBG lattice ever achieved in simple planar circuit technology [43].

Figure 8.23 shows a photograph of a 50 Ω microstrip line (24 mils wide, 1 inch long) on a ground plane with the UC-PBG lattice. The dielectric substrate used is 25 mil thick Duroid with $\varepsilon_r = 10.2$. The PBG section is 720 mils long, which corresponds to six periods. A short length (140 mils) of solid ground plane has been included at each end of the microstrip line to facilitate the connection with SMA connectors. Referring to Figure 8.22, the PBG pad and associated inductive branch

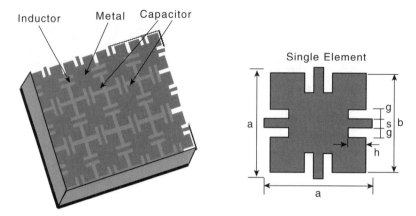

Figure 8.22. Schematic of the UC-PBG lattice and details of one single element.

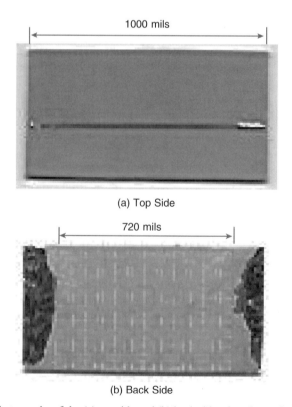

Figure 8.23. Photographs of the (a) top side and (b) back side of a microstrip line on the UC-PBG ground plane.

have the following dimensions: $a = 120$ mils, $b = 108$ mils, $s = g = 12$ mils, and $h = 30$ mils.

Figure 8.24 plots both FDTD and measured S-parameters of this test structure. For comparison the measured insertion loss of a conventional microstrip line with the same length (1 inch) has also been plotted in Figure 8.24(b). It can be seen that at lower frequencies the line insertion loss is at the same level with the reference microstrip line on a solid ground plane and with identical dimensions. The structure exhibits a deep and very broad stopband above around 10 GHz. The excellent

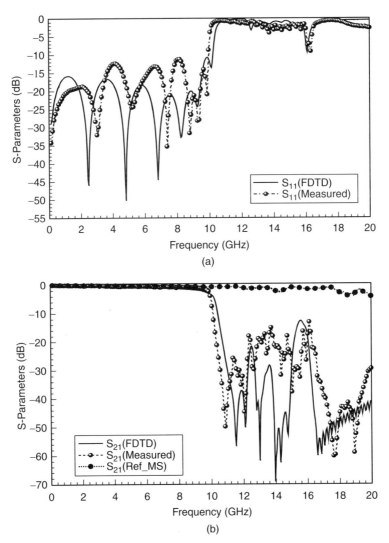

Figure 8.24. Comparison of simulation and measurement results of (a) S_{11} and (b) S_{21} of the microstrip line on the UC-PBG ground plane.

correlation between simulation and measurement indicates that this type of PBG structure can be characterized and designed with a high degree of accuracy.

Another unique feature of this PBG microstrip structure is its low-loss slow-wave effect [44]. Figure 8.25 displays the slow-wave factor (β/k_0) of the UC-PBG structure in conjunction with that of a conventional microstrip line. It can be observed that the slow-wave effect is significant even at low frequencies, and the slow-wave factor (β/k_0) has been found to be 1.2 to 2.4 times higher than that of a reference microstrip line fabricated on the same dielectric substrate. It is believed that the slow-wave factor can be further enhanced by optimizing the geometry of the UC-PBG lattice. It has also been found that the variation of slow-wave factor is less than ±5% for different alignment offsets between the microstrip and the PBG lattice, which is advantageous for practical applications.

Finally, it is interesting to observe, from the measured results of S_{21} in Figure 8.24(b), that the PBG microstrip line is well matched to the 50 Ω measurement system with SMA connectors over a wide range of frequencies, indicating that the PBG line keeps a relatively constant characteristic impedance within the passband. Since the characteristic impedance is the square root of the ratio of the inductance and capacitance per unit length of the line, this indicates that the inductance and capacitance seen by the microstrip have increased simultaneously in the UC-PBG structure. Unlike the case of other slow-wave structures, such as metal insulator semiconductor (MIS) transmission lines [45], the problem of low characteristic impedance does not exist here. The UC-PBG structure exhibits the same slow-wave effect as for a conventional microstrip line with a higher dielectric constant, but without major consequential increase in conductor loss. For example, the conductor loss of a 50 Ω line on the UC-PBG ground is 0.17 dB/in., at 6 GHz. On

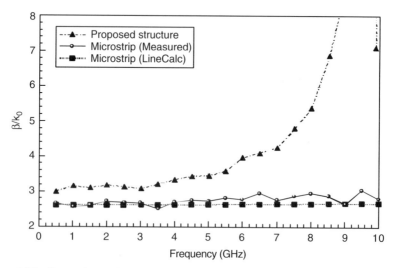

Figure 8.25. Comparison of slow-wave factor of microstrip lines on the UC-PBG and conventional solid ground plane.

the other hand, the conductor loss of a conventional microstrip line with the same impedance and propagation constant would be 0.46 dB/in., since the microstrip width has to be reduced substantially to keep the line impedance at 50 Ω due to the increase in the dielectric constant. The MIS structures with low characteristic impedances also face the necessity of using accurate photolithography for very fine features. The microstrip integrated with the UC-PBG ground plane demonstrates the advantages of low loss, moderate impedance level, and simple fabrication process, which can be exploited to build new types of slow-wave structures and phase shifters.

8.4.5 Applications of the UC-PBG Structure

The unique properties of the UC-PBG enable us to improve the performances of microwave and millimeter-wave components, both passive and active, with an unconventional approach. We now present some of the applications that take advantage of the various characteristics of the UC-PBG structure.

8.4.5.1 Spurious-Free Bandpass Filter Microstrip bandpass filters (BPFs) are very popular components in microwave integrated circuits. Conventional parallel-coupled BPFs, however, present spurious passbands at harmonic frequencies, which tend to degrade the performance of the overall RF system. Extra filters are usually required to suppress spurious transmissions, resulting in not only more complicated front-end circuitry but also increased insertion loss. The UC-PBG can be applied to construct compact microstrip bandpass filters with intrinsic spurious rejection. The wide, deep stopband of the UC-PBG structure can be employed to suppress the spurious passbands at higher harmonics. Since the stopband is intrinsic, extra filters are not required. Furthermore, the slow-wave effect reduces the physical length of the filter circuit integrated with the UC-PBG structure.

Figure 8.26 shows the photograph of a parallel-coupled bandpass filter using the UC-PBG as its ground plane. The bandpass filter is designed with four coupling

Top View   Bottom View

Figure 8.26. Spurious-free microstrip bandpass filter fabricated on the UC-PBG ground plane.

sections, a 0.5 dB equal-ripple response and a center frequency of 6 GHz. Detailed dimensions of the circuit can be found in Yang et al. [46,47]. For comparison a reference BPF using conventional microstrip lines is also designed and fabricated. Due to the slow-wave effect, the resonators in the BPF based on UC-PBG is about 20% shorter than a conventional quarter-wavelength line.

Figure 8.27 shows the measured results of the PBG bandpass filter as well as the reference one. As can be seen, the measured transmission coefficients of the conventional bandpass filter are -10 and -5 dB at 12 and 17 GHz, respectively. In comparison, the experimental result of the bandpass filter on the UC-PBG ground shows 30–40 dB suppression of the spurious response. In designing the PBG bandpass filter at a center frequency of 6 GHz, the lengths of the microstrip resonators have been scaled appropriately according to the slow-wave factor. The coupling gaps, on the other hand, were kept unchanged. This explains the increased fractional bandwidth (21.6%) of the PBG filter and a slower rolloff. The bandpass characteristics can be improved by optimizing the coupling coefficients between the resonators, in a similar manner to conventional BPF design. The minimum insertion loss of the PBG filter is 1.9 dB at 6.39 GHz, which includes the effect of two SMA connectors. The passband loss is comparable to that of the reference filter using conventional microstrip lines.

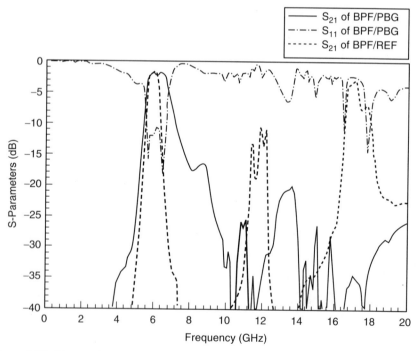

Figure 8.27. Measured S-parameters of the bandpass filter in Figure 8.26, showing effective suppression of both second and third harmonics.

HIGH Q COMPONENTS BASED ON PHOTONIC BANDGAP STRUCTURES

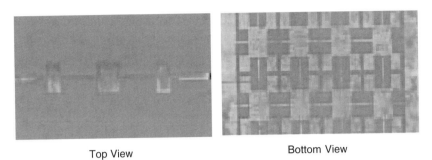

Figure 8.28. A microstrip lowpass filter fabricated on the UC-PBG ground plane.

8.4.5.2 High-Performance Lowpass Filter The UC-PBG structure can also be integrated with conventional stepped-impedance lowpass filters (LPFs) to enhance their performance [48]. The sharp cutoff in the transmission coefficient of the microstrip line as it approaches the stopband can be exploited to improve both the rolloff and the rejection level of the lowpass filter. Figure 8.28 shows a photograph of a seven-section high–low impedance LPF fabricated on the UC-PBG ground plane. Detailed dimensions of the structure can be found in Yang et al. [48]. Figure 8.29 plots the measured insertion losses of both the PBG-based lowpass filter and a reference LPF using conventional microstrip lines on a solid ground plane. As can be seen, the maximum attenuation is increased from 20 to 58 dB after applying the UC-PBG structure. In addition, the spurious passband, which appears in the reference LPF, has been completely suppressed. Meanwhile, the passband insertion loss of the PBG filter is comparable to that of the reference, indicating that applying the UC-

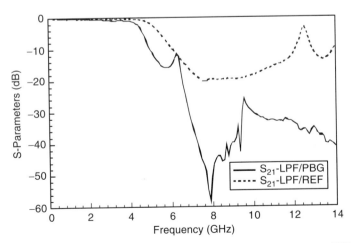

Figure 8.29. Measured insertion losses of the microstrip lowpass filter on UC-PBG and regular ground plane.

PBG does not deteriorate the performance of the LPF in the passband. The enhanced LPF was subsequently integrated with a drain mixer, where over 10 dB suppression of the LO leakage in comparison to a reference mixer with conventional filter was demonstrated.

8.4.5.3 High-Efficiency Power Amplifier

Harmonic tuning in microwave power amplifiers is an important technique to improve both the power-added efficiency (PAE) and output power level. Conventional methods for harmonic tuning include the use of single or double stubs with either open or short termination. Recently, the active integrated antenna concept has also been applied for broadband and efficient harmonic tuning of power amplifiers. Interested readers can refer to Chapter 6 of this book for a more detailed discussion on this topic.

The extremely wide stopband characteristics of the UC-PBG can be employed as an alternative technique for harmonic tuning in high-efficiency power amplifiers [49]. Figure 8.30 shows the prototype of an S-band Class AB power amplifier integrated with the UC-PBG microstrip for both second and third harmonic tuning. The UC-PBG dimensions were designed to have a cutoff frequency at 6 GHz; the measured S_{21} showed a wide stopband from 6 to 15 GHz. Figure 8.31 shows the measured PAE and output power versus input power for both the PBG and reference

Top View   Bottom View

Figure 8.30. An S-band MESFET power amplifier using UC-PBG for harmonic tuning.

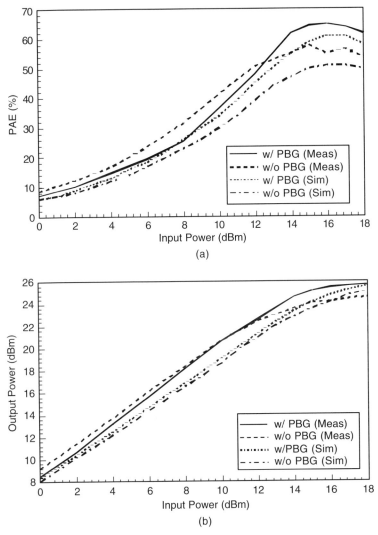

Figure 8.31. Simulation and measurement results of (a) PAE and (b) output power versus input power for the PBG and reference amplifier.

amplifiers. The measured PAE is above 50% from 3.32 to 3.62 GHz (8% bandwidth) for the PBG amplifier, and from 3.50 to 3.63 GHz (4% bandwidth) for the reference. A maximum of 10% improvement in PAE and 1.3 dB increase in output power have been achieved. Meanwhile, the output levels of the second and third harmonics were reduced from -11 and -30 dB to -48 and -60 dB, respectively, in comparison with the case of the reference amplifier terminated with standard 50 Ω load.

8.4.5.4 Leakage Suppression in CPWs and Striplines
As has been described in Section 8.2, the coplanar waveguide (CPW) is one of the most frequently employed planar circuit topologies in modern MICs and MMICs, particularly in millimeter-wave low-noise amplifier designs where minimizing the insertion loss of transmission lines is a top priority. While the original CPW does not have a conductor back, in reality it is often backed with another ground plane to increase mechanical strength, to provide heat sinking, or simply to facilitate packaging. The conductor-backed CPW (CB-CPW), however, will excite a parallel-plate mode and deteriorate the CPW performance. Several approaches have been proposed to overcome the leakage problem, such as using shorted vias to suppress the unwanted mode or using multilayered substrates to shift the dispersion curve of the parallel-plate mode. However, a planar circuit is preferable for the fabrication process of integrated circuits. The UC-PBG structure, with its unique wide stopband characteristics, can be employed to suppress the parallel plate mode, resulting in a nonleaky CB-CPW. Since the UC-PBG patterns can easily be etched in the top ground planes of the CB-CPW circuit without using any extra masks or shorted vias, there is virtually no cost addition from the fabrication viewpoint.

Figure 8.32 shows the photograph of such a nonleaky CB-CPW structure, where the UC-PBG lattice is constructed in the top ground planes [50]. The measured transmission coefficient (S_{21}) of this nonleaky CB-CPW line is plotted in Figure 8.33, together with the measured results of a conventional CPW and CB-CPW. It is found that for a conventional CB-CPW, the leakage is significant in the entire

Figure 8.32. A nonleaky CB-CPW based on the UC-PBG structure.

SUMMARY

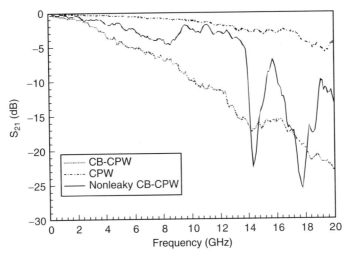

Figure 8.33. Comparison of measured insertion losses of the PBG-based nonleaky CB-CPW and conventional CPW and CB-CPW.

frequency range. Meanwhile, the insertion loss of a conventional CPW is relatively low as expected, although it increases gradually at higher frequencies due partly to reflections caused by the SMA connectors. Meanwhile, for the CB-CPW with the UC-PBG lattice, the power leakage is still present in the range between dc and 9 GHz, which is the passband of the PBG structure. However, the insertion loss is improved significantly and is comparable to that of a conventional CPW, between 9 and 14 GHz. The measurement results indicate clearly that the leakage loss has been suppressed almost completely in this frequency range, which corresponds to the stopband of the UC-PBG lattice. This novel CB-CPW structure shows great potential for applications in various types of CPW-based circuits, such as CPW-fed slot antennas.

In a similar manner, the UC-PBG structure can also be utilized to suppress the undesirable parallel-plate mode in stripline-based circuits [51]. In this case the PBG lattice can be etched on either one side or both sides of the ground planes of the stripline structure. Both simulation and measurement results have confirmed the effectiveness of this approach, which is capable of suppressing the leakage coupling in a stripline circuit by over 30 dB in the PBG stopband. Since the PBG lattice can easily be realized by standard planar processes, it provides an attractive solution to the leakage problem in stripline-based microwave circuits, in comparison with conventional approaches such as using highly populated via-hole fences.

8.5 SUMMARY

It has been recognized for a long time that integrated planar circuits cannot beat the bulky metallic waveguide systems in terms of insertion loss or quality factor. Despite this shortcoming, however, planar circuit technology has been the driving force for modern microwave and millimeter-wave application systems, because of its enormous advantages, including greatly reduced hardware size and weight, easier integration with solid-state devices, low power consumption, and easy implementation on mobile platforms. With relentless innovations in planar circuit technologies, including those described here and in other chapters of this book, we believe that high-performance microwave and millimeter-wave components compatible with modern planar fabrication processes should become more and more easily available for RF circuit designers. Together with dramatic progress in semiconductor device technology, they will provide the most cost-effective solution to modern RFICs, which are the key building blocks in all types of microwave and wireless systems being deployed rapidly around the world.

ACKNOWLEDGMENT

The authors acknowledge the contributions of Professor M. Frank Chang, Dr. Pingxi Ma, Dr. Guojin Feng, Dr. Vesna Radisic, Dr. Kuang-Ping Ma, Dr. Juno Kim, Dr. Fei-Ran Yang, and Cynthia Hang to the work described in this chapter. This work was supported by the U.S. ARO MURI program under contract DAAH04-96-1-0005.

REFERENCES

1. K. C. Gupta, R. Garg, and I. J. Bahl, *Microstrip Lines and Slotlines*, Artech House, Norwood, MA, 1979.
2. T. Itoh (ed.), *Planar Transmission Line Structures*, IEEE Press, New York, 1987.
3. T. Itoh (ed.), *Numerical Techniques for Microwave and Millimeter-Wave Passive Structures*, Wiley, New York, 1989.
4. R. A. Pucel, "Design consideration for monolithic microwave circuits," *IEEE Transactions on Microwave Theory and Techniques*, Vol. 29, 513–534 (June 1984).
5. J. Papapolymerou, J. East, and L. P. B. Katehi, "GaAs versus quartz FGC lines for MMIC applications," *IEEE Transactions on Microwave Theory and Techniques*, Vol. 46, Part 1, 1790–1793 (Nov. 1998).
6. N. H. Huynh and W. Heinrich, "FDTD analysis of submillimeter-wave CPW with finite-width ground metallization," *IEEE Microwave Guided Wave Letters*, Vol. 7, 414–416 (Dec. 1997).
7. H. Q. Tserng, P. Saunier, A. Ketterson, L. C. Witkowski, and T. Jones, "Embedded transmission-line (ETL) MMIC for low-cost high-density wireless communication applications," *IEEE Transactions on Microwave Theory and Techniques*, Vol. 45, 2540–2548 (Dec. 1997).
8. J. D. Cressler, "SiGe HBT technology: a new contender for Si-based RF and microwave circuit applications," *IEEE Transactions on Microwave Theory and Techniques*, Vol. 46, 572–589 (May 1998).
9. A. Schuppen, U. Erben, A. Gruhle, H. Kibbel, H. Schumacher, and U. Konig, "Enhanced SiGe heterojunction bipolar transistors with 160 GHz-fmax," *IEEE IEDM Digest*, 743–746 (1995).

10. J. N. Burghrtz, J.-O. Plouchart, K. A. Jenkins, C. S. Webster, and M. Soyuer, "SiGe power HBT's for low-voltage, high performance RF applications," *IEEE Electron Device Letters*, Vol. 19, No. 4, 103–105 (1998).

11. A. Schuppen, S. Gerlach, H. Dietrich, D. Wandrei, U. Seiler, and U. Konig, "1-W SiGe power HBTs for mobile communication," *IEEE Microwave and Guided Wave Letters*, Vol. 6, No. 9, 341–343 (1996).

12. B. S. Meyerson, D. L. Harame, J. Stork, E. Crabbe, J. Comfort, and G. Patton, "Silicon: germanium heterojunction bipolar transistors: from experiment to technology," in *Current Trends in Heterojunction Bipolar Transistors*, M. F. Chang (Ed.), World Scientific, River Edge, NJ, 1996, pp. 367–385.

13. P. Russer, "Si and SiGe millimeter-wave integrated circuits," *IEEE Transactions on Microwave Theory and Techniques*, Vol. 46, 590–603 (May 1998).

14. C. Warns, W. Menzel, and H. Schumacher, "Transmission lines and passive elements for multilayer coplanar circuits on silicon," *IEEE Transactions on Microwave Theory and Techniques*, Vol. 46, 616–622 (May 1998).

15. Y. Wu, H. S. Gamble, B. M. Armstrong, V. F. Fusco, and J. A. C. Stewart, "SiO_2 interface layer effects on microwave loss of high-resistivity CPW line," *IEEE Microwave Guided Wave Letters*, Vol. 9, 10–12 (Jan. 1999).

16. H. S. Gamble, B. M. Armstrong, S. J. N. Mitchell, Y. Wu, V. F. Fusco, and J. A. C. Stewart, "Low-loss CPW lines on surface stabilized high-resistivity silicon," *IEEE Microwave Guided Wave Letters*, Vol. 9, 395–397 (Oct. 1999).

17. "The one-chip radio ... search for the Holy Grail?" in *1999 IEEE MTT-S International Microwave Symposium*, Workshop WSFA, Anaheim, CA, June 1999.

18. R. J. Gutmann, "Advanced silicon IC interconnect technology and design: present trends and RF wireless implementations," *IEEE Transactions on Microwave Theory and Techniques*, Vol. 47, 667–674 (June 1999).

19. J. Kim, Y. Qian, G. Feng, P. Ma, M. F. Chang, and T. Itoh, "High-performance millimeter-wave interconnect and coupler for broadband mixed signal silicon MMICs," *2000 IEEE MTT-S International Microwave Symposium*, Boston, MA, June 2000.

20. V. Milanovic, M. Ozgur, D. C. DeGroot, J. A. Jargon, M. Gaitan, and M. E. Zaghloul, "Characterization of broad-band transmission for coplanar waveguides on CMOS silicon substrates," *IEEE Transactions on Microwave Theory and Techniques*, Vol. 46, 632–640 (May 1998).

21. T. M. Weller, L. P. B. Katehi, and G. M. Rebeiz, "High performance microshield line components," *IEEE Transactions on Microwave Theory and Techniques*, Vol. 46, 632–640 (May 1998).

22. M. F. Chang, Y. Qian, P. Ma, and T. Itoh, "Silicon/metal/polyimide (SIMPOL) interconnects for broadband mixed signal silicon MMICs," *Electronics Letters*, Vol. 34, No. 17, 1670–1671 (Aug. 1998).

23. Y. Qian, M. F. Chang, P. Ma, and T. Itoh, "Low-loss, low-crosstalk *Silicon/Metal/Pol*yimide (SIMPOL) interconnects for mixed signal silicon MMICs," in *IEEE Topical Meeting on Silicon Monolithic Integrated Circuits in RF Systems*, Ann Arbor, MI, Sept. 1998, pp. 169–172.

24. Y. Qian, M. F. Chang, and T. Itoh, "A novel high isolation interconnect for broadband mixed signal silicon MMICs," in *7th IEEE Topical Meeting on Electrical Performance of Electronic Packaging (EPEP'98)*, West Point, NY, Oct. 1998, pp. 79–82.

25. J. Kim, Y. Qian, G. Feng, P. Ma, J. Judy, M. F. Chang, and T. Itoh, "A novel low-loss low-crosstalk interconnect for broad-band mixed-signal silicon MMIC's," *IEEE Transactions on Microwave Theory and Techniques*, Vol. 47, No. 9, 1830–1835 (Sept. 1999).
26. G. E. Ponchak, D. Chen, J. G. Yook, and L. P. B. Katehi, "Characterization of plated via hole fences for isolation between stripline circuits in LTCC packages," *IEEE MTT-S International Microwave Symposium Digest*, 1831–1834 (June 1998).
27. S. Yang, Z. Hu, N. B. Buchanan, V. F. Fusco, J. A. C. Stewart, Y. Wu, B. M. Armstrong, G. A. Armstrong, and H. S. Gamble, "Characteristics of trenched coplanar waveguide for high-resistivity Si MMIC applications," *IEEE Transactions on Microwave Theory and Techniques*, Vol. 46, 623–631 (May 1998).
28. K. Nishikawa, I. Toyoda, K. Kamogawa, and T. Tokumitsu, "Three-dimensional silicon MMIC's operating up to K-band," *IEEE Transactions on Microwave Theory and Techniques*, Vol. 46, 677–684 (May 1998).
29. R. F. Drayton, R. M. Henderson, and L. P. B. Katehi, "Monolithic packaging concepts for high isolation in circuits and antennas," *IEEE Transactions on Microwave Theory and Techniques*, Vol. 46, 900–906 (July 1998).
30. Y. Qian and T. Itoh, *FDTD Analysis and Design of Microwave Circuits and Antennas—Software and Applications*, Realize Inc., Tokyo, 1999.
31. HP EEsof Series IV/PC User's Manual.
32. J. D. Joannopoulos, R. D. Meade, and J. N. Winn, *Photonic Crystals*, Princeton University Press, Princeton, NJ, 1995.
33. E. Yablonovitch, "Inhibited spontaneous emission in solid-state physics and electronics," *Physical Review Letters*, Vol. 58, 2059–2062 (May 1987).
34. Y. Qian, V. Radisic, and T. Itoh, "Simulation and experiment of photonic band-gap structures for microstrip circuits," in *1997 Asia–Pacific Microwave Conference (APMC'97)*, Hong Kong, Dec. 1997, pp. 585–588.
35. V. Radisic, Y. Qian, R. Coccioli, and T. Itoh, "Novel 2-D photonic bandgap structure for microstrip lines," *IEEE Microwave and Guided Wave Letters*, Vol. 8, 69–71 (Feb. 1998).
36. Y. Qian, D. Sievenpiper, V. Radisic, E. Yablonovitch, and T. Itoh, "A novel approach for gain and bandwidth enhancement of patch antennas," in *1998 IEEE Radio and Wireless Conference. (RAWCON'98)*, Colorado Springs, CO, Aug. 1998, pp. 221–224.
37. Y. Qian, R. Coccioli, D. Sievenpiper, R. Radisic, E. Yablonovitch, and T. Itoh, "A microstrip patch antenna using novel photonic band-gap structures," *Microwave Journal*, Vol. 42, No. 1, 66–76 (Jan. 1999).
38. F.-R. Yang, Y. Qian, and T. Itoh, "A novel high-Q image guide resonator using band-gap structures," in *1998 IEEE MTT-S International Microwave Symposium*, Baltimore, MD, June 1998, pp. 1803–1806.
39. V. Radisic, Y. Qian, and T. Itoh, "Broad-band power amplifier using dielectric photonic bandgap structure," *IEEE Microwave and Guided Wave Letters*, Vol. 8, 13–14 (Jan. 1998).
40. V. Radisic, Y. Qian, and T. Itoh, "Broadband power amplifier integrated with slot antenna and novel harmonic tuning structure," in *1998 IEEE MTT-S International Microwave Symposium*, June 1998, pp. 1895–1898.
41. V. Radisic, Y. Qian, and T. Itoh, "Novel architectures for high efficiency amplifiers for wireless applications," *IEEE Transactions on Microwave Theory and Techniques*, Vol. 46, 1901–1909 (Nov. 1998).

REFERENCES

42. D. Sievenpiper and E. Yablonovitch, "Eliminating surface currents with metallodielectric photonic crystals," in *1998 IEEE MTT-S International Microwave Symposium*, Baltimore, MD, June 1998, pp. 663–666.
43. Y. Qian, F. R. Yang, and T. Itoh, "Characteristics of microstrip lines on a uniplanar compact PBG ground plane," in *1998 Asia–Pacific Microwave Conference (APMC'98)*, Yokohama, Japan, Dec. 1998, pp. 589–592.
44. F.-R. Yang, Y. Qian, R. Coccioli, and T. Itoh, "A novel low-loss slow-wave microstrip structure," *IEEE Microwave and Guided Wave Letters*, Vol. 8, 372–374 (Nov. 1998).
45. F. Huang, "Novel slow-wave structure for narrow-band quasi-transversal filters," *IEE Proceedings on Microwaves, Antennas, and Propagation*, Vol. 142, 389–393 (Oct. 1995).
46. F.-R. Yang, Y. Qian, and T. Itoh, "A novel compact microstrip bandpass filter with intrinsic spurious suppression," in *1998 Asia–Pacific Microwave Conference (APMC'98)*, Yokohama, Japan, Dec. 1998, pp. 593–596.
47. F.-R. Yang, K.-P Ma, Y. Qian, and T. Itoh, "A uniplanar compact photonic-bandgap (UC-PBG) structure and its applications for microwave circuits," *IEEE Transactions on Microwave Theory and Techniques*, Vol. 47, 1509–1514 (Aug. 1999).
48. F.-R. Yang, Y. Qian, and T. Itoh, "A novel uniplanar compact PBG structure for filter and mixer applications," in *1999 IEEE MTT-S International Microwave Symposium*, Anaheim, CA, June 1999, pp. 919–922.
49. C. Y. Hang, V. Radisic, Y. Qian, and T. Itoh, "High efficiency power amplifier with novel PBG ground plane for harmonic tuning," in *1999 IEEE MTT-S International Microwave Symposium*, Anaheim, CA, June 1999, pp. 807–810.
50. K.-P. Ma, F.-R. Yang, Y. Qian, and T. Itoh, "Nonleaky conductor-backed CPW using a novel 2-D PBG lattice," in *1998 Asia–Pacific Microwave Conference (APMC'98)*, Yokohama, Japan, Dec. 1998, pp. 509–512.
51. K.-P. Ma, J. Kim, F.-R. Yang, Y. Qian, and T. Itoh, "Leakage suppression in stripline circuits using a 2-D photonic band-gap lattice," in *1999 IEEE MTT-S International Microwave Symposium*, Anaheim, CA, June 1999, pp. 73–76.

9

ACTIVE AND HIGH-PERFORMANCE ANTENNAS

WILLIAM R. DEAL
Malibu Networks, Inc., RF Design Center, Calabasas, CA

VESNA RADISIC
HRL Laboratories, LLC, Malibu, CA

YONGXI QIAN AND TATSUO ITOH
Department of Electrical Engineering, University of California—Los Angeles

9.1 THE ROLE OF THE ANTENNA IN NEXT-GENERATION WIRELESS SYSTEMS

Of all the radio frequency (RF) components in a microwave/wireless system, it is possible that the antenna receives the least amount of attention when next-generation wireless systems are discussed. This is probably based on a number of oversights. First, system designers often view antenna engineering as a "static" field, with little room for change or innovation. Another misconception, that the most direct path of achieving the goals for next-generation wireless products will be achieved by leveraging gains in device and processing technologies or signal processing techniques, overlooks the importance of antenna performance in a typical system. The efficiencies of the antennas in current-generation wireless handsets are often as low as 50%, with a large percentage of the radiated power often absorbed by the head and hand of the user. Therefore, it seems clear that substantial gains in system efficiency and sensitivity can be achieved by improving the performance of the antenna

---

RF Technologies for Low Power Wireless Communications
Edited by Tatsuo Itoh, George Haddad, and James Harvey
ISBN 0-471-38267-1 Copyright © 2001 by John Wiley & Sons, Inc.

platform. In this case, an integral part in the strategy to develop the next-generation low-power/low-noise products must be the development of high-performance antennas, as well as new topologies that maximize their potential benefits.

Requirements of next-generation systems will play a crucial role in the type of antenna used in these systems. Frequency congestion and the bandwidth requirements of upcoming systems such as data on demand, as well as digital video transmission, are rapidly pushing operating frequencies into the microwave and millimeter-wave regions. At these frequencies, the performances of "standard" wireless antennas such as the dipole and monopole antennas are quite poor, causing planar antennas to become the antenna of choice. These antennas have a number of advantages. First, they can be manufactured at a much lower cost than waveguide-based antenna technology and are considerably more compact and lightweight. This is essential for the many commercial applications in which planar antennas are increasingly used, such as base station or handset antennas. Second, their planar nature makes them ideal for large arrays and simplifies the integration of additional electronics, such as amplifiers and phase shifters, which are essential for electronic warfare, radar, satellite communications, or millimeter-wave imaging. Their planar nature also allows them to be applied to applications where size and shape are crucial, such as conformal printed antennas on an airplane fuselage. This myriad of applications has led to the development of a wide variety of high-performance planar antenna classes.

Another driving factor that makes planar integrated antennas so desirable is the ease of integration of these antennas with microwave or millimeter-wave circuit components. The natural extension of this is the active integrated antenna (AIA), a particular class of antenna where the antenna and active platform are directly integrated together. This results in a combined module with many unique characteristics, such as compactness, low cost, low profile, minimum power consumption, and a high degree of multiple functionality. In the decades since the approach was first introduced, the concept has been applied successfully to a wide variety of problems. A partial list of these applications includes high-efficiency transmitters and low-noise receivers, various up-converters and down-converters, as well as AIA oscillators and oscillator arrays for quasi-optical power combining. The approach is particularly successful when high performance is required or where space is at a premium, such as in active arrays.

In Section 9.2, attention is given to categorizing and detailing many of the newer planar antennas that have been developed over the last few years. Basic characteristics of many of these antennas are discussed and experimental data are presented wherever possible. Section 9.2.4 discusses a promising new twist on an old design: the quasi-Yagi antenna. This antenna features extremely broadband performance with a compact form-factor. Several simple arrays are presented that demonstrate the viability of using this antenna as an array element. In Section 9.3, active antennas are discussed from a designer's point of view. AIA circuits are categorized according to function, and design/measurement issues are discussed, such as the nonreciprocal nature of many AIA architectures. Additionally, several design examples of AIA circuits are presented that demonstrate the state of the art in this field. Since the rapidly maturing fields of planar antennas and AIA systems are a remarkably broad area, the examples have been chosen based on performance and uniqueness and to support basic concepts and do not represent a complete overview of the field. The interested reader should find the many review papers and individual papers included in the reference list a good source of additional information about specific topics.

9.2 HIGH-PERFORMANCE ANTENNAS

One of the advantages of planar antennas is the ease of integration of these antennas with standard planar transmission lines. For this reason, planar integrated antennas must be compatible with these technologies, which, at microwave and millimeter-wave frequencies, are typically microstrip or coplanar waveguide (CPW) based. These transmission lines have several advantages, including ability to integrate three-terminal devices, mechanical and heat-sinking capabilities due to metallic ground planes, as well as simplified packaging issues. Therefore, it is essential that these types of transmission lines can directly or indirectly feed the planar integrated antennas. Examples of direct feeding include the patch antenna and slot antenna, which are easily integrated with a microstrip or CPW, respectively. Indirect feeding can include transitions or various forms of electromagnetic (EM) coupling. The method of feeding is critical and can affect antenna cross-polarization, patterns and bandwidth, as well as possible array architectures.

However, the grounded dielectric slabs on which microstrip and CPW compatible antennas are fabricated support TM_0 surface waves and can propagate energy away from the antenna in the form of surface waves, thereby lowering efficiency. While the losses are small at lower frequencies, this can be a major problem at microwave and millimeter-wave frequencies, where many new applications are targeting planar antennas. The thickness of the substrate, permittivity, and frequency of operation determine the amount of surface wave losses. Several methods have been developed to reduce this, as will be briefly discussed later.

Different classes of planar antennas are capable of a broad variety of radiation characteristics. The most common classes, patch and resonant slot antennas, demonstrate broad, low-gain patterns, making them excellent for use in multi-element beam-forming arrays or for use in portable handsets. Additionally, some of these antennas can easily be modified for dual-linear or circular polarization. More sophisticated classes typically demonstrate higher gains and some are capable of frequency scanning.

9.2.1 Patch Antennas

The patch antenna has a broadside radiation pattern that allows it to be integrated into two-dimensional (2D) arrays. It has a number of additional benefits, including low profile, low cost, conformability, and ease of manufacture. Additionally, various feeding schemes can be used to achieve linear or circular polarization. The patch is readily integrated with both microstrip or coaxial feeding, and more advanced

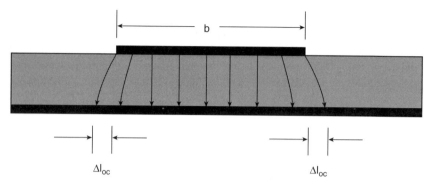

Figure 9.1. Cross section of patch antenna showing fringing fields at the edges.

schemes can be used to adapt the patch antenna to virtually any type of transmission line. Microstrip fed patches have very narrow bandwidths, almost invariably less than 5%. Other feed mechanisms have been used to increase bandwidth, including proximity coupling and aperture coupling, both of which require multilayer fabrication. A review of this technology is given by Pozar [1]. In general, the patch antenna is one of the most versatile planar antennas.

A cross section of the patch antenna is shown in Figure 9.1. A simple and intuitive technique for modeling this antenna is the transmission line model. In this model, it is assumed that the patch antenna consists of perfect magnetic conductor (PMC) walls on the sides of the patch antenna, giving rise to standing-wave-type modes inside the patch antenna cavity. The total length of the cavity is the length of the patch antenna, and an effective length at each edge due to the microstrip open-end effect. The length of the antenna will determine the resonant frequency of the patch. To first order, these occur at frequencies where the length of the antenna is a multiple of one-half of a cavity wavelength. If active circuitry, which may generate harmonics, is integrated with the patch antenna, harmonic radiation and co-site interference may occur. Several approaches for eliminating this problem are discussed in Chapter 6, Section 6.7.

An alternative technique for increasing the input return loss bandwidth of the antenna is to make the antenna substrate electrically thicker, effectively lowering the Q factor of the antenna cavity for increased bandwidth. However, high levels of TM_0 surface waves can result and therefore reduce the radiation efficiency as well as degrade the radiation pattern if the surface wave generates radiation (which can occur at the edge of finite ground antennas). The problem of an electrically thick substrate is also a common one for high-frequency antennas on high permittivity substrates such as those commonly used in microwave monolithic integrated circuits (MMICs). This can result in high amounts of TM surface waves.

Several methods have been proposed to combat this problem. One method of reducing this is to micromachine closely spaced holes into the high dielectric constant substrate in order to lower the effective permittivity in the region of the antenna and therefore reduce the electrical thickness [2,3]. Although this will increase radiation efficiency, some bandwidth may be sacrificed due to the lowered effective permittivity. More recently, the photonic bandgap (PBG) concept has been used for this purpose. A PBG crystal is a periodic structure that forbids the propagation of all electromagnetic waves with a particular frequency span. Photonic engineers commonly refer to this frequency span as the "bandgap." It is analogous to the stopband in a conventional microwave filter. In general, a periodic array of perturbations is used to suppress the undesired surface wave mode. More about this topic can be found in Chapter 8.

One particular PBG structure that has demonstrated success at suppressing surface wave effects when integrated with a patch antenna is shown in Figure 9.2 [4,5]. This 2D square lattice consists of square capacitive pads on the top plane connected to the ground plane by inductive vias. Precise geometry of the unit cell consists of a square pad of 88 mils by 88 mils (1 mil = $\frac{1}{1000}$ in.) connected to the ground plane by a via with a diameter of 8 mils at the center of the pad. The structure is on 25 mil thick Duroid with a permittivity of 10.2. Each pad is separated from its neighbors by an 8 mil gap.

The dispersion diagram shown in Figure 9.3 demonstrates the performance of the periodic structure. The three sections refer to different directions of wave propagation. For each section, the propagation vector β in the wave vector space is varied along the edges of the Brillouin zone, represented by the shaded triangle in the inset. The two dotted lines represent propagation in air and a homogenous

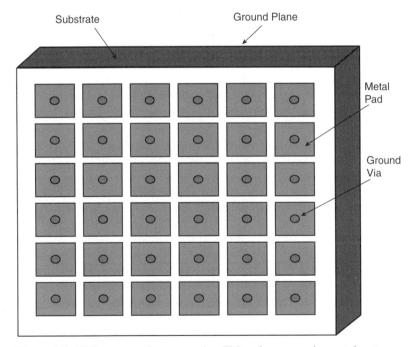

Figure 9.2. PBG structure for suppressing TM surface waves in a patch antenna.

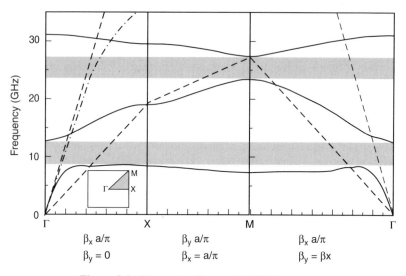

Figure 9.3. Dispersion diagram for PBG structure.

dielectric slab of permittivity 10.2. The dash-dot line on the left of the plot represents the TM_0 mode of the grounded dielectric slab without the PBG material. This is the substrate mode responsible for losses in a typical patch antenna. Finally, the three solid lines represent the propagation of the first three modes supported by the lattice. It is apparent that no modes will propagate in the lattice over two frequency spans (shaded gray), regardless of direction. Therefore, if a patch antenna operating in either of these frequency spans is incorporated into the structure, it is reasonable to expect reduced surface wave effects.

A photograph of a Ku-band patch antenna surrounded by this lattice is shown in Figure 9.4. The inset fed patch operates at approximately 14 GHz to take advantage of the surface wave suppression effect. The measured input return loss shown in Figure 9.5 demonstrates one additional benefit of the structure. The lattice, acting as a parasitic, has significantly increased the bandwidth of the structure from 1.6% to 5.4%, an unexpected side benefit.

Radiation properties of the antenna surrounded by the lattice structure were also measured. It was found that peak measured power was about 1.7 dB higher than a reference patch without the lattice structure. This measurement was taken at 14.15 GHz where both antennas had identical return loss (−9.4 dB). This indicates that the gain of the antenna has been increased. The measured E and H-plane cuts are shown in Figures 9.6(a) and 9.6(b), respectively, for both the antenna surrounded by the lattice structures and the reference antenna. Note that all copolarization patterns have been normalized to 0 dB. In general, reduced backside radiation is observed, indicating that edge radiation caused by surface wave effects has been reduced. The E-plane cut shows the largest improvement. Note that surface waves for a patch antenna are predominantly in this direction.

HIGH-PERFORMANCE ANTENNAS

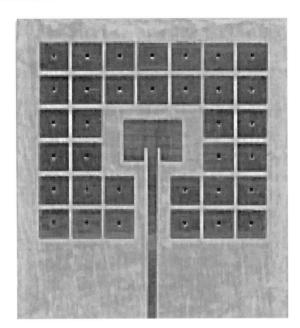

Figure 9.4. Photograph of the Ku-band patch antenna integrated with PBG for surface wave suppression.

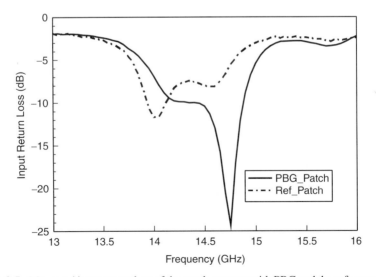

Figure 9.5. Measured input return loss of the patch antenna with PBG and the reference patch antenna.

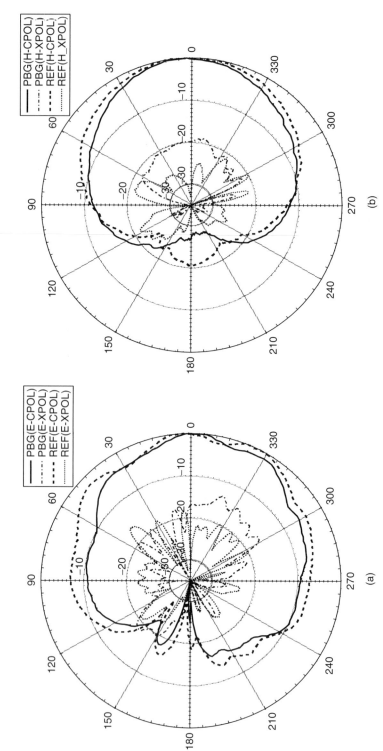

Figure 9.6. Measured radiation characteristics of the PBG patch and reference including the (a) E-plane and (b) H-plane.

Currently, research of these structures is ongoing. This type of technique, which can be implemented using standard circuit board processing, can be achieved at much lower cost than other more conventional techniques, such as the cavity-backed patch configuration.

9.2.2 Slot Antennas

The slot antenna, consisting of a narrow slit in a ground plane, is a very versatile antenna. With modification, it is amenable to waveguide, coplanar waveguide (CPW), coaxial, slotline, or microstrip feeding schemes and has found use in all aspects of wireless and radar applications. The slot antenna has been investigated since at least the 1940s [6]. Planar microstrip-fed slot antennas have been reported as early as 1972 [7].

The resonant half-wavelength slot antenna is desirable because of its compact size; but it has a large input impedance, typically larger than 300 Ω, which makes it unattractive to match to. This can be circumvented by using an offset microstrip feed or the folded-slot antenna, which stems directly from the folded dipole by Booker's relation. In this case, the slot is *folded* in upon itself. The overall length of the antenna remains approximately a half-wavelength, but increasing the number of folds reduces the radiation resistance.

The CPW version of the folded-slot has been investigated extensively [8,9]. This antenna requires no input matching, which makes it an inexpensive and compact candidate for direct integration with microwave circuits. The broad radiation pattern also makes this antenna an excellent candidate for wireless communications systems, which are currently pushing into the microwave regime.

The basic structure for the microstrip fed folded-slot antenna is shown in Figure 9.7(a) along with a finite-difference time-domain (FDTD) simulation of the aperture field distribution in Figure 9.7(b). The folded-slot is etched in the ground plane of the substrate. One of the inner metalizations of the slot is connected to the microstrip conductor on the top plane by a shorting. A 50 Ω input impedance is easily obtained for a two-fold slot on a RT/Duroid with a permittivity of 2.33 and substrate thickness of 31 mils.

The FDTD simulation allows easy visualization about the operation of the slot. From this, we can see that the feeding structure operates effectively and that the fields are directed across the slot, as desired. Perpendicular fields exist at the edges of the slot, which contribute to the cross-polarization of the radiated fields. Simulations show that reducing the area at the edges of the slot reduces the cross-polarization, but sacrifices bandwidth [10].

The bandwidth (BW) for this structure is measured to be from approximately 1.5 to 2.9 GHz (BW = 61% for $S_{11} < -10$ dB). This is shown along with the FDTD simulation in Figure 9.8. Finally, the E- and H-plane patterns of the antenna are measured at the center frequency of 2.2 GHz. The patterns are comparable to the standard slot antenna and are shown in Figure 9.9. One complication with this sort of antenna is its bidirectional pattern, which may not be suitable for most applications. This can be overcome by using a reflector at the expense of additional cost and bulk.

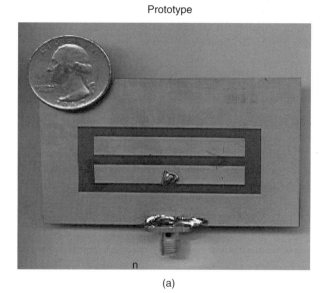

Figure 9.7. Characteristics of a microstrip folded-slot antenna including (a) a photograph and (b) the simulated aperture field distribution.

9.2.3 Planar Traveling Wave Antennas

In the previous sections, we discussed several examples of resonant-type (or standing wave) antennas including the patch and slot antennas. These types of antennas typically have narrow bandwidth and relatively simple, modest gain radiation patterns. Another class of antennas, traveling wave antennas or nonresonant antennas, is comprised of structures that radiate as the wave propagates. These structures are electrically large when compared to resonant-type structures and typically

HIGH-PERFORMANCE ANTENNAS

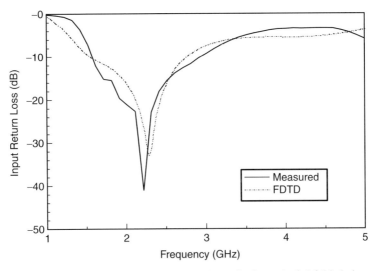

Figure 9.8. Measured and simulated input return loss of microstrip fed folded-slot antenna.

demonstrate higher gain and may demonstrate broad-bandwidth performance. Traveling wave antennas are typically divided into two classes: leaky wave antennas and surface wave antennas. The leakywave antenna will continuously radiate as the wave propagates. The surface-wave antenna will radiate at points in the structure

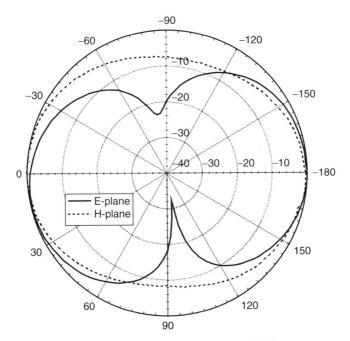

Figure 9.9. Measured folded-slot antenna E- and H-plane patterns.

where discontinuities disrupt the propagating surface wave and is typically a slow-wave structure. The ability to engineer a broad variety of planar transmission line structures and surface wave structures means that we should be able to develop excellent planar versions of these antennas. In the following two sections we give two examples of traveling wave antennas. However, we believe that work in this area is far from mature and many possibilities remain to be developed.

9.2.3.1 Microstrip Leaky Wave Antenna This type of leaky wave antenna utilizes a higher order radiative microstrip mode to form a leaky wave antenna. The radiation mode has odd symmetry and peaks at the edges of the microstrip cross-section with a null at the center. Therefore, as the signal propagates along the microstrip transmission line, energy will continuously radiate along the edges, resulting in a main radiation beam. If the end of the structure is not terminated, the remainder of the energy will be reflected from the open end and a second beam will result. For long antennas, the amount of power reflected from the open end will be small, resulting in a smaller second beam. Additionally, the direction of the main beam occurs as a function of frequency, opening the possibility that the antenna can be frequency scanned. However, the microstrip leaky wave will only radiate efficiently over a fixed bandwidth, which will in practice limit maximum scanning. Additionally, feeding and length of the antenna will also have some effect on the pattern. An aperture-coupled microstrip leaky wave antenna with a broadside-directed beam has been reported [11].

Feeding such an antenna is an important concern, and various schemes have been reported [12,13]. It is essential for the higher order leaky wave mode to be excited with minimal excitation of the fundamental (even) mode. A 5λ long prototype microstrip leakywave antenna shown in Figure 9.10 has recently been applied to the problem. A uniplanar microstrip-to-CPS transition is used to convert the quasi-TEM mode from the 50 Ω feedline into an odd mode that feeds the two edges of the leaky microstrip. The transition is inherently broadband, as demonstrated by Qian and Itoh [14]. Additionally, a periodic structure of transverse slots suppresses excitation of the fundamental even mode. A substantial amount of asymmetry in the vertical field was observed, in the cross section of the electric field profile without the mode suppressor, indicating that both the fundamental (even) and the leaky (odd) modes

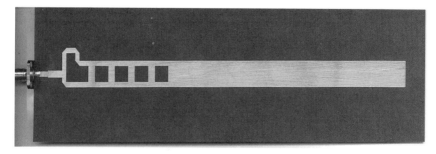

Figure 9.10. Photograph of prototype leaky wave antenna.

HIGH-PERFORMANCE ANTENNAS 317

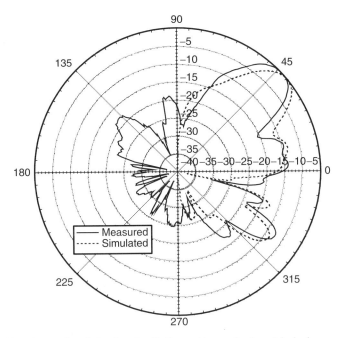

Figure 9.11. Measured and simulated radiation patterns of microstrip leaky wave antenna.

are present. After inclusion of the mode suppressor, the mode purity is significantly improved. Simulations reveal that the field null has been shifted to the center of the microstrip, indicating a pure odd mode. The antenna demonstrated a measured bandwidth of 20.2% [15]. Simulated and measured radiation patterns of the structure are shown in Figure 9.11. Note that both simulated and measured patterns indicate a substantial (−9 dB) lobe at −35°. This is due to substantial power being reflected from the open end of the 5λ long antenna. Simulation of a 10λ long version indicates that the lobe will be reduced to −13 dB.

9.2.3.2 Tapered Slot Antenna In the last decade, quite a bit of work has been conducted on tapered slot antennas (TSAs) on dielectric substrates. Extensive reviews can be found in the literature [16,17]. These antennas are completely planar, have an endfire pattern, and are capable of obtaining high directivity and/or bandwidth. Proposed applications include millimeter-wave imaging, power combining, and use as an active integrated antenna element. The TSA is etched into the metalization on one side of a dielectric substrate. The dielectric constant is usually low. Three popular configurations are shown in Figure 9.12 and include the Vivaldi (exponential taper), the linear taper (LTSA), and the constant width slot antenna (CWSA). Radiation of a particular frequency will occur where the slot is a certain width. Slot width should reach at least one-half wavelength for efficient radiation to occur. Therefore, maximum and minimum widths roughly determine the bandwidth of the structure.

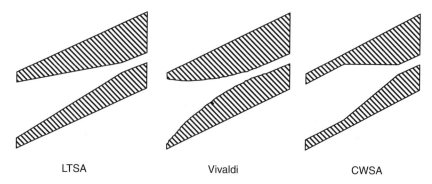

Figure 9.12. Various tapered slot antenna geometries, including linear taper, exponential taper (Vivaldi) and constant width slot antenna.

There are several important design concerns with this type of antenna. TSAs are usually built on thin, low-permittivity substrates if they are to achieve good radiation patterns and maintain good radiation efficiency, which will be reduced by TM_0 surface wave losses. As with the patch antenna, micromachining has been used to reduce the electrical thickness of millimeter-wave TSAs by reducing the effective permittivity [3]. This also allows a physically thicker substrate, which is essential for mechanical support for this kind of structure at millimeter wavelengths.

Additionally, when used as an integrated antenna, a transition from the transmission line of choice (microstrip or CPW typically) must be used. The bandwidth of the transition may limit the bandwidth of the structure. A CPW fed TSA slot with exponential taper is shown in Figure 9.13. The antenna is fabricated on a 25 mil substrate with 10.2 permittivity. The broadband CPW–slotline transition allows broadband response; measured bandwidth is greater than 70% centered at 13 GHz [18], as shown in Figure 9.14.

9.2.4 Printed Quasi-Yagi

Printed wire antennas are planar adaptations of wire antennas such as the dipole, loop, or spiral antennas. Due to a number of reasons, only limited success has been achieved at developing working versions of truly planar wire antennas. This is in part due to the popularity of microstrip and coplanar waveguide transmission lines at these frequencies, which are difficult to integrate with wire-based antennas. For instance, a printed microstrip dipole will have very small radiation resistance due to the shorting effect of the microstrip ground plane and consequently will have low efficiency when realistic estimates of losses are taken into consideration. This can be overcome by using an electrically thick substrate, at the cost of increased losses due to substrate waves and increased weight and cost.

Alternatively, the microstrip ground plane can be modified to accommodate the printed wire antenna. Recently, a new type of microstrip compatible planar endfire antenna based on the well-known Yagi–Uda antenna has been mentioned in the

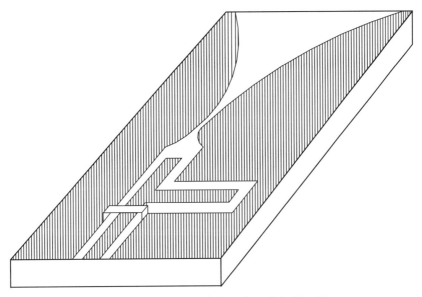

Figure 9.13. A uniplanar CPW fed version of the Vivaldi antenna.

literature [19–21]. The classic Yagi–Uda antenna consists of a driven element, a reflector, and one or more directors. The reflector and directors act as parasitic elements. The antenna has an endfire pattern and can demonstrate high gain, but typically has narrow bandwidth.

The new architecture and an X-band prototype quasi-Yagi antenna are shown in Figures 9.15 and 9.16. The antenna uses the truncated microstrip ground plane as the

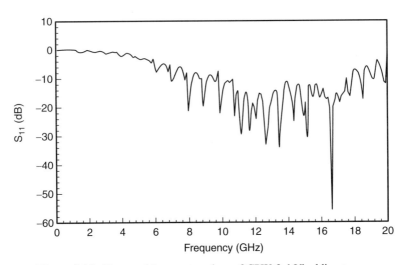

Figure 9.14. Measured input return loss of CPW fed Vivaldi antenna.

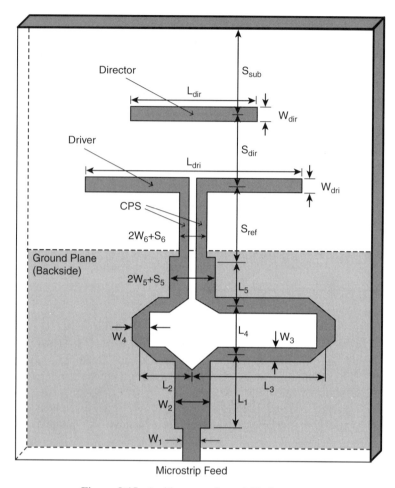

Figure 9.15. Architecture of quasi-Yagi antenna.

reflector and uses a microstrip–CPS transition as balanced feed on a single layer, high dielectric substrate (Duroid, $\varepsilon_r = 10.2$) [14]. The odd excitation of the CPS-driven dipole is clearly evident in the FDTD simulation of the current distribution shown in Figure 9.17. The X-band antenna uses 0.635 mm thick Duroid.

One of the most unique and effective features of this antenna is the use of the truncated ground plane as the reflector element. The driven printed dipole is used to generate a TE_0 surface wave with very little undesired TM_0 content, which can contribute to cross-polarization. The truncated ground plane acts as an ideal reflector for this TE_0 mode, which is completely cu toff in the grounded dielectric slab region. In fact, the cut off frequency for this mode is 39 GHz, much higher than the operating frequency of the antenna. The dipole elements of the quasi-Yagi antenna are strongly coupled by the TE_0 surface wave, which has the same polarization and same direction as the dipole radiation fields.

Figure 9.16. X-band prototype of quasi-Yagi antenna.

As with the classic Yagi–Uda antenna, proper design requires careful optimization of the driver, director, and reflector parameters, which include element spacing, length, and width. The antenna's dimensions are optimized by an in-house FDTD code to achieve broadband performance. The antenna is realized on Duroid with $\varepsilon_r = 10.2$. The antenna's dimensions are (unit: mm): $W_1 = W_3 = W_4 = W_5 = W_{dri} = W_{dir} = 0.6$, $W_2 = 1.2$, $W_6 = S_5 = S_6 = 0.3$, $L_1 = 3.3$, $L_2 = L_5 = 1.5$, $L_3 = 4.8$, $L_4 = 1.8$, $S_{ref} = 3.9$, $S_{dir} = 3.0$, $S_{sub} = 1.5$, $L_{dri} = 8.7$ and $L_{dir} = 3.3$. In this optimized quasi-Yagi design, the length of the antenna's director element is shorter than the conventional Yagi–Uda antenna design and contributes to the broadband characteristics of the antenna. The total area of the substrate is approximately $\lambda_0/2$ by $\lambda_0/2$ at the center frequency.

The antenna radiates an endfire beam, with a front-to-back ratio 15 dB and cross-polarization level below -12 dB across the entire frequency band. A very low mutual coupling level below -22 dB has been measured for a two-element array with $\lambda_0/2$ separation. Figure 9.18 shows plots for both FDTD simulation and measurement results for the input return loss. The simulated and measured bandwidths (VSWR<2) are 43% and 48%, respectively, which covers the entire X-band. The antenna is at least two orders smaller in volume than a standard horn antenna for the same frequency coverage. Radiation patterns of the endfire antenna are shown in Figure 9.19.

We have also found that the quasi-Yagi antenna can be scaled linearly to any frequency band of interest while retaining its wideband characteristics. In fact, a C-band prototype, which has been simulated and fabricated on 1.27 mm thick

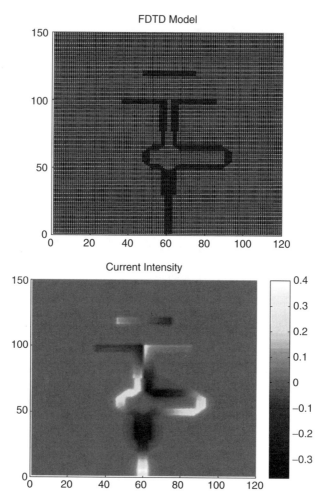

Figure 9.17. Simulated current distribution of quasi-Yagi antenna. Note successful odd-mode excitation of the driver dipole by the broadband microstrip–CPS transition.

Duroid ($\varepsilon_r = 10.2$), measured 50% frequency bandwidth (4.17–6.94 GHz). FDTD simulation for a millimeter-wave version indicates that a single quasi-Yagi antenna will function from 41.6 to 70.1 GHz (51% bandwidth), which covers part of the Q-band and most of the V-band.

This antenna has several additional advantages over the tapered slot, including narrower width, direct microstrip feeding, and the ability to design on a high-permittivity substrate. Additionally, radiation efficiency may be improved because of the truncated ground plane. This will eliminate losses due to the TM_0 mode in the antenna region that may be a significant loss mechanism with the TSA.

The quasi-Yagi should prove to be an excellent array element and should find application in any architecture that has been proposed for tapered slot antennas such

HIGH-PERFORMANCE ANTENNAS

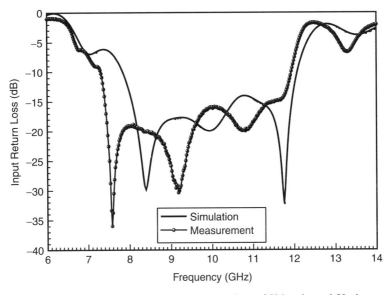

Figure 9.18. Measured and simulated input return loss of X-band quasi-Yagi antenna.

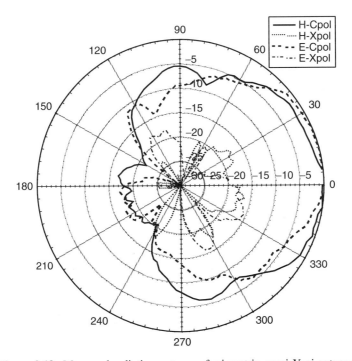

Figure 9.19. Measured radiation patterns of microstrip quasi-Yagi antenna.

as imaging arrays and power combining. An economic method of implementing full-scale 2D arrays based on this antenna is a "card" based approach, where individual 1D cards are stacked to form a fully functional 2D array for radar, communications, and quasi-optical power combining applications. This is conceptually shown in Figure 9.20. The 1D design of each card allows inclusion of phaseshifters and amplifiers with minimal difficulties in routing and heat sinking, making this scheme extremely desirable. The low mutual coupling between adjacent quasi-Yagi elements and the narrow physical width of the antenna gives great flexibility in array spacing not available with other endfire antennas. Several simple 1D cards have recently demonstrated the viability of this approach [22,23].

A simple equal amplitude eight-element linear array is used to demonstrate the viability of the quasi-Yagi as an array antenna. Three arrays are built, the first with the main beam at endfire, and the second with microstrip delay lines so that the main beam is tilted to 12°. The third array incorporates amplifiers at each antenna feed to achieve higher output and realized gain. Each eight-element linear array is fabricated on a 25 mil substrate with permittivity of 10.2 and utilizes the X-band quasi-Yagi antenna as the radiating element. Photographs of the active and tilted beam array are shown in Figures 9.21(a) and 9.21(b), respectively. The amplifiers of the active array are clearly visible and are integrated directly at the antenna feeds for maximized power combining efficiency. For simplicity, matched gain blocks are used for the amplifiers.

Measured S-parameters of the passive and active arrays are shown in Figures 9.22(a) and 9.22(b), respectively. In the case of both the broadside and tilted beam passive arrays, the bandwidth (VSWR <2.0) is approximately 50%. The bandwidth

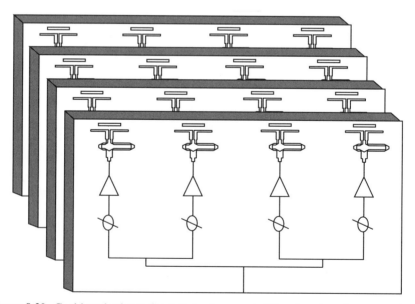

Figure 9.20. Card-based scheme for realizing large-scale 2D active phased array topology.

HIGH-PERFORMANCE ANTENNAS

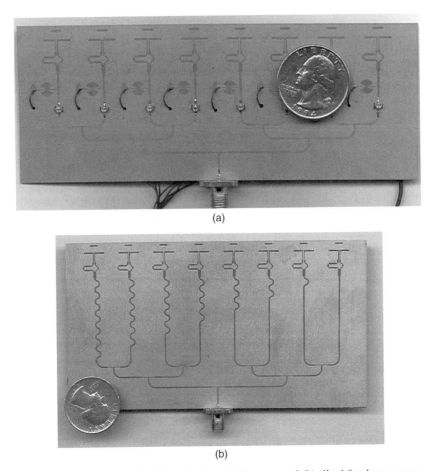

Figure 9.21. Photograph of (a) active broadside array and (b) tilted fan-beam array.

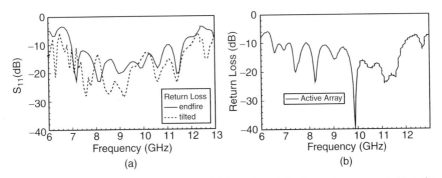

Figure 9.22. Measured input return loss of (a) passive eight-element arrays and (b) active eight-element array.

of the active array is even more impressive, a measured 60%. This is an effect of the broadband internally matched amplifiers. However, this increased VSWR bandwidth probably does not translate into increased usable bandwidth in terms of radiation characteristics of the array. The radiation characteristics of the quasi-Yagi element are optimized for in band performance and tend to deteriorate at the outer edges of the VSWR bandwidth. Therefore, the radiation properties in the region of increased VSWR bandwidth are not optimal.

The radiation patterns of each antenna have been measured across the operating bandwidth of the quasi-Yagi element. Radiation patterns for the endfire array at 9 GHz are shown in Figure 9.23(a). Note that only the radiation patterns for the passive array are shown. The normalized patterns for the active broadside array were virtually identical. The front-to-back ratio of the endfire array is better than 20 dB. Cross-polarization in the main beam is better than -15 dB. The E-plane of the tilted beam antenna is shown in Figure 9.23(b) at 8, 10, and 11.7 GHz. The 3 dB beamwidth ranges from 10 to 17° in this frequency range. A well-defined pattern dominated by the array factor is clearly observable.

Finally, the efficiency of the array has been estimated from gain (10–12 dB in the operating bandwidth) and directivity measurements of the endfire array. From this, it is estimated that the efficiency of the passive array is greater than 65% and the efficiency of the single element is greater than 85% after deembedding the losses of the feed network. Losses of the feed network were determined by fabricating a back-to-back corporate feed identical to the one used in the array. Half the measured losses for the back-to-back feed are used in the efficiency calculation.

9.3 ACTIVE INTEGRATED ANTENNAS

The active integrated antenna (AIA) approach combines the antenna and active platforms into one highly integrated system. AIA systems feature several unique features [24,25]. The inclusion of active, nonreciprocal devices means that the AIA system is typically nonreciprocal, unlike passive antennas. Therefore, AIAs often function as either transmitting or receiving class antennas. Several classes, notably the recently developed AIA retrodirective array, defy classification and function simultaneously as a transmit and a receive antenna. While this may seem an obvious or trivial fact, it does have some important ramifications, most notably in AIA architectures and measurement methodology.

In this section, we first discuss the design methodology and measurement issues for AIA systems. Since AIA measurements require the merging of active and antenna measurement techniques, detail is given to this issue. Two examples designed to highlight some of the issues involved are given. Next, a review of several new AIA architectures is given. Basic features of the design are illustrated as well as suggested applications and measured performance of these systems.

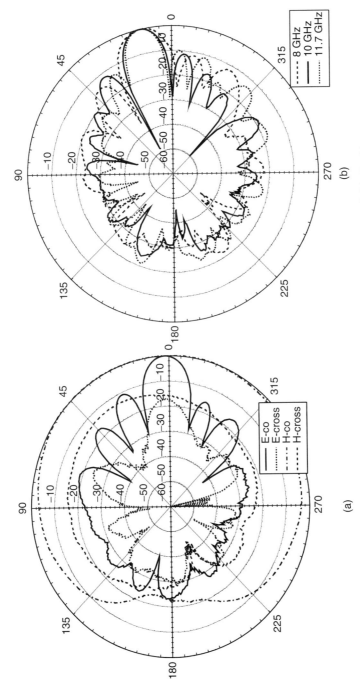

Figure 9.23. Measured radiation pattern of (a) eight-element broadside array and (b) tilted beam array.

9.3.1 Design Methodology

With the current abilities of microwave computer-aided design (CAD) tools and EM simulators, single-pass design success is feasible when designing highly integrated AIA modules. This is highly desirable for more than the obvious reason that it speeds up the design and prototyping process. Another less obvious reason is the difficulty in performing manual tuning of AIA modules. Since the antenna and active circuitry are tightly integrated together, the designer must test the module as a whole in an anechoic chamber, rather than testing the active circuitry separately from the antenna. In general, antenna measurements can be time consuming. If an entire set of measurements must be repeated as the active circuitry is "tweaked," this can be a prohibitive task. Therefore, it is essential that the design be as close to first-pass success as practically feasible.

While a number of successes have been reported in the literature on global simulation of microwave circuits [26,27] (note that global simulations combine both active circuitry and full-wave analysis in one concurrent simulation), this technique is still prohibitively slow on today's personal computers. Ultimately, it would be highly desirable that the microwave/millimeter-wave designer could use one global simulation CAD tool that would intrinsically treat such EM phenomena as radiation and coupling, as well as nonlinear effects such as gain and mixing. It would be necessary that this task could be performed in a short enough time frame that circuit optimization would be feasible.

Currently, the most effective technique for AIA design is to first optimize the antenna structure for the traits deemed beneficial for the total module (radiation properties, input impedance). Then, either measured or simulated S-parameter measurements of the optimized antenna structure are incorporated into a commercial microwave CAD tool that can accurately predict nonlinear effects, such as Agilent's Series IV. At this point, the designer can optimize performance with the same efficiency as achieved by a more conventional topology. This process may require additional iterations of antenna design to achieve required specifications for the module. It should be noted that this technique is not essentially different from that used by millimeter-wave designers to verify their designs. The biggest difference is that millimeter-wave designers are typically verifying that EM effects have not irretrievably compromised their designs, where the AIA designer relies on EM effects for the fundamental performance of the circuit. Note that for many AIA applications, accurate S-parameters are required over a broad bandwidth. In this case, time-domain simulation techniques are preferred if EM simulation is used, such as finite-difference time-domain (FDTD), which can accurately and efficiently predict the entire frequency response. A thru-return-line (TRL) calibration may be necessary if experimental data are used. This calibrates out connector and feedline effects.

9.3.2 Measurement Issues

To perform AIA measurements, all of the usual equipment for testing active topologies must be incorporated into an anechoic chamber. Therefore, testing an

ACTIVE INTEGRATED ANTENNAS

AIA module can be quite a bit more tedious than testing either active circuitry or antennas. However, the basic technique is straightforward and consists of the logical merging of the typical active measurement setup with standard antenna measurement techniques. To illustrate these issues, we will demonstrate several techniques for performing calculations on AIA power amplifier modules in terms of directly measurable quantities. In the first example, power-added efficiency (PAE) is calculated using measured power and radiation patterns. The second example uses a second passive version of the AIA module to calibrate out some of the potential sources for error that were present in Example 9.1. The benefits and limitations of each technique will also be discussed in the context of the examples. The fundamental concepts can easily be modified to other types of AIA modules, such as AIA oscillators, mixers, and receivers. For calculations, one of the most useful relations is the well-known Friis transmission formula:

$$P_{rec} = P_{trans} \frac{G_{trans} G_{rec} \lambda_2}{(4\pi R)^2}. \tag{9.1}$$

This form assumes that the antennas are pointed and polarized for maximum response and are impedance matched.

A typical experimental setup for measuring a power amplifier AIA transmitter is shown in Figure 9.24. Measurement parameters are defined below. Note that the

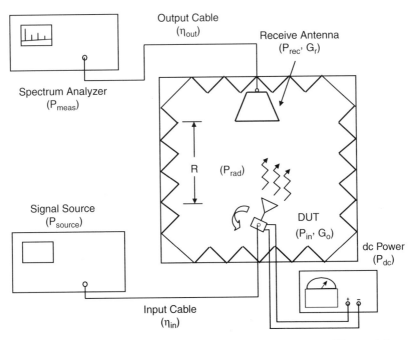

Figure 9.24. Experimental setup for measuring typical AIA amplifier module.

superscript refers to the measurement performed with either the AIA module (a) or a passive reference antenna (p) if both are used, as in Example 9.2.

P_{source} Power supplied by the source
η_{in} Efficiency of the input cable; determined by its losses
P_{in} Calibrated input power to the DUT
$P_{rad}^{a,p}$ Total radiated power for a given P_{in}
$P_{rec}^{a,p}$ Power received by receive antenna
η_{out} Efficiency of the output cable; determined by its losses
$P_{meas}^{a,p}$ Measured power for a given P_{in}
Γ_a Reflection coefficient of AIA module; assumed to be zero
Γ_a Reflection coefficient of passive standard
G_0 Gain of passive standard (same as the *antenna* gain of AIA module)
G_r Gain of the receive antenna
G_p Power gain of the amplifier
R Distance between transmit and receive antennas

Since large-signal operation is of primary interest in power amplifier measurements, the network analyzer typically found in anechoic chambers is replaced with a source capable of powering the AIA and a spectrum analyzer. A dc power supply must also be included to power the module. In the following examples, there may be two devices under test (DUT)—the AIA module and a passive reference antenna. The passive reference antenna is identical to the antenna used in the AIA module with a 50 Ω feed.

Example 9.1. Given $P_{source}=20.0$ dBm, $P_{meas}^a=-10.0$ dBm, $\eta_{in}=\eta_{out}=-3.0$ dB, $P_{dc}=1.0$ W, $G_r=10$ dB, and $R=69$ cm, determine an expression for the amplifier PAE in terms of these parameters and determine the PAE of this particular AIA amplifier module. The operating frequency is 9.5 GHz. A quasi-Yagi antenna is being used as the radiating element. The measured -3 dB beamwidth for the E- and H-planes, Θ_E and Θ_H, of the antenna are $\Theta_E=150°$ and $\Theta_H=90°$. Assume the radiated pattern of the quasi-Yagi antenna contains only negligible power in the sidelobes. Additionally, the efficiency of the quasi-Yagi antenna has previously been measured to be 93% at the design frequency.

Solution: When free space is used as a load, the standard definition for PAE is given by:

$$\eta_{PAE} \equiv \frac{P_{rad}-P_{in}}{P_{dc}}. \tag{9.2}$$

This can be rewritten in terms of measurable parameters as

$$\eta_{PAE} \equiv \frac{P_{meas}}{P_{dc}} \frac{1}{G_0 G_r \eta_{out}} \left(\frac{4\pi R}{\lambda}\right)^2 - \frac{P_{source}}{P_{dc}} \eta_{in}. \tag{9.3}$$

ACTIVE INTEGRATED ANTENNAS

To use this equation, the antenna gain of the AIA module, G_0, must be known. The hint about the -3 dB beamwidth of the radiation pattern provides an approximate way of calculating the antenna gain. To first order, the gain of an antenna with a single main beam and negligible power in the sidelobes can often be approximated as [28]

$$G \cong \eta_{\text{rad}} \frac{41,253}{\Theta_E \Theta_H}. \tag{9.4}$$

With the given information, the approximate gain of the antenna is calculated to be 4.54 dB. After first converting all parameters into a linear scale, the PAE is calculated to be 47.9%.

Example 9.2. Given $P_{\text{source}} = 20.0$ dBm, $P_{\text{meas}}^{\text{a}} = -10.0$ dBm, $P_{\text{meas}}^{\text{p}} = -20.0$ dBm, $\eta_{\text{in}} = -3.0$ dB, and $P_{\text{dc}} = 1.0$ W, determine the amplifier PAE in terms of these parameters and determine the PAE of this particular AIA amplifier module. Assume that $P_{\text{meas}}^{\text{a}}$ and $P_{\text{meas}}^{\text{p}}$ were both measured with the DUT oriented for maximum received power and that both have the same normalized radiated patterns. Assume the input impedance of the passive reference is 30 Ω.

Solution: The PAE may be rewritten in terms of the given measurable parameters as

$$\eta_{\text{PAE}} \equiv \frac{\eta_{\text{in}} P_{\text{source}}}{P_{\text{dc}}} \left[\frac{P_{\text{meas}}^{\text{a}}}{P_{\text{meas}}^{\text{p}} (1 - |\Gamma_p|^2)} - 1 \right].$$

Note that we have used the standard definition of PAE and the Friis transmission equation to obtain this result. Additionally, for a fair comparison, mismatch loss of the passive antenna is compensated for. However, since the AIA is assumed to be a complete module, no mismatch loss is compensated. Of particular interest is the fact that the PAE is a function only of standard circuit parameters. No antenna parameters are explicitly required to perform the calculation other than consecutively measuring the AIA and a passive antenna gain reference. After converting all of the given parameters to a linear scale from a log scale, the PAE of the module is found to be 48.3%.

From Example 9.2 we see that by using a passive reference antenna and performing successive measurements, antenna gain is calibrated out of the equation. This is analogous to the gain substitution technique commonly used in antenna measurements to find the absolute gain of an antenna. However, in our case we are interested in the relative gain between the two measurements and no gain standard is required. The final result yields the proper PAE of the amplifier itself. Of primary importance to this technique is repeatability of the measurement, which will rely on the quality of the measurement setup and the ability to precisely position the reference antenna and AIA module so that their patterns are perfectly aligned during the two measurements. The simplest technique is to orient the main beam of each DUT directly toward the receive antenna. Then, each antenna is adjusted so that maximum received power is obtained. This technique also assumes the patterns of the AIA and passive antennas are identical, which is easily verified. Note that this might not be the case if a finite ground plane antenna is used at lower frequencies. In this case, radiation from the edges of the ground plane can perturb the measurements.

Finally, a comment should be made on the PAE calculated in each example, which was found to be 47.9% and 48.3%, respectively. The given parameters of the examples were chosen to be self-consistent. The values for Example 9.2 were calculated using measured gain data for the quasi-Yagi antenna, which was experimentally found to be 4.5 dB at 9.5 GHz. These values were then used as the basis of Example 9.1. However, Example 9.1 called for the calculation of antenna gain from measured efficiency and radiation parameters using the *approximate* expression given in Eq. (9.4). Readers will find that if they use the measured value of 4.5 dB for Example 9.1, they will find exactly the answer of Example 9.2. In either case, the two calculations agree quite well in this case, demonstrating the utility of using such approximations as Eq. (9.4).

9.3.3 AIA Examples

In this section, the value of the AIA approach is demonstrated by several examples from the literature. These examples are chosen to illustrate the potential that this approach brings to the design table. In the first example, the AIA approach is extended to the push–pull power amplifier (PA). It is demonstrated that dual-feed antennas can be used to achieve an extremely compact high-efficiency transmitter front-end. The next example highlights a 60 GHz AIA receiver module. A 60 GHz self-oscillating mixer (SOM) AIA realizes an extremely compact design with minimal component count. This type of architecture is intended to provide a cost-effective alternative to high-rate data transmission at millimeter-wave frequencies. An AIA noncontact ID card for application in keyless entry systems is discussed. This compact transponder provides a flexible and low-cost solution at microwave frequencies. Finally, the AIA retrodirective array is introduced.

9.3.3.1 Push–Pull AIA Power Amplifier Output power is another crucial factor related to cost in transmitter systems. Higher power commercial devices are often considerably more expensive than devices with only slightly lower output power. This is because these devices either approach the limit of current manufacturing techniques or require additional engineering to obtain a useful product, and are therefore available at a premium cost. A cost-effective solution to this problem is to use two lower power devices and combine the output power. One of the most common ways of doing this is by using the push–pull PA architecture. In this scheme, a hybrid is used to provide -3 dB of input signal to two Class B field effect transistors (FETs) with a 180° phase shift between the two inputs. The output at the drain of each FET consists of the fundamental component and a series of even harmonics. The fundamental components at the output of each device are antiphase with respect to each other while the even harmonics are in phase. Therefore, when the power is combined through a second 180° hybrid, the fundamental component will combine at the Σ port and the harmonics are dissipated through a dummy load at the Δ port. Disadvantages to this scheme are the insertion losses of the output hybrid and the harmonic power dissipated through the dummy load.

Recently, the AIA concept has been applied to this problem [29–32]. In order to reduce the output losses, the output power is combined through a dual-feed planar antenna, as shown in Figure 9.25. This limits the output losses to that of the antenna. For proper power combining, it is essential that the feeding of the antenna be done in such a way that signals add constructively. For the push–pull, the feeds of the antenna must be placed so that the radiation mode is antiphase. One particular example of this is the dual-feed patch antenna shown in Figure 9.26(a), along with its corresponding mode profile for push–pull excitation. From this, we see that the power from the two antiphase driven ends will power combine properly. Of additional interest are the measured S-parameters of the dual-feed antenna, shown in Figure 9.26(b). If the operating frequency is chosen at the first resonance, about 2.5 GHz, the second harmonic will be almost totally reflected. By properly adjusting the phase of the reflected harmonic power, this power can be converted to fundamental power for increased PAE. This process is known as harmonic tuning. More about AIA harmonic tuning can be found in Chapter 6, Section 6.7. A prototype AIA utilizing this antenna is shown in Figure 9.27. A measured PAE of 55% at 2.5 GHz was obtained, as shown in Figure 9.28. While significant, this figure could be improved by adding vias to short even harmonics, as with the patch with pins AIA

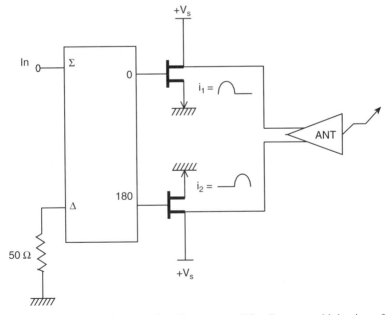

Figure 9.25. Architecture of AIA push–pull power amplifier. Power combining is performed using a dual-feed antenna.

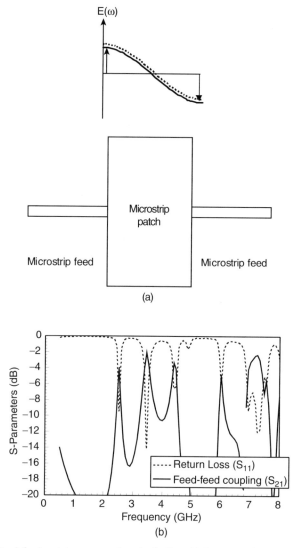

Figure 9.26. Dual-feed patch antenna data including (a) layout and radiation mode profile and (b) measured two-port S-parameters.

amplifier of Section 6.7 in Chapter 6. Since the push–pull has only even harmonics, this could be used to suppress all harmonics and increase the PAE through harmonic tuning.

A fully tuned slot-antenna-based push-pull PA based on this concept has also been demonstrated. The configuration of the dual-feed slot is shown in Figure 9.29(a) for proper power combining when antiphase excitation is applied. The structure uses $\lambda/4$ microstrip stubs for coupling to the slot. This scheme has the added benefit of

ACTIVE INTEGRATED ANTENNAS

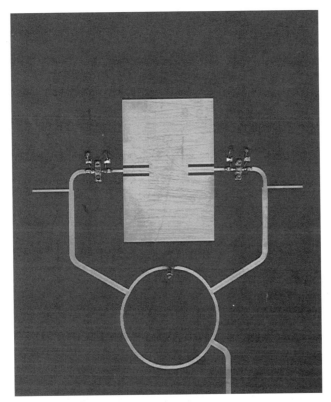

Figure 9.27. Photograph of patch-antenna based AIA push–pull power amplifier module.

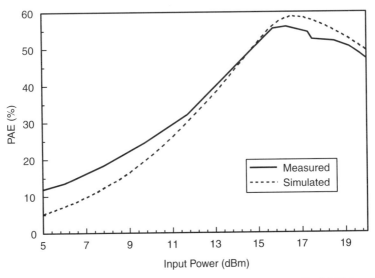

Figure 9.28. Measured PAE of patch-based push–pull. Maximum PAE of 56% is achieved.

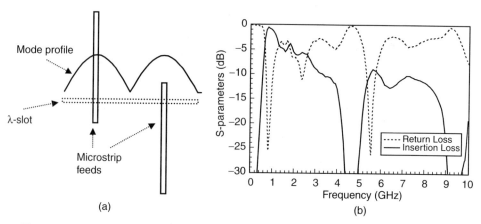

Figure 9.29. Dual-feed patch antenna data including (a) layout and radiation mode profile and (b) measured two-port S-parameters.

reflecting all even harmonics, as is seen from the input return loss shown in Figure 9.29(b). Therefore, this structure has a built-in mechanism for harmonic tuning. This amplifier reached a maximum PAE of 63% and yielded PAE better than 55% over an 8% bandwidth centered at 2.46 GHz. The PAE characteristic is shown in Figure 9.30(a). This demonstrates that integrated antenna front-ends are capable of reaching extremely high efficiencies with minimal component count and cost.

Also of interest is the harmonic suppression level of the AIA module. Using the antenna to reflect the power back to the device not only increases the PAE through harmonic tuning, but also lowers undesired harmonic radiation. This can be observed by measuring the radiation patterns at the fundamental and higher harmonics. Note that the proper wavelength for each frequency must be used in the Friis transmission formula to accurately calibrate the power in each harmonic. Shown in Figure 9.30(b) is the harmonic suppression level for the dual-feed slot push–pull AIA. All harmonics are at least 30 dB below peak fundamental power for all angles in the measured E-plane.

The module has also been tested for linearity. At maximum PAE of 63%, the amplifier fails the IS98 linearity specification. However, when backed off so that the ACPR is -42 dBc, the amplifier yields 48% PAE, indicating that linearity is not compromised by this approach.

9.3.3.2 60 GHz AIA SOM Receiver The first example featured an AIA transmitter, which benefits from increased efficiency/range by reducing the output losses of the transmitter front-end. Conversely, AIA receivers benefit from increased sensitivity by eliminating losses between the antenna and receiver electronics. In the present example, a 60 GHz AIA receiver module is discussed. Due to restrictions on available gain from active devices at millimeter-wave frequencies, these bands are particularly attractive for application of AIA systems.

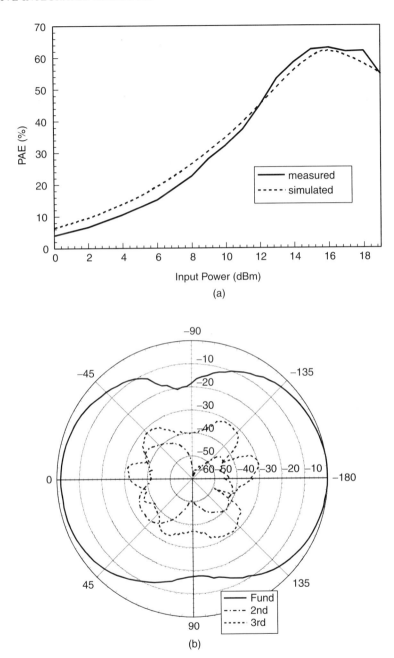

Figure 9.30. Measured data for slot-antenna based push–pull. In (a), a maximum PAE of 63% is achieved. In (b), harmonic suppression is demonstrated to be −30 dB below the fundamental in all planes. Note that the harmonics are calibrated to the fundamental power using the Friis transmission equation.

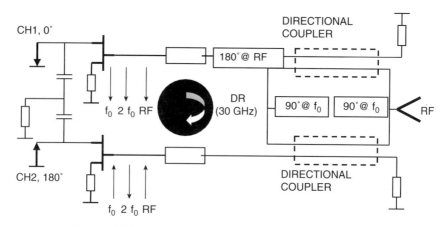

Figure 9.31. Architecture of 60 GHz second harmonic SOM.

The architecture of the AIA receiver front-end is shown in Figure 9.31. The structure uses a second harmonic self-oscillating mixer (SOM). A SOM, a mixer that generates its own local oscillator (LO), is attractive because of reduced component count. A second harmonic SOM allows the use of lower frequency devices for further cost reduction. These features give the potential for a low-cost design suitable for a short-range communications link at 60 GHz. The high RF frequency allows high data bandwidth, a desirable feature for such next-generation applications as wireless data or real-time digital video transmission.

In the design, two high electron mobility transistor (HEMT) devices are used in a push–push configuration to achieve balanced mixing at the second harmonic frequency. Using a dielectric resonator coupled to the gates of both HEMTs, shown at the center of Figure 9.31, forms a loop for the fundamental current and stabilizes oscillation performance. The DR is placed on a quartz support in a cylindrical copper cavity. The $\lambda/4$ stubs in the RF feed network ensure that all fundamental current flow to the gates of the devices is restricted to the DR. The RF is received by a patch antenna and split by a coupler network and fed to the gate of each device. One of the ends undergoes a 180° delay to achieve proper phasing at the mixer. Intermediate frequencies (IFs) are extracted through the two feeds above and below the coupler network. Absorber is used to eliminate any RF or LO leakage. For optimal operation, the devices are biased near pinch-off. A photograph of the completed structure is shown in Figure 9.32.

From Sironen et al. [33], the RF to IF conversion efficiency was measured by illuminating the circuit with a horn antenna placed at broadside and combining the IF channels with an external hybrid. IF power is referenced to available RF power calculated by the Friis transmission formula. The RF to IF conversion efficiency was measured over a 1 GHz bandwidth with two different heights for the DR support, 25 mils and 30 mils. With the 25 mil case, the fundamental oscillation frequency was 29.97 GHz and a conversion efficiency of -21 dB was achieved over the measured

ACTIVE INTEGRATED ANTENNAS 339

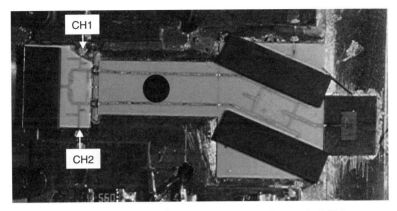

Figure 9.32. Photograph of 60 GHz second harmonic SOM.

band. The IF phase noise of -65 dBc/Hz at 10 kHz corresponded to a loaded Q of 900. With the 30 mil case, the fundamental oscillation frequency was 29.95 GHz and a conversion efficiency of -22 dB was achieved over the measured band. The IF phase noise of -68 dBc/Hz at 10 kHz corresponded to a loaded Q of 1300. The RF–IF conversion efficiency is shown in Figure 9.33 for the 30 mil case, given by the difference channel (CH1 $-$ CH2) for this balanced design. Also shown is the sum channel (CH1 $+$ CH2), giving an idea of the channel balance. Note that while the phase noise is superior with the 30 mil case, the channel balance was approximately 5 dB worse. The measured performance is sufficient for several meters of video transfer with typical solid state sources.

9.3.3.3 Noncontact ID The first two examples featured transmit and receive AIAs. The following examples combine both functions into a single module. The

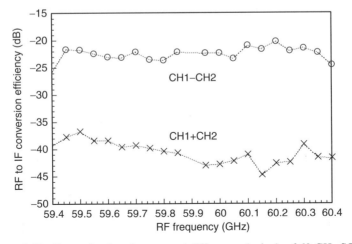

Figure 9.33. Conversion loss for sum and difference channels of 60 GHz SOM.

AIA is an ideal candidate for designing compact transceivers and transponders for current- and next-generation wireless systems in the microwave/millimeter-wave frequency range. This allows the exciting possibility of building the entire RF subsystem, including oscillator, amplifier, antenna, and mixer, onto a single dielectric substrate.

One example of this is the noncontact RF ID transponder card shown in Figure 9.34. A transponder is a circuit that, when triggered by an external interrogation source at a predefined frequency, will transmit a response signal to the interrogator, preferably at a different frequency for the purpose of interference reduction. The use of small low-cost microwave "tag" transponders for noncontact identification such as entry systems, toll collection, and inventory control has received much interest in recent years. The transponder in Figure 9.34 is based on a subharmonically pumped quasi-optical mixer using a broadband bowtie antenna [34].

A schematic diagram of the noncontact RF ID card is shown in Figure 9.35. When illuminated by a 6GHz interrogating signal, it transmits a unique digital identification code at a response frequency near 12 GHz. In the case of the schematic shown in Figure 9.34, the response frequency will shifted from the second harmonic frequency by an amount equal to the modulation rate of 10 MHz. Due to the frequency shift, the returned signal can easily be distinguished from spurious second harmonic radiation, a benefit of this system. By upconverting the digital ID code to variable frequencies, multiple cards can be read simultaneously. Another unique feature is the use of a slot antenna coupled with a rectifying diode to trigger the card from its standby state. The slot antenna, and the transmitting bowtie antenna are clearly visible in Figure 9.34. The standby and operating currents are 25 μA and 1 mA, respectively, from a +1.5 V silver oxide battery. Reducing the standby current

Figure 9.34. Photograph of noncontact RF ID card.

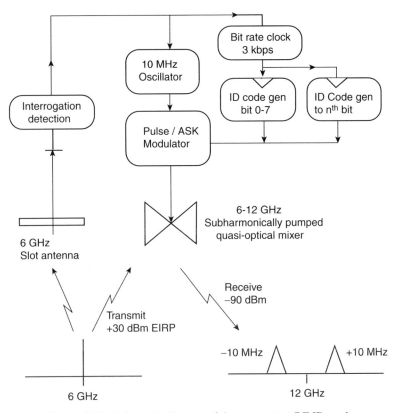

Figure 9.35. Schematic diagram of the noncontact RF ID card.

to 1 μA could increase the current one-to-two year battery life to approximately ten years, assuming that the card is interrogated daily for 1 minute.

9.3.3.4 Retrodirective Array Retrodirective antennas represent a special class of antennas, which have the specific characteristic that they reradiate any incident signal back toward the source direction with no prior knowledge of the source direction. This unique characteristic allows retrodirective antennas to simultaneously provide a broad range of coverage while maintaining a high level of gain. This trait has potential application in many areas, including self-steering antennas, radar transponders, search and rescue, and communication nodes.

Retrodirective performance is achieved by requiring that each element in the array radiate an outward-going wave whose phase is the conjugate of the incoming signal relative to a common reference [35]. Early designs were based on the Van Atta array. In this case, conjugation is achieved by using a symmetric array with conjugate elements connected by transmission lines of equal length [36]. To overcome the requirement of array symmetry and uniformity of the phase front in the Van Atta array, a more general approach of phase conjugation based on heterodyne mixing has been proposed [37–38]. This technique uses a local oscillator at twice the received incoming RF frequency. The resulting lower sideband product is at the same frequency as the received signal, but with inverted phase. Therefore, phase inversion occurs at each element, thus allowing irregular spacing or nonplanar arrays. This heterodyne-type retrodirective array is extremely useful for conformal arrays such as on an airplane fuselage. As will be seen, this scheme is also naturally suited for the AIA approach.

Shown in Figure 9.36 is one such AIA self-conjugating element that can be used in a retrodirective array. A dualfeed is connected to a modified rat race mixer by a lowpass filter designed to filter out any LO leakage. The rat race mixer design provides excellent RF/IF isolation. This is essential because any RF leakage would

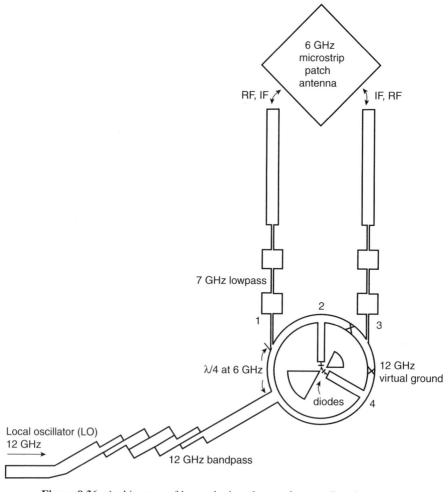

Figure 9.36. Architecture of heterodyning element for retrodirective arrays.

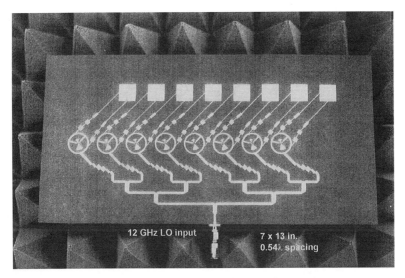

Figure 9.37. Eight-element retrodirective array.

reradiate, perturbing the radiation patterns. This element has been integrated into a linear eight-element array [39], as shown in Figure 9.37. A 4 × 4 retrodirective array utilizing ring slot antennas and FET resistive mixers demonstrates the feasibility of low-cost, large-scale 2D retrodirective arrays using the AIA concept [40]. The array featured a remarkable scan range greater than 100° in both azimuth and elevation planes.

Recently, another advance has been achieved in AIA retrodirective arrays. By using active mixers, an array has been developed that features not only retrodirective performance, but also demonstrates conversion gain [41]. This is vital if a retrodirective array is to be used in a communications link. The newly proposed phase conjugating element is shown in Figure 9.38(a). The element is based on a balanced quasi-optical mixer with an integrated patch antenna. The FET gate mixer is used to provide conversion gain. However, since the LO frequency is twice the RF, special consideration is needed to efficiently inject both the LO and RF signals for maximal conversion gain. A dual-frequency rat race coupler with a common output port is adopted for this purpose. More than 30 dB isolation between the RF and LO ports has been achieved for this particular hybrid. After optimization, the coupling between the input and output ports is −3.0 dB at each port. Since both the RF and LO signals applied to each FET are out of phase, the leakage is canceled when combined at the output port, while the in-phase IF signals are added.

A four-element prototype array based on this mixer design has been implemented using 6 GHz patch antennas with dual orthogonal feeds for transmission and reception and is shown in Figure 9.38(b). The orthogonal feeds cause the transmitted retrodirective signal to be orthogonal to the received signal. This is beneficial because it allows easy discrimination between the transmit and receive modes.

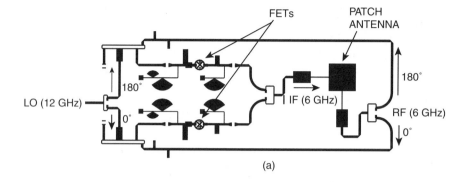

Figure 9.38. Details of active retrodirective array with conversion gain including (a) layout of active mixer and (b) photograph of prototype four-element array.

Measured isolation between the orthogonal modes of the patch antenna was determined to be 30 dB. The array spacing is $0.68\lambda_0$ to prevent grating lobes. To test the array, a horn antenna at 6.0 GHz placed 5 feet away is used to illuminate the structure. The LO frequency is at 12.05 GHz, giving a slight frequency shift between the RF and IF. This allows the IF signal to be distinguished from room backscatter of the RF signal. A second horn antenna is used as a receiving antenna. In this manner, the bistatic RCS of the structure is determined with the source placed at broadside and at 30°, as shown in Figure 9.39. The retrodirective performance of the array can clearly be seen.

9.4 CONCLUSION

In this chapter, a variety of state-of-the art planar antennas have been presented, as well as several new active architectures that take full advantage of the high performance, low profile, low cost, and low weight of this still developing field of antennas. As communications bands become increasingly cluttered and more commercial and military systems switch to microwave and millimeter-wave frequency bands, planar antennas will be increasingly used because of their qualities and flexible nature. Additionally, the AIA approach provides a key component in the

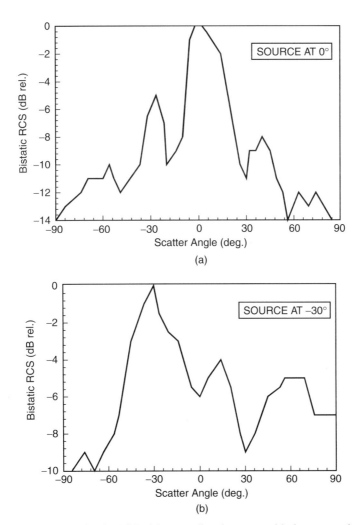

Figure 9.39. Measured bistatic RCS of the retrodirective array with the source placed at (a) 0° (broadside) and (b) 30°.

search for the low-noise receivers and low-power transmitters of next-generation systems through tight integration of the active and antenna platforms. While the broad scope of research that has been done in the areas of planar antennas and AIA systems in the last few decades prevents a comprehensive overview, it is hoped that the examples and issues discussed in this chapter provide a sufficient "flavor" of the field to show its true potential.

REFERENCES

1. D. M. Pozar, "Microstrip antennas," *Proceedings of the IEEE*, Vol. 80, No. 1, 79–81 (Jan. 1992).
2. G. M. Rebeiz, "Millimeter-wave and terahertz integrated circuit antennas," *Proceedings of the IEEE*, Vol. 80, No. 11, 1748–1770 (Nov. 1992).
3. G. P. Gauthier, A. C. Courtay, and G. M. Rebeiz, "Microstrip antennas on synthesized low dielectric-constant substrates," *IEEE Transactions on Antennas and Propagation*, Vol. 45, No. 8, 1310–1314 (Aug. 1997).
4. Y. Qian, R. Coccioli, D. Sievenpiper, V. Radisic, E. Yablonovitch, and T. Itoh, "A novel approach for gain and bandwidth enhancement of patch antennas," in *1998 IEEE Radio and Wireless Conference (RAWCON'98) Digest*, Colorado Springs, CO, Aug. 1998, p. 221.
5. Y. Qian, R. Coccioli, D. Sievenpiper, V. Radisic, E. Yablonovitch, and T. Itoh, "A microstrip patch antenna using novel photonic band-gap structures," *Microwave Journal*, Vol. 42, No. 1, 42 (Jan. 1999).
6. H. G. Booker, "Slot aerials and their relation to complementary wire aerials," *Journal of the IEE (London)*, Vol. 93, Part IIIA, 620–626 (1946).
7. Y. Yoshimura, "A microstripline slot antenna," *IEEE Transactions on Microwave Theory and Techniques*, Vol. 20, 760–762 (Nov. 1972).
8. T. M. Welller, L. P. B. Katehi, and G. M. Rebeiz, "Single and double folded-slot antennas on semi-infinite substrates," *IEEE Transactions on Antennas and Propagation*, Vol. 53, No. 12, 1423–1428 (Dec. 1995).
9. H. S. Tsai, M. J. W. Rodwell, and R. A. York, "Planar amplifier array with improved bandwidth using folded-slots," *IEEE Microwave and Guided Wave Letters*, Vol. 4, No. 4, 112–114 (Apr. 1994).
10. W. R. Deal, V. Radisic, Y. Qian, and T. Itoh, "A broadband microstrip-fed slot antenna," in *1999 IEEE MTT-S Topical Symposium Digest*, pp. 202–212.
11. T. L Chen and Y. D. Lin, "Aperture-coupled microstrip line leaky wave antenna with broadside mainbeam," *Electronics Letters*, Vol. 34, No. 14, 1366–1367 (July 1998).
12. Y. D. Lin, J. W. Sheen, and C. K. C. Tzuang, "Analysis and design of feeding structures for microstrip leaky wave antenna," *IEEE Transactions on Microwave Theory and Techniques*, Vol. 44, No. 9, 149–152 (Sept. 1996).
13. Y. D. Lin and J. W. Sheen, "Mode distinction and radiation-efficiency analysis of planar leaky-wave line sources," *IEEE Transactions on Microwave Theory and Techniques*, Vol. 45, No. 10, 1672–1680 (Oct. 1997).
14. Y. Qian and T. Itoh, "A broadband uniplanar microstrip-to–CPS transition," in *1997 Asia–Pacific Microwave Conference Digest*, pp. 609–612 (Dec. 1997).

15. Y. Qian, B. C. C. Chang, T. Itoh, K. C. Chen, and C. K. C. Tzuang, "High efficiency and broadband excitation of leaky mode in microstrip structures," in *1999 IEEE MTT-S Symposium Digest*, pp. 1419–1422.

16. K. S. Yngvesson, D. H. Schaubert, T. L. Korzeniowski, E. L. Kollberg, T. Thungren, and J. F. Johansson, "Endfire tapered slot antennas on dielectric substrates," *IEEE Transactions on Antennas and Propagation*, Vol. 33, No. 12, 1392–1400, Dec. 1985.

17. K. S. Yngvesson, T. L. Korzeniowski, Y. K. Kim, E. L. Kollberg, and J. F. Johansson, "The tapered slot antenna—a new integrated element for millimeter-wave applications," *IEEE Transactions on Microwave Theory and Techniques*, Vol. 37, No. 2, 365–374 (Feb. 1989).

18. K. P. Ma, Y. Qian, and T. Itoh, "Analysis and application of a new CPW–slotline transition," *IEEE Transactions on Microwave Theory and Techniques*, Vol. 47, No. 4, 426–432 (Apr. 1999).

19. N. Kaneda, Y. Qian, and T. Itoh, "A novel Yagi–Uda dipole array fed by a microstrip-to-CPS transition," in *1998 Asia–Pacific Microwaves Conference Digest*, pp. 1413–1416.

20. Y. Qian, W. R. Deal, N. Kaneda, and T. Itoh, "Microstrip-fed quasi-Yagi antenna with broadband characteristics," *Electronics Letters*, Vol. 34, No. 23, 2194–2196 (Nov. 1998).

21. Y. Qian, W. R. Deal, N. Kaneda, and T. Itoh, "A uniplanar quasi-Yagi antenna with wide bandwidth and low mutual coupling characteristics," in *1999 IEEE AP-S*, pp. 924–927.

22. W. R. Deal, J. Sor, Y. Qian, and T. Itoh, "A broadband uniplanar quasi-Yagi array for power combining," in *Proceedings of RAWCON'99*, pp. 231–234.

23. J. Sor, W. R. Deal, Y. Qian, and T. Itoh, "A broadband quasi-Yagi antenna array," in *European Microwave Conference*, pp. 255–258, 1999.

24. J. Lin and T. Itoh, "Active integrated antennas," *IEEE Transactions on Microwave Theory and Techniques*, Vol. MTT-42, 2186–2194 (Dec. 1994).

25. Y. Qian and T. Itoh, "Progress in active integrated antennas and their applications," *IEEE Transactions on Microwave Theory and Techniques*, Vol. 46, No. 11, 1891–1900 (Nov. 1998).

26. C. N. Kuo, B. Houshmond, and T. Itoh, "Full-wave analysis of packaged microwave circuits with active and nonlinear devices: an FDTD approach," *IEEE Transactions on Microwave Theory and Techniques*, Vol. 45, 819–826 (May 1997).

27. K. P. Ma, M. Chen, B. Houshand, Y. Qian, and T. Itoh, "Global time-domain full-wave analysis of microwave circuits involving highly nonlinear phenomena and EMC effects," *IEEE Transactions on Microwave Theory and Techniques*, Vol. 47, 859–866 (June 1999).

28. C. A. Balanis, *Antenna Theory: Analysis and Design*, Wiley, New York, 1982.

29. W. R. Deal, V. Radisic, Y. Qian, and T. Itoh, "Novel push-pull integrated antenna transmitter front-end," *IEEE Microwave Guided Wave Letters*, Vol. 8, 405–407 (Nov. 1998).

30. W. R. Deal, V. Radisic, Y. Qian, and T. Itoh, "A high efficiency slot antenna push–pull power amplifier," in *1999 IEEE MTT-S International Microwave Symposium*, Anaheim, CA, pp. 659–662 (June 1999).

31. W. R. Deal, V. Radisic, Y. Qian, and T. Itoh, "Integrated-antenna push–pull power amplifiers," *IEEE Transactions on Microwave Theory and Techniques*, Vol. 47, No. 8, 1901–1909 (Aug. 1999).

32. C. Y. Hang, W. R. Deal, Y. Qian, and T. Itoh, "Push–pull power amplifier integrated with microstrip leaky-wave antenna," *Electronics Letters*, Vol. 35, 1891–1892 (Oct. 1999).
33. M. Sironen, Y. Qian, and T. Itoh, "A dielectric resonator balanced second harmonic quasi-optical self oscillating mixer for 60 GHz applications," in *1999 IEEE MTT-S International Microwave Symposium*, Anaheim, CA, pp. 139–142.
34. C. W. Pobanz and T. Itoh, "A microwave noncontact identification transponder using subharmonic interrogation," *IEEE Transactions on Microwave Theory and Techniques*, Vol. 43, 1673–1679 (July 1995).
35. M. I. Skolnik and D. D. King, "Self-phasing array antennas," *IEEE Transactions on Antennas and Propagation*, Vol. AP–12, 142–149 (Mar. 1964).
36. E. D. Sharp and M. A. Diab, "Van Atta reflector array," *IRE Transactions on Antennas and Propagation*, Vol. AP–8, 436–438 (July 1960).
37. C. C. Culter, R. Kompfner, and L. C. Tilotson, "A self-steering array epeater," *Bell System Technical Journal*, Vol. 52, 2013–2027 (Sept. 1963).
38. C. Y. Pon, "Retrodirective array using the heterodyne technique," *IEEE Transactions on Antennas and Propagation*, Vol. AP–12, 176–180 (Mar. 1964).
39. C. W. Pobanz and T. Itoh, "A conformal retrodirective array for radar applications using a heterodyne phased scattering element," in *IEEE MTT-S International Microwave Symposium Digest*, Orlando, FL, May 1995, pp. 905–908.
40. C. W. Pobanz and T. Itoh, "A two-dimensional retrodirective array using slot ring FET mixer," in *Proceedings of the 26th European Microwave Conference*, Prague, Czech Republic, Sept. 1996, pp. 217–220.
41. R. Y Miyamoto, Y. Qian, and T. Itoh, "A retrodirective array using balanced quasi-optical FET mixers with conversion gain," in *1999 IEEE MTT-S International Microwave Symposium Digest*, pp. 655–658.

10

MICROELECTROMECHANICAL SWITCHES FOR RF APPLICATIONS

SERGIO P. PACHECO AND LINDA P. B. KATEHI
The Radiation Laboratory, The University of Michigan, Ann Arbor, MI

10.1 INTRODUCTION

The coming generation of tetherless (terrestrial wireless and satellite) communications technology promises a leap forward in information accessibility with an attendant increase in economic yield similar to that stimulated by the Internet in the 1990s. Researchers face many technical challenges, but data rates of tens of megabits per second are apparently a realizable goal early in the 21st century. Typical communications satellites (Fig. 10.1) employ traditional waveguide-based front-end architectures due to excellent electrical performance and high reliability. However, these systems are extremely massive and utilize large volume mostly attributed to the low-insertion-loss waveguide switch, diplexer, and waveguide-packaged solid-state power amplifier. These three components are interconnected by waveguide sections for compatibility and result in subsystems with dimensions of 18 cm × 40.6 cm × 10.5 cm.

Despite the large volume and mass, the metallic waveguide has been the transmission medium of choice in space applications due to the low-loss requirements. Its use in communications systems has resulted in overall system loss of less than 2 dB. Replacement of the waveguide components by micromachined ones without substantially affecting electrical performance can lead to a breakthrough in wireless communications [1,2].

RF Technologies for Low Power Wireless Communications
Edited by Tatsuo Itoh, George Haddad, and James Harvey
ISBN 0-471-38267-1   Copyright © 2001 by John Wiley & Sons, Inc.

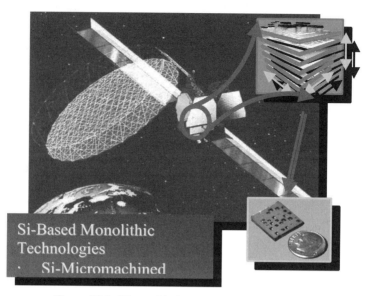

Figure 10.1. Waveguide-based satellite RF front end.

Communications via satellites and, in general, wireless communications require optimum high frequency performance, lightweight hardware, advanced packaging, high-density interconnect technology, and high reliability. In addition, future communications needs require increasing the system functionality to meet performance expectations of new highly integrated sensors and instruments commonly planned for the new generation of satellites. To maximize data transfer and minimize operation cost, future communications systems are forced to move to higher frequencies. The use of electronics in space poses interesting but large challenges. It also provides the opportunity for using revolutionary concepts in circuit design, fabrication, and implementation to achieve what is considered by today's standards as ultimate performance, minimal volume, and very low cost. Circuit optimization methods applied to existing technologies cannot meet the specifications, but critical advancements based on new concepts at fundamental levels of circuit design and diagnostics are needed. The capability to integrate the RF system on a single chip while preserving electrical performance will provide a breakthrough in communications systems for any type of wireless applications. Circuit miniaturization can be achieved by implementing a number of enabling technologies for next generation wireless systems. This includes new circuit and component architectures, low-cost adaptive and reconfigurable phased array antennas, millimeter-wave devices, advanced materials and novel integration techniques for systems on a chip, and new packaging methods for high-frequency electronics. RF micromachining and microelectromechanical systems (MEMS) technologies promise to provide an innovative approach in the development of effective and low-cost circuits and systems

INTRODUCTION

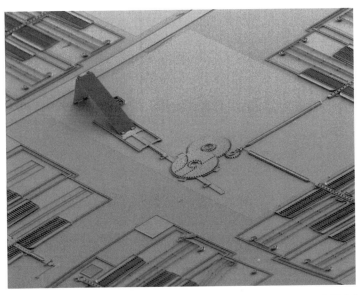

Figure 10.2. Sandia's ultraplanar multilevel MEMS technologies (SUMMIT) based on surface micromachining for complex actuators and optical systems [3].

and are expected to have significant application in the development of low-cost antenna arrays and reconfigurable apertures (see Fig. 10.2).

RF MEMS has been identified as an area that has the potential to provide a major impact on existing RF architectures in sensors (radar) and communications by reducing weight, cost, size, and power dissipation. The impact of this technology on communications system cost, size, and volume will be a few orders of magnitude. Key MEMS devices for current RF architectures are switches and microrelays in radar systems (see Fig. 10.3) and filters in communications systems. From a research standpoint, the RF MEMS is enabling new RF system architectures, such as a radar simultaneous transmit and receive system (STAR). Like a fully duplexed communications systems, this will allow radar and communications to occur through a common aperture and common electronics. In communications, a new architecture is that of "cognitive radio." In essence this will put a spectrum analyzer at the front of the otherwise digital radio, thereby allowing the radio to adaptively select frequencies based on the RF environment. For any RF system requiring electronic beam steering, RF MEMS will enable new antenna technologies based on a concept called "reconfigurable aperture." The idea is to use embedded MEMS switches to change the radiating topology of an aperture antenna by changing the current distribution. From a practical viewpoint this will allow the consolidation of several apertures into a single common aperture [5].

Passive components currently take up $\sim 90\%$ of the total transceiver board area. Via orders of magnitude reduction in the size of high-Q passive components,

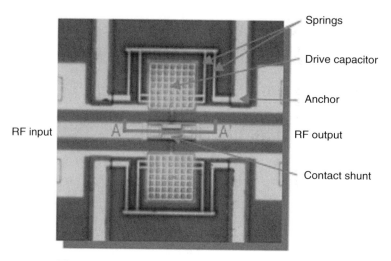

Figure 10.3. Rockwell Science Center's microrelay [4].

MEMS technology stands to greatly reduce this percentage. The circuit board may in fact be eliminated entirely if micromechanical passive components can be integrated alongside transistors in a single planar process, or using a wafer-level bonding technology. Due to their tiny size, zero dc power consumption, and batch fabrication technology, MEMS passive components can be used in large numbers, and this can revolutionize the system architectures used for transceivers. In particular, with MEMS passive components, architectures that presently minimize the use of high-Q components, in order to reduce size, may give way to architectures that emphasize their use, trading high Q (or selectivity) for power—hence greatly reducing the power consumption of communications transceivers. Both transceiver size and battery size stand to benefit from such novel communications architectures.

With more selectivity provided by MEMS, transceivers will now be able to function in more hostile environments, since co-site and other interferers will be much less of a problem to the new architectures. The use of high-Q passive components in transceivers allows one to relax the design specifications for the surrounding transistor circuits. In particular, dynamic range and phase noise specifications can greatly be relaxed, allowing not only substantial power reduction (as described above), but also making possible the use of less expensive technologies for certain functions. For example, with relaxed power requirements attained via MEMS, low-noise and low-power amplifiers may now be achievable in silicon, rather than GaAs, thus reducing the cost of the total system. In addition, in a fully integrated system using a merged circuits/MEMS technology, packaging costs can also be reduced, since board-level assembly may no longer be needed. The key enabler is, again, the outstanding performance of RF MEMS devices, particularly switches and filters.

INTRODUCTION   353

Another key enabler is the ultimate ability of these devices to be fabricated on a variety of heterogeneous materials such as plastics (kapton), thereby allowing flexibility and configurability and reducing the cost per unit area of RF systems and components well below Si, which costs approximately \$10/ in.^{2} (CMOS VLSI) in production [5].

The advent of bulk and surface micromachining techniques during the 1970s enabled the emergence of microelectromechanical systems (MEMS). MEMS fabrication allows the coupling of mechanical and electronic functionality in a single device. Moreover, MEMS devices are designed and fabricated by techniques similar to those of very large-scale integration (VLSI), thus using traditional batch-processing methods. The use of MEMS was most readily noticed in the arena of sensors and actuators such as accelerometers, gyroscopes, pressure and chemical sensors, and relays. From the original concept of a MEMS microwave switch by Petersen in 1979, there has been a constant push to develop MEMS relays and/or switches for applications requiring low power and higher frequencies [6]. At first this push was driven by the development of low-frequency MEMS devices such as resonators and filters. The major advantage of these devices was their low-power operation and ease of integration with CMOS circuitry. However, the great majority of these devices relied on polysilicon structures, which at high frequencies are extremely lossy. Due to this fact most researchers in the area of RF MEMS devices have used metal structures driven by electrostatic actuation.

In the last ten years, MEMS as applied to microwave and millimeter-wave circuits have experienced an exponential growth. In 1991, Larson et al. [7,8] described rotary MEMS switches with good performance at RF frequencies. C. T. Nguyen demonstrated the successful development of MEMS HF filters in 1993–1994 [9] and Yao and Chang [10] demonstrated a surface micromachined series switch for telecommunications applications in 1995. Recently, shunt microwave switches have been developed in the X to K/Ka band [11,12,14]. These switches are usually electrostatic in nature and commonly driven by bias voltages in the 30–80 V range. Most recently, low actuation voltage switches requiring 9 V of dc actuation have been demonstrated [13,15,16] and have opened new directions in system implementation and development of communications systems architectures. These MEMS devices are designed primarily for low-loss applications that do not require fast rates such as airborne and/or deep space communication.

The advantage of MEMS switches over their solid state counterparts such as FETs or PIN diodes is their extremely low series resistance and low drive power requirements. In addition, since MEMS switches do not contain a semiconductor junction, they exhibit negligible intermodulation distortion and, as a result, they provide better linearity at RF, microwave, and millimeter-wave frequencies. The above properties of the MEMS switches have created a lot of excitement and anticipation and have driven many research efforts in groups in the United States, Europe, and Japan. A variety of RF MEMS switch architectures have been introduced and successfully demonstrated; however, only a small number of them will be featured in this chapter. Further details on the remaining architectures may be found in the literature.

There are a number of ways to classify various switch architectures. The classification followed in this chapter places switches in three main groups: *fixed–fixed beam switches*, *cantilever beam switches*, and *compliant beam switches*. Each of the above architectures satisfies specific requirements related to the type of switch, *series versus shunt*, the type of contact, *metal-to-metal versus capacitive*, and the actuation voltage required for the switch operation. A discussion on switch characteristics in each of the above groups is provided in detail below.

10.2 FIXED-FIXED BEAM SWITCHES

All of the demonstrated fixed–fixed beam architectures are of capacitive type but they may operate in a series or shunt form. One of the very first demonstrations was by Goldsmith et al. [11,12,14]. These switches are made of a metal membrane, as shown in Figures 10.4 and 10.5, which moves between the up and down positions with the use of an externally applied dc voltage between 30 and 55 V. This membrane switch is a shunt type and operates on a coplanar waveguide (CPW) interconnect environment. As evident from the figures, the switch consists of a lower electrode fabricated on the surface of the integrated circuit (IC) and a thin aluminum membrane suspended over the electrode. The membrane is connected directly to the grounds of the CPW line, while a thin dielectric covers the lower electrode in order to avoid direct metal-to-metal contact. The air-gap between the two conductors determines the on-to-off capacitance ratio and specifies the switch isolation in the frequency range of interest.

With no applied actuation voltage, the residual intrinsic stress, keeps the membrane in the upper position and effectively makes it an electrical airbridge, which may require some appropriate compensation for effective RF operation (see Fig. 10.4). When a dc voltage is applied, the resulting electrostatic field causes the

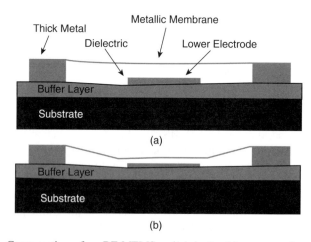

Figure 10.4. Cross section of an RF MEMS switch in the (a) unactuated and (b) actuated positions.

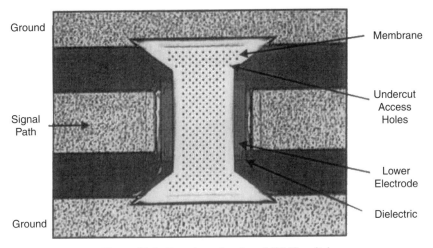

Figure 10.5. Top view of a shunt MEMS switch.

formation of positive and negative charges on the electrode and membrane conductor surfaces and as a result applies an attractive force, which when strong enough forces the membrane to collapse on the surface of the lower electrode (see Fig. 10.5). A similar MEMS device architecture can be used to develop MEMS capacitors. Details related to MEMS device design, fabrication, and RF characterization can be found in Goldsmith et al. [12,14] and Pacheco et al. [13]. Most recently, airbridge MEMS switches have been developed successfully at the University of Michigan. The RF performance of these switches is very similar to the membrane switches as extensively described in Muldavin and Rebeiz [17,18]. A picture of this switch is shown in Figure 10.6.

10.3 CANTILEVER BEAM SWITCHES

A number of efforts have been reported on the development of cantilever beam switches [10,16,19]. A schematic of such a switch is shown on Figure 10.7. In all these efforts the cantilevered structure was chosen to minimize the effects of residual stress on the mechanical characteristics of the switch. In this design architecture, the gradient stress is relaxed in a fully released structure, which is supported on one end only by a fixed beam.

As reported in the literature [19], the residual intrinsic stress can be controlled by process development and temperature control of the switch itself under operating conditions. As will be described in detail in later sections, stress gauges can be utilized effectively to specify the necessary fabrication conditions for the minimization of vertical stress gradients. In addition, to minimize this stress gradient, the trilayer structure shown in Figure 10.7(b) has been developed [10,16,19]. In this structure, plated or evaporated gold is sandwiched between two layers of dielectric,

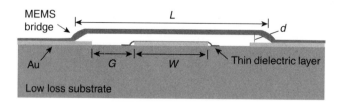
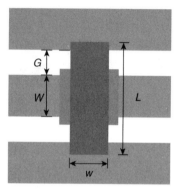

Figure 10.6. Typical shunt capacitive airbridge switch over a CPW shown in cross section in up position and top view.

Si₃N₄ or SiO₂, with appropriately designed thickness. This arrangement improves the thermal stability of the MEMS devices and reduces the stress mismatch between the materials. The primary difference between this structure and the fixed–fixed beam switches presented earlier is in the use of the dielectric layers to support the beam and provide isolation between the dc and RF electrodes, thus allowing for a metal-to-metal contact in a series or shunt switch architecture.

The RF performance of a series cantilever beam switch with a direct metal-to-metal contact, measured and fabricated, is shown in Figure 10.8 and indicates the ability of this switch to operate over a very broad frequency range from high-frequency to a few gigahertz. To extend the switch performance at higher frequencies while keeping the isolation below -30 dB, combinations of resonant switches are required, as described in later sections.

As shown by these measurements, the series switch has very low insertion loss (0.2 dB) up to 40 GHz and isolation that starts at -40 dB at 500 MHz and reduces to -15 dB at 40 GHz. A detailed description of the design, fabrication, and experimental verification for this switch are provided in Yao and Chang [10], Cai and Katehi [16], and Hyman et al. [19].

10.4 COMPLIANT BEAM SWITCHES

One of the disadvantages of the switches presented in the earlier sections is the need to operate under relatively high dc voltages varying from as low as 30 V to as high as

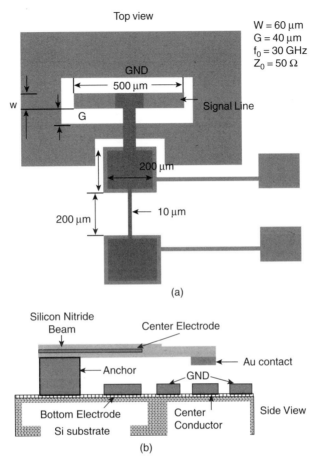

Figure 10.7. (a) Top view of the cantilever beam switch [11]. (b) Side view of the cantilever beam switch.

120 V. For MEMS switches to become appropriate for wireless handheld communications systems, actuation voltages less than 6 V are required. The primary goal of this section is to demonstrate the design, fabrication, and dc as well as RF characterization of low-actuation-voltage electrostatic shunt microwave switches. The ultra-low-loss, low actuation voltage RF MEMS switches presented herein are appropriate for use with finite ground coplanar waveguides, but the design methodology and fabrication approach are very general and can be utilized in the development of any low-actuation-voltage switch architecture.

This section is divided into three parts. In the first part, the mechanical design and properties of the switch such as stress and switching speed are discussed. The second part introduces the RF design, analysis, and measured performance of the RF MEMS switch. Finally, the last part discusses the use of specifically designed RF MEMS resonant switches for high isolation.

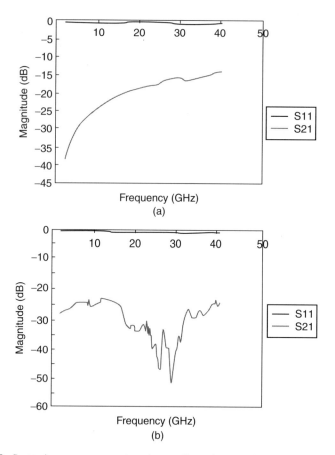

Figure 10.8. Scattering parameters when the cantilever beam switch is in the (a) up position and (b) down position.

10.4.1 Mechanical Design

As described in many books on fundamental mechanics, the actuation or pull-in voltage in a fixed beam switch is given by [20,21]

$$V_{pi} = \sqrt{\frac{8K_z g_0^3}{27\varepsilon_0 A}}, \qquad (10.1)$$

where K_z is the spring constant in the z-direction, g_0 is the initial gap between the switch and the bottom electrode, A is the area of the actuation pads, and ε_0 is the permittivity of air. In order to lower the pull-in voltage of the structure, three different design goals may be pursued: (1) increasing the area of actuation, (2) diminishing the gap between the switch and bottom electrode, and (3) designing a structure with low spring constant. In the first case, the area can only be increased by so much

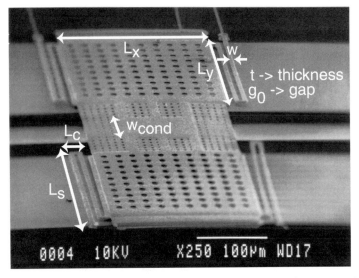

Figure 10.9. Scanning electron micrograph (SEM) of MEMS microwave switch.

before compactness becomes a prevailing issue. In the second case, the return loss associated with the RF signal restricts the gap. The third design goal is the one with the most flexibility, since the design of the springs does not considerably impact the size, weight, and/or RF performance of the circuit. Figure 10.9 is a scanning electron micrograph (SEM) of a shunt switch design appropriate for operation with a finite ground CPW (FGCPW) interconnect.

It consists of two folded suspensions of serpentine type attached to two square plates that provide the electrostatic area of actuation. Attached to the actuation plates is the plate that provides the parallel capacitance at the center conductor of the FGCPW line. The entire structure is anchored at the ends of the folded suspensions and is composed of electroplated metal. Table 10.1 lists the physical dimensions of the MEMS microwave switch. In the next section, a detailed analysis of the mechanical suspension and its design will be discussed.

10.4.1.1 Mechanical Suspension Mechanical suspension design is one of the most important factors in the design of a low actuation voltage switch. The folded suspension design shown in Figure 10.10 was chosen due to its ability to provide

TABLE 10.1. Physical Dimensions of MEMS Microwave Switch Shown in Figure 10.9

L_s	220 μm	L_x	220 μm
L_c	18 μm	L_y	220 μm
t	2 μm	w_{cond}	60 μm
w	6 μm	g_0	3 μm

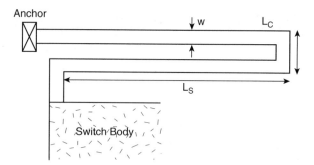

Figure 10.10. Diagram of folded suspension.

very low values of spring constant in a compact area. In addition, this design provides high cross-axis sensitivity between vertical and lateral dimensions. The folded suspension is made of metal (gold or nickel) of uniform thickness t, span beam length L_s, connector beam length L_c, and uniform width w. The spring constant in the z-direction, k_z, for the suspension is given by Eq. (10.2) [20,21]:

$$k_z = \frac{\left(\frac{Ew}{2}\right)\left(\frac{t}{L_c}\right)^3}{1+\frac{L_s}{L_c}\left(\left(\frac{L_s}{L_c}\right)^2 + 12\frac{1+\nu}{1+(w/t)^2}\right)}, \qquad (10.2)$$

where E and ν are the Young's modulus and Poisson's ratio for the metal. The total spring constant is the sum of all four suspensions attached to the structure and is given by

$$K_z = 4k_z. \qquad (10.3)$$

In similar fashion, the spring constants in the x- and y-directions can be determined from [22, 23]

$$K_x = \frac{8Etw^3}{L_c^3}\left(\frac{1+L_s/L_c}{1+4L_s/L_c}\right), \qquad K_y = \frac{8Etw^3}{L_s^3}\left(\frac{1+L_s/L_c}{1+4L_s/L_c}\right). \qquad (10.4)$$

Table 10.2 lists the physical dimensions of the serpentine spring used in the MEMS switch of Figure 10.9. Since $L_s/L_c \gg 1$, Eq. (10.3) and (10.4) can be

TABLE 10.2. Physical Parameters for Serpentine Spring

L_s	220 μm	E_{Au}	80 GPa
L_c	18 μm	ν_{Au}	0.22
t	2 μm	E_{Ni}	207 GPa
w	6 μm	ν_{Ni}	0.31

approximated by

$$K_z = 2Ew\left(\frac{t}{L_s}\right)^3,$$

$$K_x = 2Et\left(\frac{w}{L_c}\right)^3, \quad (10.5)$$

$$K_y = 2Et\left(\frac{w}{L_s}\right)^3.$$

Note that the spring constants given by Eq. (10.5) are equivalent to the spring constants of cantilever beams of length $2L_s$ for K_z and K_y and $2L_c$ for K_x. Table 10.3 compares the calculated spring constants for the folded suspension and the cantilever beam approximations. From Eq. (10.5) the cross-axis sensitivity can be determined to be

$$\frac{K_x}{K_z} = \left(\frac{w}{t}\right)^2 \left(\frac{L_s}{L_c}\right)^3,$$

$$\frac{K_y}{K_z} = \left(\frac{w}{t}\right)^2, \quad (10.6)$$

$$\frac{K_x}{K_y} = \left(\frac{L_s}{L_c}\right)^3.$$

The cross-axis sensitivity is determined by the ratios of the lengths L_s/L_c and w/t with w the width and t the thickness of the beam. Thus the proper choice of these physical parameters leads to a design of maximal cross-axis sensitivity. As can be seen, the cantilever beam equations are excellent predictors in the regime defined by $L_s/L_c \gg 1$. In addition, they are simple and can easily be calculated, giving the designer an immediate understanding of how mechanical parameters change with L_s, L_c, w, and t. In the next section, these equations will be used to study the effect of tensile axial stress on the mechanical parameters of the folded suspension design.

TABLE 10.3. Spring Constants

Constant	Folded Suspension Equation	Cantilever Beam Approximation	% Difference
K_z^{Ni}	1.85 N/m	1.86 N/m	1.1%
K_x^{Ni}	3.25×10^4 N/m	3.07×10^4 N/m	5.7%
K_y^{Ni}	17.8 N/m	16.8 N/m	5.7%
K_z^{Au}	0.71 N/m	0.72 N/m	1.0%
K_x^{Au}	1.26×10^4 N/m	1.19×10^4 N/m	6.0%
K_y^{Au}	6.88 N/m	6.49 N/m	6.0%

10.4.1.2 Residual Axial Stress Residual stress present on the structural material of the switches after fabrication can have a profound impact on mechanical parameters, such as the spring constant K_z. The analysis presented here assumes that the operation of the folded suspension is limited to small deflections and, therefore, the mechanical behavior can be modeled using a linear spring constant. Since $L_s/L_c \gg 1$, a cantilever beam under simultaneous axial tension and transverse loading is used to approximate the folded suspension behavior. Figure 10.11 shows the free-body diagram of a cantilever beam with the right end fixed and the left end guided. M_A is the reaction moment at the left end, R_A is the vertical reaction at the left end, l is the length of the beam, and a is the distance from the left end where the concentrated vertical loading W is located. The vertical deflection as a function of load position is given by [24]

$$z = z_A + \frac{M_A}{P} F_3,$$

$$z_A = \frac{W}{\gamma P}\left(\frac{C_3 C_{a3}}{C_2} - C_{a4}\right), \quad (10.7)$$

$$M_A = \frac{W}{\gamma} \frac{C_{a3}}{C_2},$$

where

$$\gamma = \sqrt{\frac{P}{EI}} \quad \text{and} \quad I = \frac{wt^3}{12}. \quad (10.8)$$

In the above equations, P is the axial tensile load, I is the moment of inertia for a rectangular cross section (with w the width and t the thickness of the beam), E is the Young's modulus, and C_2, C_3, C_{a3}, C_{a4}, and F_3 are hyperbolic coefficients. The spring constant is then determined by the ratio of concentrated loading W at the left

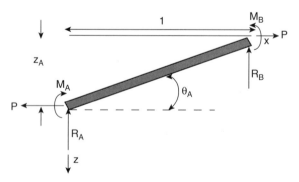

Figure 10.11. Free-body diagram of cantilever beam under simultaneous axial tension and concentrated transverse loading.

end of the beam ($a = 0$) over the deflection z, resulting in the following equation:

$$K_z = \frac{\gamma P \sinh(\gamma l)}{2[1 - \cosh(\gamma l)] + \gamma l \sinh(\gamma l)}.\qquad(10.9)$$

Using the physical parameters from Table 10.2 for nickel suspensions, and assuming an axial tensile loading of 1 Pa, Eq. (10.9) results in $K_z = 1.86$ N/m. As expected, this value is identical to the cantilever approximation for K_z in Table 10.3.

Figure 10.12 demonstrates the variation in spring constant K_z and pull-in voltage V_{pi} as a function of residual axial tensile stress from 0 to 300 MPa. As can be seen, the residual stress can cause an increase of about 20 times in the spring constant and consequently a fourfold increase in pull-in voltage. Therefore accurate control of the axial residual stress during fabrication is of extreme importance in the design of low actuation voltage switches.

In order to evaluate the axial stress present in the switch structures, and consequently their actuation voltage, micromechanical strain gauges developed by Lin et al. [25] were fabricated and measured. Figure 10.13 shows the layout of the sample containing strain gauges to measure compressive or tensile axial stress and cantilever beams of different lengths to measure the stress gradient across the thickness t of the beam. The stress gradient measurement consists of measuring the deflection ζ of the tip of the cantilever beam in the $+z$ (positive stress gradient) or $-z$ (negative stress gradient) direction.

The stress gradient (in MPa/μm) is then calculated from [26, 27]

$$\frac{d\sigma}{dt} = \frac{2E\zeta}{(1-\nu)l^2},\qquad(10.10)$$

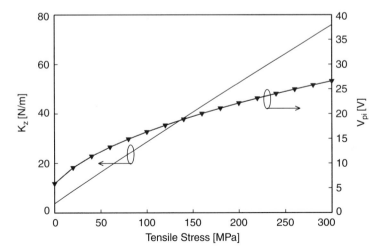

Figure 10.12. Variation of K_z and V_{pi} as a function of residual axial tensile stress.

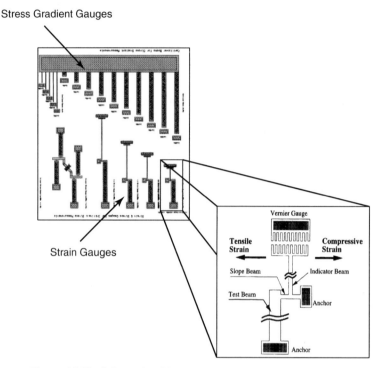

Figure 10.13. Schematic of layout and diagram of strain gauges.

where l is the length of the beam in microns and v is the Poisson ratio. The axial residual stress is measured by detecting the amount of lateral displacement δ of the strain gauge. The direction of deflection indicates if the stress is compressive or tensile, and the length of the indicator beam determines the amount of mechanical amplification and resolution of the gauge. The residual axial stress is given by the formula [25]

$$\sigma = \frac{2EL_{sb}\delta}{3(1-v)L_{ib}L_{tb}C}, \qquad (10.11)$$

where C is a correction factor to account for the presence of the indicator beam, and L_{sb}, L_{ib}, and L_{tb} are the lengths of the slope, indicator, and test beams in microns respectively.

TABLE 10.4. Electroplating Conditions for Strain Gauge Samples

Sample #	Current	Current Density	Time	Temperature
1	3 mA	1.5 mA/cm^2	40 min	49 °C
2	6 mA	3 mA/cm^2	20 min	49 °C
3	12 mA	6 mA/cm^2	10 min	50 °C

TABLE 10.5. Measured Stress Gradient and Residual Axial Stress

Sample #	Stress Gradient	Residual Axial Stress
1	86.9 MPa/μm	187.1 MPa (T)
2	26.9 MPa/μm	157.4 MPa (T)
3	−70.3 MPa/μm	60.8 MPa (T)

Three strain gauge samples were put through the same fabrication process as the switches with resulting structures that were 2 μm of electroplated nickel. Table 10.4 shows the conditions and variations in current and current density for the three strain gauge samples. Table 10.5 shows the measured stress gradient and residual axial stress for each sample (T indicates tensile stress). Sample 2 had the smallest stress gradient and indicated virtually no deflection of the cantilever beams up to 280 μm. Sample 3 resulted in a negative stress gradient and the tips of all cantilever beams 160 μm long or above touched the substrate. The electroplating conditions for sample 2 were used to fabricate the MEMS switches since it provided a structure with minimal warping of the folded suspensions. The tensile axial loading of 157.4 MPa determined by the strain gauges was input in Eq. (10.9) and resulted in a spring constant K_z of 42.41 N/m and a pull-in voltage V_{pi} of 19.90 V.

Figure 10.14 shows the measured dc response of the MEMS switch. The measurement was taken with an HP 4285A LCR meter at 100 kHz. The pull-in voltage of 20 V has excellent agreement with the analytical value of 19.90 V. The off capacitance (C_{off}) and on capacitance (C_{on}) were measured to be 39 fF and 800 fF,

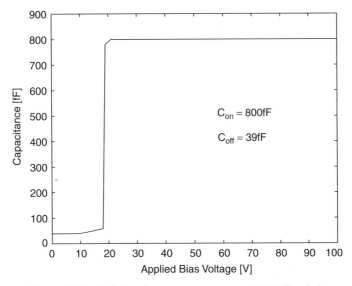

Figure 10.14. Pull-in voltage measurements of MEMS switch.

respectively. The value of C_{on} was approximately half that of the theoretically expected result in an on-to-off ratio of 20.5. This difference is due to the roughening of the circuit metal during processing, which prevents an appropriate intimate contact between the switch and the line and has deleterious effects on reliability. A more detailed discussion on this matter will be included in Section 10.4.2.5.

10.4.1.3 Switching Speed The switching speed can be modeled using the differential equation of motion given by [28]

$$m\frac{d^2z}{dt^2} + b\frac{dz}{dt} + k_z z = -\frac{\varepsilon_0 A V^2}{2(g-z)^2}, \qquad (10.12)$$

where m is the mass, k is the spring constant, and b is the damping coefficient. Microstructures operating at atmospheric pressure experience squeeze-film damping due to the vertical motion of the actuation electrode over the substrate. Continuum fluid mechanics can be used to analyze squeeze-film damping if the gap g_0 is much larger than the mean free path λ of the air molecules [29]. The mean free path of a gas is given by

$$\lambda = \frac{1}{\sqrt{2}\pi d_0^2 n}, \qquad (10.13)$$

where πd_0^2 is the collision cross section of the gas molecules, and n is the molecular density, which, for an ideal gas is given by

$$n = \frac{P}{k_B T}. \qquad (10.14)$$

In the above equation, P is the pressure of the squeeze film, k_B is Boltzmann's constant, and T is absolute temperature. Assuming atmospheric pressure and room temperature, the mean free path of air is 650 Å. The original 3 μm gap is approximately 46 times greater than the mean free path and, therefore, the air in the gap can be modeled as a viscous fluid. Using the viscous fluid approximation and one-dimensional analysis, the damping coefficient for square plates of length l and gap g_0 from the substrate can be derived to be [30]

$$b = 0.4217\mu \frac{l^4}{g_0^3}, \qquad (10.15)$$

where $\mu = 1.79 \times 10^{-5}$ Pa·s is the viscosity of air at atmospheric pressure and $T = 288$ K.

The MEMS switch design, however, contains plates with access square holes for faster etching of the sacrificial layer. These holes also serve a second purpose of diminishing the squeeze-film damping significantly. McNeil et al. [31] have

TABLE 10.6. Damping Coefficient, Damping Factor, and Switching Speed for MEMS Switch with and Without Perforations

	No Perforations	With Perforations
b	1.31×10^{-3} Pa·s	2.15×10^{-6} Pa·s
ξ	2.32	3.81×10^{-3}
t_s	80 μs	10.5 μs

modeled damping in a perforated plate as an ensemble of N smaller plates acting independently of each other, as shown below:

$$b = 0.4217 N\mu \frac{l_h^4}{g_0^3}, \qquad (10.16)$$

where l_h is the size of the smaller square plates. Differential equation (10.12) was solved numerically using MATHCAD for varying values of bias voltage. Table 10.6 shows the results for the damping coefficient b, damping factor $\xi = b/2\sqrt{km}$, and switching speed t_s for MEMS switches without perforations and with square holes 10 μm in length. In both cases, $k = 40.65$ N/m, $V_{pi} = 19.9$ V, and the applied bias voltage $V_{dc} = 20$ V. The damping factor reduces by a factor of 600 due to the inclusion of the perforations, and the system evolves from an overdamped to an underdamped case. In this configuration, the switching speed reduces almost tenfold.

Figure 10.15 shows the modeled switching speed results for the MEMS switch versus varying applied bias voltage. As expected, switching speed decreases with

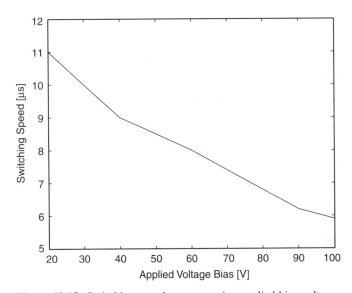

Figure 10.15. Switching speed versus varying applied bias voltage.

increasing bias voltage due to the increase of the force between the plates. It should be noted here that the inclusion of perforations does not significantly affect the capacitance between electrodes because of the presence of fringing fields [30]. Therefore the electrostatic force acting on the switch remains nearly constant, as in the case without perforations, while the damping decreases appreciably.

10.4.2 RF Microwave Design

Figure 10.9 shows a micrograph of the MEMS switch over a FGCPW line. At the time of switching, the necessary pull-in voltage is applied between the ground plane of the FGCPW line and the switch actuation pads. The resulting electrostatic force causes the switch to clamp down over the line. The high capacitance present at the center conductor provides a virtual short at RF. Since the switch is confined completely above the FGCPW line, it is very compact and only minimal real estate is needed for bias lines. In addition, the shunt configuration together with low actuation voltages results in negligible dc power consumption.

10.4.2.1 Fabrication A five mask batch process is used to fabricate the micromechanical switches. The substrate is composed of high resistivity silicon 400 μm thick with a passivation layer of SiO$_2$ approximately 1000 Å thick. Figure 10.16 shows the fabrication process flowchart. In (a) 500/7500 Å of Ti/Au is deposited and

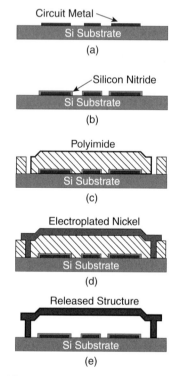

Figure 10.16. Fabrication process.

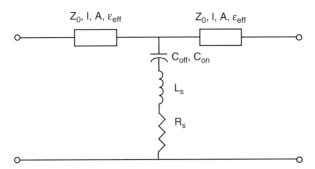

Figure 10.17. Libra lumped element model for single MEMS switch.

the circuit metal layer is defined via a liftoff process; (b) 2000 Å of plasma enhanced chemical vapor deposition (PECVD) silicon nitride (Si_3N_4) is patterned over the locations of switch actuation; (c) a sacrificial layer 3 μm thick of polyimide Du Pont PI2545 is spun cast, soft baked, and patterned for anchor points; (d) 2 μm of metal is electroplated to define the switch structure; (e) sacrificial etching of the polyimide layer is followed by supercritical CO_2 drying and release of the switch structure.

10.4.2.2 Circuit Modeling The circuits are modeled using HP EESof's Libra. The model shown in Figure 10.17 was developed for one single MEMS switch. The model consists of two sections of physical transmission line to represent the input and output sections of the FGCPW line and a capacitor–inductor series combination shunted across the transmission line to represent the MEMS switch. In this approach, the line impedance Z_0, line length l, and the effective dielectric constant ε_{eff}, are determined from physical dimensions of the FGCPW line. The unloaded line attenuation, α, switch capacitances on, C_{on}, and off, C_{off}, inductance, L_s, and series resistance, R_s, are all varied to fit the model to the measured data. The attenuation in the physical transmission line model is specified at 20 GHz and then follows a $\sqrt{f}$ variation. Once the model parameters are determined for the single MEMS switch, more complex circuits can be modeled by placing the switch section in the appropriate locations within the circuits.

The RF characteristics of a thru line without a MEMS switch was measured and a physical transmission line model was fitted to the data. Table 10.7 shows the fitted model parameters for the thru line. These parameters were used in conjunction with the capacitor–inductor–resistor series lumped element for the MEMS switch to fit

TABLE 10.7. Fitted Model Parameters for a Physical Transmission Thru Line

Z_0	54.8 Ω
l	200 μm
ε_{eff}	6.313
α (20 GHz)	1.2 dB/cm

TABLE 10.8. Fitted Model Parameters for Capacitor–Inductor–Resistor Series Lumped Element for the MEMS Switch

C_{off}	39 fF
C_{on}	800 fF
L_s	1 pH
R_s	0.3 Ω

the data for the measured response with the switch on and off. An analytical static model was used as a first order guess for the values of on and off capacitances:

$$C_{off} = \frac{\varepsilon_r \varepsilon_0 A}{\varepsilon_r g_0 + d_d},$$
$$C_{on} = \frac{\varepsilon_r \varepsilon_0 A}{d_d},$$
(10.17)

where ε_0 is the permittivity of air, ε_r is the relative permittivity of silicon nitride, A is the switch center pad area over the FGCPW center conductor, g_0 is the air gap under the switch, and d_d is the dielectric thickness. Table 10.8 shows the fitted model parameters for the MEMS switch of Figure 10.18. The C_{off} corresponded to the analytical value and, as expected, the inductance and resistance are approximately negligible.

However, the fitted value of the on capacitance was approximately half of the analytical value. The cause for this discrepancy will be discussed in more detail in Section 10.4.2.5. The fitted parameters for the physical transmission line and MEMS

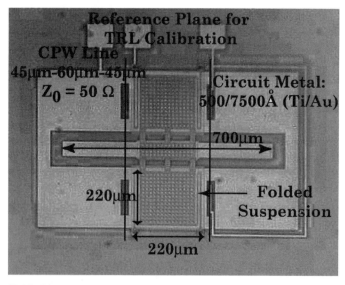

Figure 10.18. Photograph of MEMS switch and reference plane for TRL calibration.

switch were then used to model the RF response of the measured circuits as shown in the next section.

10.4.2.3 Switch Measurements The RF response of the switch is measured using a 8510C Vector Network Analyzer, Alessi Probe Station, and GGB Picoprobe 150 micron pitch coplanar probes. A TRL calibration software, MULTICAL, developed at NIST [32,33], is used to deembed the effects of the probe tips and feedlines from the measurement response, thus extending the reference plane as shown in Figure 10.18. The plot in Figure 10.19 shows the measured and modeled responses of the FGCPW line with a switch in the off position. Note the increase of the return loss at higher frequencies due to the capacitance introduced by the switch. Figure 10.20 compares the measured and modeled insertion losses between a plain FGCPW line and a line with the switch in the off position. There is almost no appreciable loss due to the switch up to 20 GHz with a loss of 0.16 dB at 40 GHz. The RF measured and theoretically modeled characteristics with the switch in the on position is shown in Figure 10.21 and demonstrates an isolation of approximately −16 dB at 40 GHz. The isolation can be further improved by decreasing the thickness of the dielectric between the RF pad of the switch and the line center conductor (see Fig. 10.16), thus increasing the on capacitance. However, care must be taken in order to make sure that the thin film thickness is sufficient not to cause dielectric breakdown and current flow.

10.4.2.4 Power Handling Measurements The power handling capabilities of the MEMS switches can be limited either by the current density on the transmission line causing excessive heating due to ohmic losses or by the self-actuation of the switches due to the average RF voltage on the FGCPW line (denoted as "self-biasing"). Since

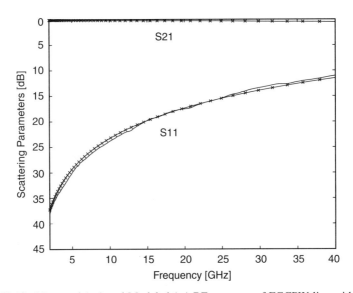

Figure 10.19. Measured (—) and Modeled (×) RF response of FGCPW line with switch.

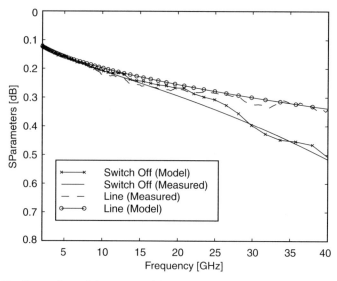

Figure 10.20. Comparison of the measured (—) and modeled (×) insertion loss of a FGCPW line with no switch and a FGCPW line with switch off.

the electrostatic force acting on the switch can derive from either a positive or negative voltage, the average voltage level of the rectified sine wave due to the RF power on the FGCPW line causes attractive forces to be applied between the RF line and RF electrode of the switch.

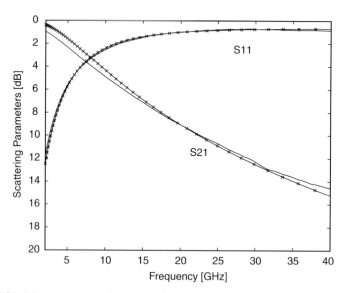

Figure 10.21. Measured (—) and modeled (×) RF response of FGCPW line with switch on.

The average voltage of a rectified sine wave is given by

$$V_{\text{avg}} = \frac{1}{T} \int_0^{T/2} V_p \sin(\omega t)\, dt = \frac{V_p}{\pi}, \qquad (10.18)$$

where $T = 2\pi/\omega$ and V_p is the peak RF voltage. The RF power in terms of peak voltage V_p is given by

$$P = \frac{V_p^2}{2Z_0}, \qquad (10.19)$$

where Z_0 is the characteristic impedance of the FGCPW line. Using Eq. (10.18), the RF power on the FGCPW line can be written as

$$P = \frac{\pi^2 V_{\text{avg}}^2}{2Z_0}. \qquad (10.20)$$

Figure 10.22 is a diagram of the power measurement setup. The RF power is generated by an X-band (8–12 GHz) TWT amplifier and the variable attenuator is used to vary the output power from 1 mW up to the maximum level of 6.6 W. The output power from the MEMS switch is coupled to a 20 dB attenuator and fed into the power meter. From Eq. (10.20) the predicted average voltage for the maximum power level of 6.6 W is 8.56 V, not enough to cause the switch to actuate. In fact, since the RF power is delivered to the center conductor region of the FGCPW line,

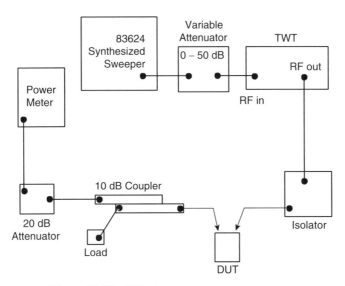

Figure 10.22. X-Band power measurement set up.

the area of electrostatic attraction is much smaller and requires a pull-in voltage of 53.89 V. This pull-in voltage corresponds to a RF power level of 261.5 W. As expected, the MEMS switch with the dimensions given in Table 10.1 did not "self-bias" during the duration of the high-power test. Figure 10.23 shows a plot of the power readings for the cases of the switch in the off and on positions. The switches used in this power experiment were actuated using bias voltages from 20 to 90 V at power levels from 1 mW to 6.6 W. The difference of 3–3.5 dB in power levels corresponds to the isolation of the switches at 10 GHz (see Fig. 10.21). There was no observed catastrophic failure of the switches and/or dielectric film. The dielectric strength for a PECVD Si_3N_4 thin film is 5×10^6 V/cm [34], corresponding to a breakdown voltage of 100 V for a 2000 Å thick film. There was a slight bending of the folded suspensions at power levels above 4.5 W with the switch in the on position. The suspensions reverted back to their original position once the power was decreased or turned off. This warping was due to heating of the switches as the power levels increased.

10.4.2.5 Silicon Nitride Thin Film Dielectric The characterization of the thin film dielectric over the FGCPW lines is of extreme importance, since it limits the on capacitance of the MEMS switch. As mentioned in Section 10.4.2.1, the Si_3N_4 film is deposited via a PECVD process. The quality of this dielectric is not as good as a low pressure chemical vapor deposition (LPCVD) nitride. However, the LPCVD process occurs at high temperatures (600–700 °C), while the PECVD nitride is a low temperature process (400 °C). In addition, the chamber configuration in the PECVD allows the deposition over metals. Still, the film contains pinholes as well as

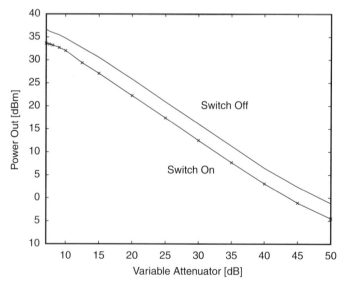

Figure 10.23. Measured power levels for switch off (—) and switch on (×).

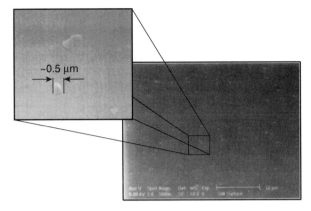

Figure 10.24. Scanning electron micrograph (SEM) of dielectric layer.

roughening of the metal during deposition, thus preventing an intimate contact between switch and dielectric. Figure 10.24 also shows that after fabrication inclusions approximately 0.5 μm in size are present over the dielectric. These inclusions are a result of the metal roughening due to the deposition temperature (400 °C) being above the eutectic temperature of the gold lines (372 °C). Using a lower deposition temperature, the inclusion problem can be alleviated. These inclusions prevent direct contact of the switch and dielectric, creating an effective air gap, thus considerably decreasing the on capacitance. Figure 10.25 is a schematic of the gap due to the inclusions present on the Si_3N_4 layer.

The resulting on capacitance is derived from the effective air gap capacitance and dielectric capacitance in series:

$$\begin{aligned}
\frac{1}{C_{on}} &= \frac{1}{C_{gap}} + \frac{1}{C_d} \\
&= \frac{g_{eff}}{\varepsilon_0 A} + \frac{d_d}{\varepsilon_r \varepsilon_0} \\
&= \frac{\varepsilon_r g_{eff} + d_d}{\varepsilon_r \varepsilon_0 A}, \\
C_{on} &= \frac{\varepsilon_r \varepsilon_0 A}{\varepsilon_r g_{eff} + d_d}.
\end{aligned} \quad (10.21)$$

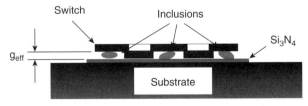

Figure 10.25. Schematic diagram of the effective air gap due to inclusions on the Si_3N_4 layer.

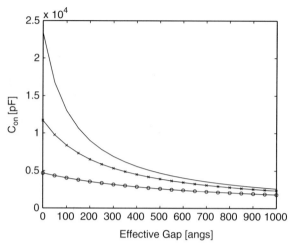

Figure 10.26. Plot of C_{on} versus g_{eff} for Si_3N_4 thicknesses of 500 Å (—), 1000 Å (×), and 2500 Å (○).

Using Eq. (10.21) and solving for g_{eff}, it is determined that the inclusions account for an effective air gap of 836 Å and decrease the on capacitance from 1.87 to 0.8 pF in the MEMS switch. Figure 10.26 shows calculated C_{on} versus g_{eff} for dielectric thicknesses of 500Å, 1000Å, and 2500Å. In order to attain a high value of on capacitance, the effective air gap must be very small; an effective air gap of 300 Å will decrease the on capacitance by almost half. Furthermore, the effective air gap limits the maximum value of on capacitance. Using Eq. (10.21) and solving for d_d,

$$d_d = \varepsilon_r \left(\frac{\varepsilon_0 A}{C_{on}} - g_{eff} \right). \tag{10.22}$$

Equation (10.22) provides physically acceptable solutions, only if

$$\frac{\varepsilon_0 A}{C_{on}} > g_{eff}. \tag{10.23}$$

Equating the two sides in Eq. (10.23) results in a $d_d = 0$ and determines the maximum achievable C_{on} for a given g_{eff}:

$$C_{on,max} = \frac{\varepsilon_0 A}{g_{eff}}. \tag{10.24}$$

Using Eq. (10.24) and assuming the effective air gap of 836 Å, a maximum on capacitance of 1.398 pF is calculated. Note that the upper limit on the on capacitance is independent of the type of dielectric film and inherent to the effective air gap.

COMPLIANT BEAM SWITCHES

Therefore it is crucial to optimize the thin film material growth and fabrication process to achieve a smooth, defect-free dielectric surface.

10.4.3 High Isolation Switching

The typical MEMS shunt switch uses the variation in the capacitance ratio between the on and off stages to provide an RF switching operation. To provide the correct modeling, we need to take into account the parasitic inductance and resistance of the MEMS switch. Muldavin and Rebeiz [17] particularly show that the small parasitic inductance can considerably affect the isolation of the switch. This section demonstrates that the appropriate design of switch inductance and capacitance can provide switches with very special characteristics. It is shown that the parasitic inductance can be controlled and used to increase the isolation in the down state. In particular, single shunt switches as well as combinations of switches connected in parallel with very high isolation at a specified frequency can be realized. In addition, modifications in the switch geometry and mechanical design can change its frequency and bandwidth, thus giving design goal flexibility.

10.4.3.1 Resonant Switch Design The resonant shunt switch has the same basic design as the RF MEMS switches discussed in Section 10.4. The main difference is in the geometry of the beams connecting the large actuation pads to the center capacitive pad of the RF MEMS switch. These connecting beams have a crucial role in the isolation of the switch because they effectively form a small parasitic inductance in series with the RF switch capacitor. A SEM of four connected switches is shown in Figure 10.27.

The value of the inductors mentioned above is determined mainly by the geometry of the connecting beams and can either be found by a full wave electromagnetic solver [35] or from measured S-parameters. Since the up-state switch

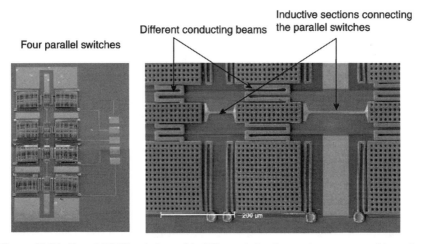

Figure 10.27. Four MEMS switches with different inductive sections connected in series.

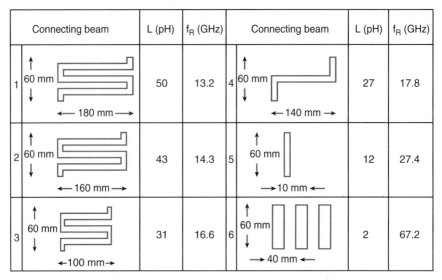

Figure 10.28. Inductances and resonant frequencies for six different connecting beam geometries.

capacitance is usually very small, the response of the switch in the up-state is almost independent of the inductive connecting beams. Therefore S-parameters showing the switch isolation can be used to accurately extract the inductance value. Nevertheless, theoretical and experimental data are in close agreement.

The RF MEMS switch discussed in Section 10.4 has an effective inductance formed by the connecting beams of approximately 1 pH, while the new designs have inductances between 12 and 50 pH. Six single switches, each with a different connecting beam, have been fabricated and measured. Each connecting beam yields a different response and resonant frequency. Figure 10.28 gives inductance values extracted from the measured S-parameters and Figure 10.29 presents the measured MEMS switch performance in the down position.

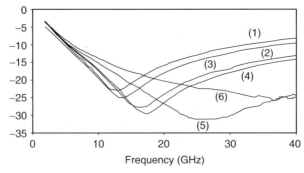

Figure 10.29. Isolation in the down state for single switches in relation to their connecting beam geometries (the numbers correspond to the geometries in Fig. 10.28).

COMPLIANT BEAM SWITCHES

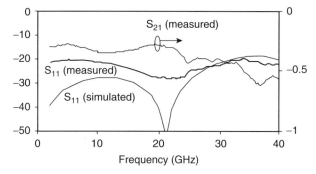

Figure 10.30. S-parameters for the four-switch network of Figure 10.27 with all switches off.

The measured on-to-off capacitance ratio was found to be approximately 70:1 (on capacitance of 2.6 pF and off capacitance of 37 fF).

This design implies that both the resonant frequency and the bandwidth of the switch can be changed. For example, a more selective switch can be designed by increasing the inductance and reducing the capacitance while keeping the resonant frequency the same.

10.4.3.2 Resonant Switches in Series The previous method of forming a resonance can be applied to other designs where higher isolation is required and therefore connecting switches in parallel is necessary. One of the advantages of this design is that the inductance and, consequently, the resonant frequency of each switch can be used as design parameters and can be optimized for desired performance. Therefore, instead of using switches that are all tuned at one specific frequency, they are designed to resonate at different frequencies, thus increasing the isolation and bandwidth of the final circuit.

By applying the previous technique, it is possible to have several poles at different frequencies, all in one compact design. Using similar techniques to those used in a bandstop filter design, the final four-switch network in Figure 10.27 has been fabricated. Its S-parameters are shown in Figures 10.30 and 10.31.

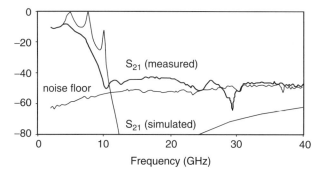

Figure 10.31. S-parameters for the four-switch network of Figure 10.27 with all switches on.

The equivalent switch capacitance and inductances for every switch are

$C_{down} = 2.9$ pF (Common)

$L_1 = 50$ pH, $L_2 = 43$ pH, $L_3 = 31$ pH, $L_4 = 27$ pH

Although the theoretical isolation is very high and has a maximum value of 95 dB, it is not seen in the measurements because of the noise floor, which is also shown in Figure 10.31. It is also seen that the return loss for this structure is better than -20 dB over a bandwidth of almost 40 GHz. This is achieved by the two resonances shown (around 22 and 42 GHz), which come from the combination of the up-state capacitances and the short, high impedance CPW lines (180 and 100 µm). Finally, when the switches are down the return loss is less than -0.5 dB.

10.5 CONCLUSIONS

The micromechanical switch is without doubt the paradigm RF MEMS device. Within the last decade the RF community has experienced a growing plethora of MEMS switch designs. These designs have achieved a high level of RF performance, while maintaining ultra-low-power dissipation and large-scale integration. Due to the above characteristics, such RF MEMS devices should enable a wide variety of new system capabilities. Barker and Rebeiz [36] have demonstrated the use of periodically loaded transmission lines with MEMS switches in order to achieve phase shifts of up to 120° with minimal insertion loss. More recently, X-band distributed MEMS transmission lines (DTML) have yielded one- and two-bit phase shifters with 180°/dB of insertion loss [37]. Research efforts have also been reported in the arena of quasi-optical components and reconfigurable antenna arrays [38]. MEMS switches are used to alter the functionality as well as the topology of such systems. In addition, design tools and manufacturing capabilities are increasing at an astounding pace. Therefore it is expected that the market potential in RF MEMS will grow substantially in the coming decade. According to research conducted by Ernst & Young Entrepreneurs Conseil, Paris, France, estimates indicate that by 2002 the MEMS market will have grown to $34 billion for devices and $96 billion for systems [39]. With the ability to support any combination of electrical and micromechanical devices on a single chip, RF MEMS open an endless horizon of tantalizing possibilities for future designs and systems. In addition, the continual research on new design tools and methodologies ensures that MEMS hold a tremendous promise in contributing heavily to the next generation of RF-based applications.

REFERENCES

1. L. P. B. Katehi et al., "Si micromachining in high-frequency applications," in *CRC Handbook on Si Micromachining*, 1995.
2. L. P. B. Katehi, "Novel transmission lines for the submillimeter-wave region," *IEEE Proceedings*, Vol. 80, No. 11, pp. 1771–1787, Nov. 1992.

REFERENCES

3. B. D. Staple, "Surface micromachining fabrication for the 21st century," *NSF Workshop on RF Micromachining and MEMS Technology for Wireless Communications Systems*, Arlington, MD, December 1999.
4. E. Sovero, "RF MEMS switches," *NSF Workshop on RF Micromachining and MEMS Technology for Wireless Communications Systems*, Arlington, MD, December 1999.
5. E. R. Brown, "Architectural issues in RF-MEMS based reconfigurable apertures," *NSF Workshop on RF Micromachining and MEMS Technology for Wireless Communications Systems*, Arlington, MD, December 1999.
6. K. E. Petersen, "Micromechanical membrane switches on silicon," *IBM Journal of Research and Development*, Vol. 23, 376–385, (July 1971).
7. L. E. Larson, R. H. Hackett, and R. F. Lohr, "Micromachined microwave actuator (MIMAC) technology—a new tuning approach for microwave integrated circuits," in *IEEE Microwave and Millimeter-Wave Monolithic Circuit Symposium Proceedings*, pp. 27–30, 1991.
8. R. H. Hackett, L. E. Larson, and M. Melendes, "The integration of micro-machine fabrication with electronic device fabrication on III–V semiconductor materials," in *Transducers Proceedings*, pp. 51–54, 1991.
9. C. T.-C. Nguyen, *Micromechanical Signal Processors*. Ph.D. Thesis, University of California at Berkeley, 1994.
10. J. J. Yao and M. F. Chang, "A surface micromachined miniature switch for telecommunications applications with signal frequencies from dc up to 4 Ghz," in *The 8th International Conference on Solid-State Sensors and Actuators Digest*, pp. 384–387, 1995.
11. C. Goldsmith, T. H. Lin, B. Powers, W. R. Wu, and B. Norvell, "Micromechanical membrane switches for microwave applications," *IEEE MTT-S International Microwave Symposium Digest*, pp. 91–94, 1995.
12. C. Goldsmith, J. Randall, S. Eshelman, and T. H. Lin, "Characteristics of micromachined switches at microwave frequencies," *IEEE MTT-S International Microwave Symposium Digest*, pp. 1141–1144, 1996.
13. S. Pacheco, C. T. Nguyen, and L. P. B. Katehi, "Micromechanical electrostatic K-band switches," *IEEE MTT-S International Microwave Symposium Digest*, pp. 1569–1572, 1998.
14. C. L. Goldsmith, A. Malczewski, Z. J. Yao, S. Chen, J. Ehmke, and D. H. Hinzel, "RF MEMS variable capacitors for tunable filters," Special Issue on RF Applications of MEMS Technology, *RF and Microwave Computer-Aided Engineering*, Vol. 9, No. 4, 362–374 (July 1999).
15. S. P. Pacheco, L. P. B. Katehi, and C. T.-C. Nguyen, "Design of low actuation voltage RF MEMS switch," *IEEE MTT-S International Microwave Symposium Digest*, pp. 165–168, 2000.
16. Y. Cai and L. P. B. Katehi, "Wide band series switch fabricated using metal as sacrificial layer," *European Microwave Conference*, Paris, France, Oct. 2000.
17. J. B. Muldavin and G. M. Rebeiz, "30 GHz tuned MEMS switches," *IEEE MTT-S International Microwave Symposium Digest*, pp. 1511–1514, 1999.
18. J. B. Muldavin and G. M. Rebeiz, "High-isolation CPW MEMS switches—part 2: design," *IEEE Transactions on Microwave Theory and Techniques*, Vol. 48, No. 6, 1053–1056 (June 2000).

19. D. Hyman et al., "Surface-micromachined RF MEMS switches on GaAs substrates," Special Issue on RF Applications of MEMS Technology, *RF and Microwave Computer-Aided Engineering*, Vol. 9, No. 4, 348–361 (July 1999).
20. E. P. Popov, *Introduction to Mechanics of Solids*, Prentice-Hall, Englewood Cliffs, NJ, 1968.
21. J. E. Shigley and L. D. Mitchell, *Mechanical Engineering Design*, 4th ed. McGraw-Hill, New York, 1983.
22. A. P. Pisano and Y.-H. Cho, "Mechanical design issues in laterally- driven microstructures," *Sensors and Actuators*, Vol. A21-A23, 1060–1064 (1990).
23. Y. -H. Cho and A. P. Pisano, "Optimum structural design of micromechanical crab-leg flexures with microfabrication constraints," *Symposium on Micromechanical Systems*, Nov. 1990.
24. R. J. Roark and W. C. Young, *Formulas for Stress and Strain*, 6th ed. McGraw-Hill, New York, 1989.
25. L. Lin, A. P. Pisano, and R. T. Howe, "A micro strain gauge with mechanical amplifier," *Journal of Microelectromechanical Systems*, Vol. 6, 313–321 (Dec. 1997).
26. F. Ericson, S. Greek, J. Söderkvist, and J.-Å. Schweitz, "High-sensitivity surface micromachined structures for internal stress and stress gradient evaluation," *Journal of Micromechanics and Microengineering*, Vol. 7, 30–36 (Mar. 1997).
27. W. Fang and J. A. Wickert, "Determining mean and gradient residual stresses in thin films using micromachined devices," *Journal of Micromechanics and Microengineering*, Vol. 6, pp. 301–309, Sept. 1996.
28. N. S. Barker, *Distributed MEMS Transmission Lines*. Ph.D. Thesis, University of Michigan, 1999.
29. R. T. Howe and R. S. Muller, "Resonant microbridge vapor sensor," *IEEE Transactions on Electron Devices*, Vol. ED-33, pp. 499–506, Apr. 1986.
30. G. K. Fedder, *Simulation of Microelectromechanical Systems*. Ph.D. Thesis, University of California at Berkeley, 1994.
31. V. M. McNeil, M. J. Novack, and M. A. Schmidt, "Design and fabrication of thin film microaccelerometers using wafer bonding," *Technical Digest of the 7th International Conference on Solid-State Sensors and Actuators*, pp. 822–825, June 1993.
32. R. B. Marks and D. F. Williams, *Program Multical*, Tech. Rep., NIST, Aug. 1995. Rev. 1.
33. R. B. Marks, "A multiline method of network analyzer calibration," *IEEE Transactions on Microwave Theory and Techniques*, Vol. 39, pp. 1205–1215, July 1991.
34. M. Madou. *Fundamentals of Microfabrication*, CRC Press, New York, 1997.
35. Zeland's IE3D, Release 6, 1999.
36. N. S. Barker and G. M. Rebeiz, "Distributed MEMS true-time delay phase shifters and wide-band switches," *IEEE Transactions on Microwave Theory and Techniques*, Vol. 46, No. 11, 1881–1890 (Nov. 1998).
37. J. S. Hayden and G. M. Rebeiz, "One and two-bit low-loss cascadable MEMS distributed X-band phase shifters," *IEEE MTT-S International Microwave Conference Digest*, pp. 161–164, 2000.
38. E. R. Brown, "RF-MEMS switches for reconfigurable integrated circuits," *IEEE Transactions on Microwave Theory and Techniques*, Vol. 45, No. 11, pp. 1868–1880, Nov. 1998.
39. J. M. Karam, "Automation mainstream MEMS design," *Electronic Design*, Vol. 46, No. 16, pp. 67–71 (July 1998).

11

MICROMACHINED K-BAND HIGH-Q RESONATORS, FILTERS, AND LOW PHASE NOISE OSCILLATORS

ANDREW R. BROWN AND GABRIEL M. REBEIZ
Department of Electrical Engineering and Computer Science
University of Michigan, Ann Arbor

11.1 INTRODUCTION

Microwave and millimeter-wave communications systems are expanding rapidly because they offer many advantages over conventional wireless links. They allow the use of very wideband radio links suitable for intersatellite and interpersonal communications. Commercially available systems are under development at 28 GHz for the local multipoint distribution system (LMDS) [1], the PCS networks at 38 GHz [2,3], and also a new short-range telecommunication band at 60 GHz. Commercial systems demand high yield and the ability to fabricate large volumes of systems using low-cost techniques.

Current millimeter-wave wireless front-end transceivers use a hybrid approach with a combination of waveguide components, solid state devices, and dielectric resonators (Fig. 11.1). All of the active components low-noise amplifier, power amplifier, mixer) are based on solid state technology and are implemented using planar microwave monolithic integrated circuits (MMICs) or hybrid devices. The MMICs or devices are placed automatically on the carrier substrate using pick-and-place machines and are connected to the transceiver circuit using either bond wires

---

RF Technologies for Low Power Wireless Communications
Edited by Tatsuo Itoh, George Haddad, and James Harvey
ISBN 0-471-38267-1   Copyright © 2001 by John Wiley & Sons, Inc.

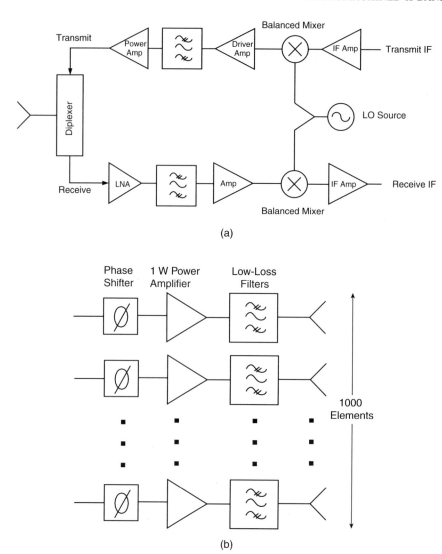

Figure 11.1. Typical millimeter-wave (a) transceiver front-end block diagram and (b) new generation K-band phased array for satellite communications.

or using flip-chip technology. This technology has advanced considerably in the past ten years and does not contribute significantly to the total cost of the transceiver.

However, the diplexer and other filters and the local oscillator tank are expensive to manufacture at Ka-band to K-band applications. These components are implemented using high-Q structures such as resonant waveguide cavities or dielectric resonators. The high-Q resonators are needed in order to have a low insertion loss, high out-of-band rejection, and high channel-to-channel isolation in the case of

INTRODUCTION

filters, and for achieving low phase noise in the case of oscillators. However, they are expensive to assemble and require manual placement and tuning on the carrier substrate (for dielectric resonators) and waveguide-to-microstrip transitions (for waveguide cavities).

The theoretical insertion loss of a filter is given by [4]

$$\Delta L_A(\text{dB}) \approx 8.686 \frac{c_n}{\bar{\omega} Q_u}, \qquad (11.1)$$

where $\Delta L_A(\text{dB})$ is the in-band insertion loss, c_n is the filter prototype coefficient, which is a function of the number of poles and passband ripple of the filter, $\bar{\omega}$ is the filter fractional bandwidth, and Q_u is the unloaded quality factor of the resonator. Figure 11.2 shows the impact of the quality factor on the insertion loss of a 3% and an 8% bandwidth filter. It is evident that in order to obtain high performance, a resonator unloaded Q of 500 or higher should be used for a narrowband diplexer design.

The local oscillator should also exhibit a very low phase noise performance, which is strongly dependent on the resonator quality factor. The oscillator phase noise, using a linear approximation, is given by

$$N_{\text{pn}}(f_m) = \frac{FkT}{2P_{\text{avs}}}\left[1 + \left(\frac{1}{2Q_L}\right)^2 \left(\frac{f_0}{f_m}\right)^2\right], \qquad (11.2)$$

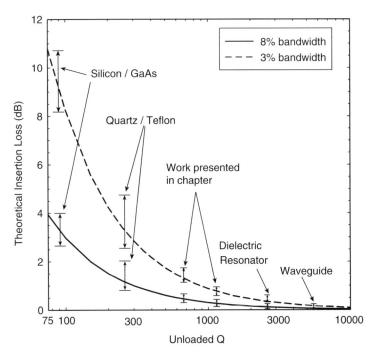

Figure 11.2. Theoretical insertion loss for 3% and 8% filters. The estimated Q is based on data from 30 to 60 GHz.

where F is the noise figure of the active circuit with the positive feedback removed, k is the Boltzmann constant, T is the temperature, P_{avs} is the available signal power, Q_L is the resonator loaded Q, f_0 is the oscillation frequency, and f_m is the frequency offset from the carrier where the noise spectral power is measured. For frequencies near the carrier, the phase noise is a function of $1/Q^2$. Millimeter-wave oscillators are typically fabricated with either a waveguide cavity, such as the case of Gunn or IMPATT diodes [5], or a dielectric resonator, as in the case of high electron mobility transistor (HEMT) or heterojunction bipolar transistor (HBT) devices [6–8]. These resonators exhibit a Q of 1000–3000 at 30–60 GHz and result in excellent phase noise performance. However, excellent performance is also achieved with a micromachined resonator with a Q of 500–1000.

This chapter discusses alternative methods for obtaining *planar* microwave and millimeter-wave high-Q resonators by using micromachining techniques to alter the geometry of a silicon substrate. The micromachined resonators are fully compatible with MMIC technology and, most importantly, do not require low-loss millimeter-wave transitions, which are necessary in waveguide technology. Also, micromachined resonators are lithographically defined and therefore do not require exact placement and manual tuning such as in dielectric resonator designs. It is expected that micromachining techniques will drastically lower the cost of future K-band to V-band communications and phased array systems.

11.2 HIGH-Q MICROMACHINED SUSPENDED MICROSTRIP RESONATORS

For the case of a distributed transmission line resonator, such as a microstrip or coplanar waveguide (CPW) resonator, the quality factor can be found by

$$Q = \frac{\pi}{\lambda \alpha}, \qquad (11.3)$$

where λ is the guided wavelength and α is the attenuation (in Np/m) including attenuation by radiation, substrate loss, and ohmic loss.

Distributed resonators using microstrip structures on conventional substrates have been used extensively for applications at X-band and below with reasonable values of Q (150–300 on quartz/Teflon). However, with increasing frequency, thinner substrates must be used to reduce radiation loss in substrate modes. This results in narrow line dimensions for a given impedance, which greatly increases the ohmic loss and drastically reduces the resonator Q. Micromachining techniques are used to produce a micropackaged, air dielectric line with wide transverse dimensions resulting in high-Q resonators at millimeter-wave frequencies.

Membrane-supported microstrip structures are formed by removing the silicon substrate and suspending a microstrip line on a thin (1–1.5 μm) dielectric membrane. A ground plane is formed by another micromachined substrate and attached to the top of the circuit. The bottom is also shielded with a third substrate (Fig. 11.3). For this structure, dielectric loss is eliminated with the air dielectric, the radiation loss is

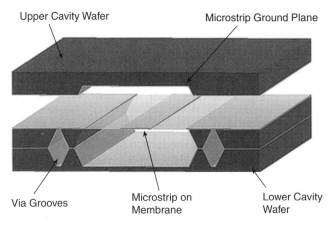

Figure 11.3. Transverse section of the microstrip structure.

minimized by shielding the structure on all sides using thick via grooves to limit substrate modes, and ohmic loss is greatly reduced by allowing for very wide transverse microstrip geometries. Silicon has been used for its low-cost anisotropic etching properties. Other materials can also be used, such as InP and GaAs. In the past, micromachining techniques have successfully been applied to K- and W-band membrane-supported microstrip lines and filters [9,10].

11.2.1 Fabrication of Membrane-Suspended Microstrip Resonators

The micromachined suspended microstrip transmission line is based on a three-wafer process (Fig. 11.3). The top wafer will be referred to as the ground plane wafer, the middle wafer as the circuit wafer, and the bottom wafer as the shielding wafer. For the circuit wafer, a stress-compensated 1.4 μm membrane layer consisting of $SiO_2/Si_3N_4/SiO_2$ (7000 Å/4000 Å/3000 Å) is deposited on a high-resistivity 525 μm thick silicon substrate using thermal oxidation and low-pressure chemical vapor deposition (LPCVD). This process deposits the thin film on both sides of the silicon wafer, allowing for a membrane on the top side of the wafer and a good etch mask for the silicon removal on the back side. The thicknesses of the SiO_2/Si_3N_4 layers are optimized to balance the net stress, leaving the membrane in slight tension to result in flat and rigid membranes. After the membrane is deposited, the circuit is patterned on the top side of the wafer using either a standard 2 μm gold electroplating technique or a 1 μm evaporated gold and lift-off procedure depending on the operating frequency of the resonator. Other circuit components such as thin film capacitors, resistors, or air bridges can also be included at this time. Next, an opening is defined on the back side of the wafer under the resonators and areas where via holes are to be formed by etching the back side membrane with a reactive ion etching (RIE) machine. The silicon is then completely etched under the circuit to the dielectric membrane, which acts as an excellent etch stop. The etchant used in this work was a solution of 12.5% tetramethyl ammonium hydroxide (TMAH) and water [11]. This solution has a 1.1 µm/min etch rate for the $\langle 100 \rangle$ crystal plane with a $\langle 100 \rangle : \langle 111 \rangle$ selectivity of 25 : 1.

The bottom shielding wafer and the top ground plane wafer are 525 µm thick, low-resistivity wafers with a 1.2 µm layer of thermal oxide that acts as an etch mask. The bottom wafer is etched down 400 µm and metalized with gold to prevent radiation.

The ground plane wafer requires a two-step etch from both the top and bottom side of the wafer for the formation of the ground plane plus access windows for on-wafer probing. The process flow is shown in Figure 11.4. Alignment marks are deposited on the front side of the wafer to allow for alignment from front to back sides. Photoresist is spun on the front side of the wafer and hardbaked to prevent etching of the silicon dioxide layer on the front of the wafer. Photoresist is then spun on the back side of the wafer and patterned with the probe window openings. The

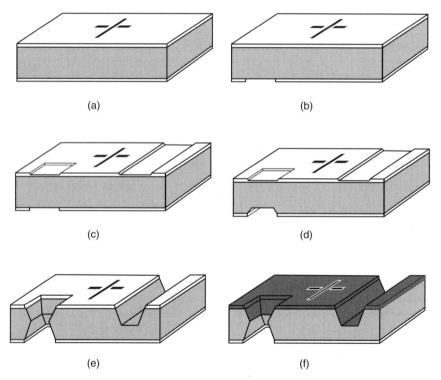

Figure 11.4. Fabrication of the ground plane wafer. (a) Alignment marks are deposited on a low-resistivity oxide wafer (oxide thickness = 1.2 µm). (b) The back side is patterned and the oxide is etched in the pattern openings. (c) The front side of the wafer is patterned and 9000 Å of oxide is etched. (d) The silicon is etched from the back of the wafer using the oxide as an etch mask. (e) Now 3000 Å of oxide is etched from the entire sample, exposing the silicon on the front side. The sample is then etched until the groud plane is at the specified depth. (f) The oxide is stripped off the sample and the sample is metalized.

silicon dioxide is completely removed from the openings, exposing the bare silicon. Similarly, the back side of the wafer is protected with photoresist and the front side is patterned with the ground plane pattern plus the probe window openings. Only 9000 Å of silicon dioxide is etched from the pattern openings, leaving 3000 Å of oxide to mask the etch. The sample is then placed in a solution of 12.5% TMAH and water to begin the etch of the probe window openings. The wafers used for the ground plane wafer are only polished on one side, resulting in a rough surface on the back side of the wafer. The rough surface has a much slower etch rate due to the formation of small hillocks in the crystal plane [12]. The etch rate for the rough, unpolished surface is approximately 0.7 times the etch rate for polished surfaces. The backside of the substrate is etched to a depth given by

$$t_{back} = t_{sub} - 1.7 t_{ground}, \qquad (11.4)$$

where t_{back} is the etch depth of the initial etching of the probe window openings, t_{sub} is the thickness of the substrate, and t_{ground} is the total depth of the ground plane. For the case of a 525 μm wafer with a 250 μm ground plane, the initial etch depth is 100 μm. The oxide layer is then etched uniformly by 3000 Å, completely exposing the silicon patterns on the front side of the sample. The oxide layer is thick enough that reducing it by 3000 Å in the unpatterned areas still acts as a good etch mask. The sample is placed back in the silicon etchant, where the silicon is etched from both sides of the sample. When the ground plane has been etched to the specified depth, the sample is removed from the etchant and all of the oxide masking layer is stripped off to provide a smoother surface for bonding. The ground plane is then metalized with a 500 Å/2 μm layer of titanium/gold using a static deposition with a sputtering tool. The three wafers are finally assembled together under a microscope using EPO-TEK H20E (product of Epoxy Technology, Inc., Billerica, MA) conductive silver epoxy to form the complete resonator.

11.2.2 Simulation and Measurement

Several resonators were constructed using both the bandpass and the bandstop configurations. In all cases, the resonators are $\lambda_0/2$ microstrip lines that are weakly coupled by capacitive coupling gaps or magnetically coupled by parallel coupled lines.

A single resonator was fabricated at 29 GHz in a bandstop configuration (Fig. 11.5). The ground plane height is 250 μm with an 800 μm wide conductor. The metal thickness of the conductor is 2 μm of electroplated gold (4 skin depths at 29 GHz). The shielding cavity is 800 μm away from the resonator. The length of the resonator was adjusted for the correct resonant frequency to 29.0 GHz by using a $2\frac{1}{2}$D moment method package IE3D [13]. The measured resonance was at 28.7 GHz, showing a 1% shift in the resonant frequency. This is due to the fringing capacitance to the sidewalls of the cavity that was not modeled with the full-wave analysis technique.

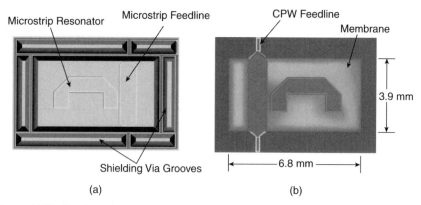

Figure 11.5. Circuit wafer of a 29 GHz microstrip resonator in bandstop configuration: (a) bottom view and (b) top view.

The suspended microstrip feedline is first measured without the resonator to study the effect of the CPW to microstrip transition. The microstrip line is 1025 μm wide, resulting in an impedance of 50 Ω over a 250 μm ground plane. The transition can be modeled with a series inductor to account for the extra inductance between the CPW and microstrip ground planes. Figure 11.6 shows the measured and modeled S_{11}, and excellent agreement is achieved with a 0.05 nH series inductance. Figure 11.6 also shows the S_{11} response with the resonator present. As shown, the microstrip line length is chosen to result in a good match at 27–32 GHz with the effects of the

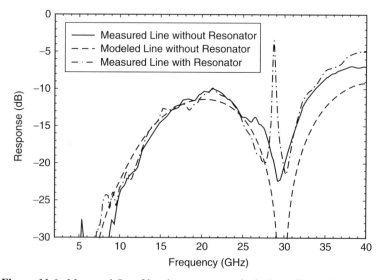

Figure 11.6. Measured S_{11} of bandstop resonator including effects of transition.

transition discontinuity. The measured loaded Q was 190 with a coupling -4.6 dB, giving an extracted unloaded Q of 460 at 28.7 GHz.

Two different resonators were also fabricated at 60 GHz. Both resonators have a ground plane height of 250 µm with widths of 500 and 700 µm. The resonator thickness is 1 µm of evaporated gold (3 skin depths at 60 GHz). The 500 µm lines had an extracted unloaded Q of 454 and the 700 µm lines had an extracted unloaded Q of 503. The repeatability of the measurements are to within 1% in Q and to within 0.03% in frequency.

Design curves for the micromachined resonator Q factor were generated using a code based on the surface impedance method [14] called Simian [15] (Fig. 11.7) for both the 30 GHz ($t = 2$ µm gold) and 60 GHz ($t = 1$ µm gold) resonators. The conductivity of the gold lines was assumed to be $3.9 \times 10^7 \, \Omega^{-1} \text{m}^{-1}$ in the simulations. Simian models the loss very accurately and is within 5% of the measured values. For the 60 GHz calculations, increasing the metal thickness to 2 µm will result in a 2-3% improvement in Q. The measured Q_u values at 30–60 GHz are 10× larger than microstrip resonators on GaAs or silicon, and 2× larger than corresponding resonators on low dielectric constant substrates such as Teflon [16].

The value of the resonator Q is sensitive to the ground plane height. Since the cavity sidewalls are far from the line, the structure is insensitive to $X - Y$ alignment. However, with increasing height ($h > 250$ µm), the CPW to membrane transition becomes more difficult with added transition inductance from the CPW ground to the microstrip ground plane. Also, increasing the width of the lines past 800–900 µm results in very wide transmission lines and makes circuit modeling and design more difficult (Fig. 11.8). Therefore, practical considerations will limit the Q of such resonators to 550–700 for most applications.

11.2.3 Effects of Thermal Expansion on the Resonant Frequency

Changes in temperature have the effect of thermal expansion and contraction of the metallic transmission line, the membrane, and the silicon substrate, resulting in a change in resonant frequency. Since the membrane is large, is relatively elastic, and has low net stress, small changes in the geometry of the silicon carrier have little effect on the geometry of the microstrip line. The temperature coefficient of linear expansion, α, is defined as

$$\alpha = \frac{\Delta l}{l_0 T}, \tag{11.5}$$

where l_0 is the length of the structure at 0°C, Δl is the change in length, and T is the temperature in degrees Celsius. The temperature coefficients for linear expansion for gold, silicon dioxide, and silicon nitride are [17]

$$\alpha_{Au} = 14.2 \times 10^{-6} \, \text{C}^{-1}, \tag{11.6}$$

$$\alpha_{SiO_2} = 0.5 \times 10^{-6} \, \text{C}^{-1}, \tag{11.7}$$

$$\alpha_{Si_3N_4} = 4.0 \times 10^{-6} \, \text{C}^{-1}. \tag{11.8}$$

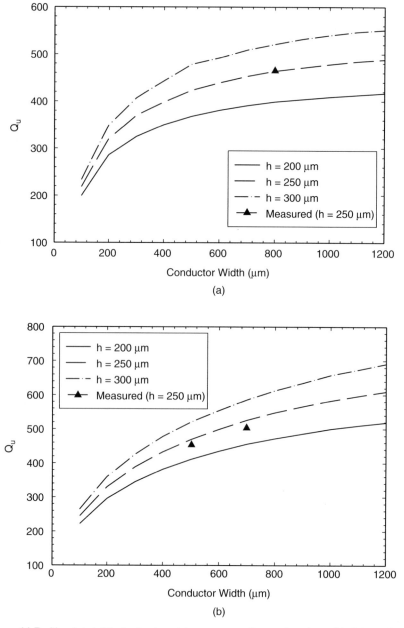

Figure 11.7. Simulated (Simian) microstrip resonator Q as a function of height and strip width (a) at 30 GHz with 2 μm thick lines and (b) at 60 GHz with 1 μm thick lines.

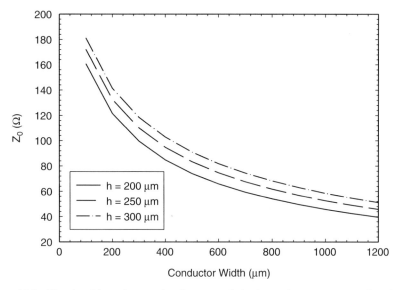

Figure 11.8. Simulated impedance value for suspended microstrip resonator as a function of height and strip width.

The coefficient of expansion for gold is much greater than that of the silicon dioxide and the silicon nitride and will dominate the expansion of the microstrip line length. The resonant frequency is directly proportional to the physical length of the transmission line. Due to the layered structure, this is a complicated problem to solve; however, one can safely assume that the percent change in resonant frequency as a function of temperature of the microstrip line is due to the thermal expansion of nitride and gold and is approximated by

$$f'_0 = \frac{f_0}{1 + [(6 \text{ to } 10) \times 10^{-6} \text{C}^{-1}]T}, \quad (11.9)$$

where f'_0 is the resonant frequency including temperature fluctuations. The fractional temperature coefficient (TC_F) is then given by

$$TC_F = \frac{1}{f'_0} \frac{\partial f'_0}{\partial T} = \frac{-(6 \text{ to } 10) \times 10^{-6}}{1 + (6 \text{ to } 10) \times 10^{-6}T}. \quad (11.10)$$

This results in a fractional temperature coefficient for the resonant frequency of −6 to 10 ppm/°C. Temperature-compensated dielectric resonators are of the order of −4 to 6 ppm/°C and since the above result is a conservative estimation, one can safely say that the micromachined resonators offer a similar thermal performance as state-of-the-art dielectric resonators at 30 GHz.

11.3 MICROMACHINED LOW-LOSS LMDS FILTERS AND DIPLEXERS

The LMDS filter and diplexer topology is based on a membrane-supported, capacitively coupled bandpass filter [18], as shown in Figure 11.9(a). The resonators are $\lambda/2$ lengths of capacitively end-coupled microstrip transmission lines. The feedlines are coplanar waveguide lines on silicon with an abrupt transition to microstrip at the edge of the membrane. The ground plane wafer is etched to conform to the membrane area and has small "mouseholes" over the CPW lines. The edge of the mouseholes contact the CPW ground planes and act as a via from the CPW ground plane to the microstrip ground plane at the silicon/membrane interface. Due to the anisotropic etching properties of the etchant used for fabrication, these ground vias are 1 mm from the signal conductor and add an inductance (0.05–0.06 nH) in the CPW-to-microstrip ground plane transition.

The resonators used in the filters and diplexer consist of 800 μm wide lines with a ground plane height of 250 μm (etch depth of the top wafer) and a shielding cavity

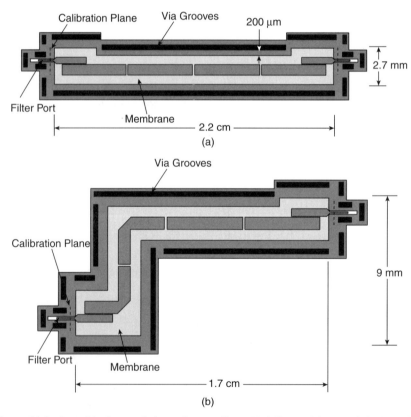

Figure 11.9. Capacitively coupled membrance filter: (a) inline and bent and (b) bent filter method.

height of 800 μm (400 μm for the thickness of the middle wafer and 400 μm for the etch depth of the bottom shielding wafer). The resonator is composed of 2 μm thick electroplated gold. The distance from the edge of the conductor to the sidewall of the micromachined channel is 700 μm. This results in a microstrip line with characteristic impedance of 62 Ω. A half-wavelength resonator constructed from this geometry has a measured unloaded quality factor of 460 at 29 GHz with a corresponding attenuation coefficient of 0.057 dB/cm (see above) [19].

11.3.1 Filter Packaging and Out-of-Band Rejection

A disadvantage of the filter shown in Figure 11.9(a) is that the out-of-band rejection is very susceptible to degradation by parasitic modes of the micromachined structure. These may result in an increase in coupling due to the evanescent modes in the cavity structure and also in propagating modes in the silicon surrounding the cavity. This is greatly improved by bending the filter structure to disturb any possible modes that are formed (Fig. 11.9(b)). The lengths of the resonators are adjusted to compensate for the bends using IE3D [13], a $2\frac{1}{2}$D commercially available software package based on the method of moments. The length of the 28 GHz straight resonator is found to be 5 mm, while the bent resonator consists of two 2.2 mm lines connected to an optimally mitered corner.

Figure 11.10 shows a comparison of measured results for the inline filter from Figure 11.9(a), the bent filter from Figure 11.9(b)), and a full-wave simulation for a four-pole capacitively coupled filter with 5.5% bandwidth at a center frequency of 28.0 GHz. The measured results are calibrated to the reference plane (Fig. 11.9) using a TRL calibration method [20] and include the CPW-to-microstrip transition. The simulated results were done using IE3D. The simulation conditions neglect the effects of the CPW-to-microstrip transition and assume a 1.4 μm thick dielectric membrane ($\epsilon_r = 5$) on a 250 μm air layer with an 800 μm high shielding cover that is infinite in extent. The simulation is done for an inline filter and has an identical response to a bent filter.

The measured insertion loss is 0.85 and 1.00 dB for the inline and bent filters, respectively, which is close to the simulated loss of 0.75 dB. Note that the bent topology has a large improvement in out-of-band rejection compared to the inline filter (better than -75 dB for the bent as compared to -35 dB for the inline filter) with only a 0.15 dB increase in insertion loss due to the mitered resonator bends. The -75 to -80 dB filter rejection is competitive with "stand alone" filters used in multisubstrate packages. This is due to the micropackaging technique used in this design: the via holes around the filter structure, the ground plane on top of the filter, and the bottom shield plane contribute to make a completely closed micromachined "box" around the filter.

11.3.2 Power Handling Capabilities of Membrane-Supported Filter

Power handling has been a concern for membrane-suspended filters since the resonators are isolated from the substrate and the only real path for thermal conduction is

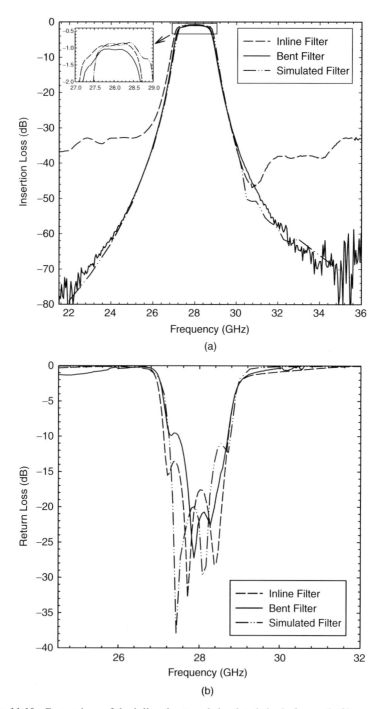

Figure 11.10. Comparison of the inline, bent, and simulated single four-pole filter response: (a) insertion loss and (b) return loss.

through air. The probable failure mechanism due to power loading would be the resonators peeling up from the membrane due to differences in expansion coefficients of the metal and the membrane. The membrane-suspended filter was measured under high power using a Hughes 8001H traveling-wave tube amplifier (TWTA). The total incident power on the filter was approximately 0.7 W with no apparent increase in insertion loss or filter failure. More testing is needed to determine the maximum allowable value of power handling, but an estimated value is 2–5 W. It is evident that the maximum power will depend on the gold thickness, resonator width, ground plane height, and the filter insertion loss.

11.3.3 Diplexer Design Details

The diplexer consists of two bent capacitively coupled bandpass filters with one port shared between the filters (see Fig. 11.11). The two channels for the diplexer were chosen to correspond roughly to the commercial local multipoint distribution system (LMDS). The receive band filter is designed using a four-pole Chebyshev prototype with a center frequency of 28.2 GHz, relative bandwidth of 5.5%, and a ripple of 0.1 dB. The transmit band filter is a three-pole Chebyshev filter with a center frequency of 31.75 GHz, relative bandwidth of 5.5%, and a ripple of 0.1 dB.

The design of the inter-resonator couplings follows a standard procedure [18, 21–23]. The inter-resonator couplings are calculated using Sonnet EM [24], a commercially available method of moments package. The effects of the micromachined sidewalls are modeled using a perfect conducting vertical wall. The coupling coefficients and associated end gaps are given in Table 11.1, where $k_{i,i+1}$ and $g_{i,i+1}$ are the coupling coefficients and corresponding microstrip series gaps between resonator i and $i+1$, respectively.

The output coupling (noncommon port to the two filters) was set by the required external Q as determined by the low-frequency prototype. The receive and transmit

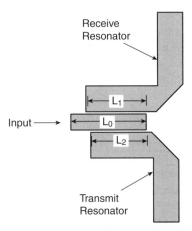

Figure 11.11. Coupling structure of the common port of the diplexer.

TABLE 11.1. Summary of Inter-resonator Couplings

i	Receive Band		Transmit Band	
	$k_{i,i+1}$	$g_{i,i+1}$	$k_{i,i+1}$	$g_{i,i+1}$
1	0.042	140 µm	0.034	160 µm
2	0.033	190 µm	0.034	160 µm
3	0.042	140 µm		

band filters have an external Q of 22 and 34, respectively. This is implemented by a section of asymmetric coupled microstrip line. The length of the 500 µm wide feedline and the length and gap of the coupled section were optimized using Sonnet EM [24]. The optimization included only the microstrip mode on a dielectric membrane and neglects the effects of the CPW-on-silicon to the microstrip-on-membrane transition. For the receive (transmit) band filter, the length of the feed line is 2500 µm (2350 µm) with a coupled section 1900 µm (1700 µm) long and a coupling gap of 60 µm (70 µm).

The coupling structure of the common port of the diplexer is a three-line coupled microstrip geometry. As with the other ports, the feed line is 500 µm wide. In order to reduce receive/transmit filter interactions, the reactance from the rejection of one filter is compensated in the other filter. This was done by extending the length of the feedline (L_0 in Fig. 11.11) to 2375 µm, increasing the length of the coupling section of the receive filter, L_1, to 1950 µm, and decreasing the length of the coupled section for the transmit filter, L_2, to 1650 µm. The coupling gap remained at 60 µm and 70 µm for the receive and transmit filters, respectively.

11.3.4 Integrated Diplexer Measurements

The micromachined diplexer was fabricated in silicon using a MMIC compatible process (Fig. 11.12). The diplexer outer dimensions are 1.5 cm by 1.6 cm and the diplexer is only 1.4 mm thick. A considerable amount of the space in the outer dimensions is not used for the actual diplexer and can fit other micromachined or planar circuitry low noise amplifier, power amplifier, biasing circuits, etc.) (Fig. 11.13).

The diplexer response was measured with an HP8510C network analyzer calibrated using a TRL calibration method [20,25] with 150 µm pitch coplanar Picoprobes (a product of GGB Industries, Inc., Naples, FL). Since this is a two-port measurement system, the third port was always terminated by another probe with an HP901C 2.4 mm broadband load. On all measurements, the system was calibrated with a reference plane located 500 µm from the silicon-to-membrane transition in order to take into account the discontinuity of the transition (see Fig. 11.9).

The measured results are summarized in Figure 11.14. The measured insertion loss of the four-pole receive filter is 1.4 dB including all transition effects, the effects of the loss in the rejection of the transmit filter, and the 500 µm long CPW feedlines on silicon at both ports. This is in very good agreement with the simulated value of

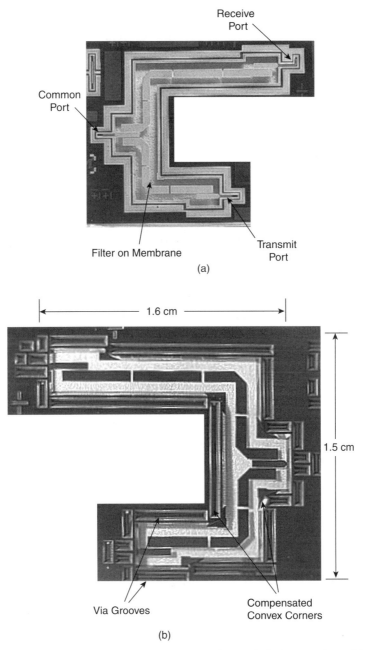

Figure 11.12. Fabricated K-band diplexer: (a) top view and (b) bottom view without the ground plane or shielding cavity.

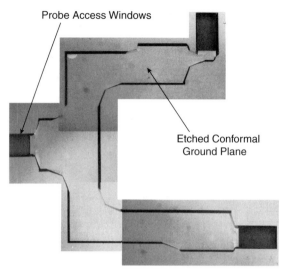

Figure 11.13. Fabricated diplexer conformal ground plane (bottom view).

0.9 dB, which neglects these losses (Table 11.2). The measured insertion loss of the three-pole transmit filter is only 0.9 dB compared to a simulated value of 0.65 dB due to the additional losses (transition, losses of CPW feedlines on silicon), which are not accounted for in the simulation. In both cases, the reflection coefficient in the passband is less than -10 dB.

The measured transmit-to-receive port isolation is better than -35 dB across the receive band (-40 dB in the center of the band) and better than -50 dB in the transmit band. The isolation closely follows the rejection of the individual filters. Not that the leakage into the substrate is below -70 to -80 dB, which is competitive with hybrid-cavity packaging techniques.

11.4 MICROMACHINED LOW PHASE NOISE OSCILLATOR

This section presents one possible alternative method for building low phase noise oscillators in an integrated, planar structure while maintaining the basic oscillator design of a DRO. Rather than using an external high-Q resonator, micromachining techniques are used to produce a micropackaged, air dielectric line with wide transverse dimensions resulting in high-Q resonators at millimeter-wave frequencies (Fig. 11.15). The 28 GHz micromachined resonator outline above with an unloaded Q of 460 is used in this design.

11.4.1 Oscillator Design

The transistor used was a commercially available Fujitsu FHR20X GaAs HEMT. The device has a minimum noise figure of 1.5 dB at 28 GHz with an associated gain

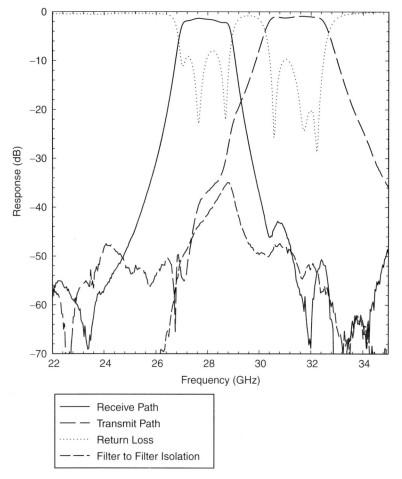

Figure 11.14. Summary of measured diplexer.

of 8 dB. A nonlinear model for this device was not available, so the design was based solely on the manufacturer supplied S-parameter file. The S-parameter file included the effects of the wirebonding of the chip HEMT to the carrier substrate. The oscillator topology follows the design of the series feedback DRO (outlined in Wilson and Carver [26]). The design is based on CPW transmission lines for everything but the micromachined resonator section. All lines are 50 Ω with a conductor

TABLE 11.2. Measured Diplexer Performance

Characteristic	Receive Filter	Transmit Filter
Insertion loss	1.4 dB	0.9 dB
Relative bandwidth	5%	6.5%
Adjacent band isolation	> 35 dB	> 50 dB

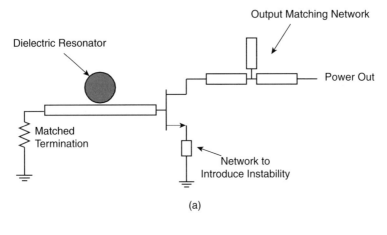

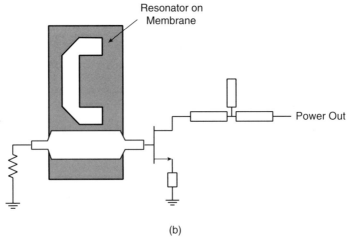

Figure 11.15. Series feedback: (a) dielectric resonator oscillator and (b) membrane-supported microstrip topology.

width of 100 μm and a gap width of 65 μm. The process includes air bridges placed at major discontinuities. The oscillator is biased with external bias-tees.

The source of the transistor was first inductively loaded with shorted stubs to increase instability. The amount of inductance added sets the source plane stability circles to coincide with the magnitude of the reflection coefficient of the resonator at resonance. The stubs are 50 Ω sections of 650 μm long CPW lines. This is a dual-source device and the stubs are symmetrical on both sides of the device.

The resonator is an 800 μm wide microstrip line that has been bent in a U-shape using optimal miters to conserve membrane space. The length of the resonator was adjusted to give a resonance at 28.7 GHz using a commercially available method of moments software package [13]. The coupling gap is 200 μm wide and 900 μm long

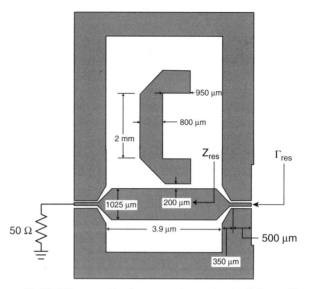

Figure 11.16. Micromachined resonant input circuit of the oscillator.

(Fig. 11.16). The measured loaded Q was 190 with a coupling of -4.6 dB ($n^2R = 144$ Ω), giving an extracted unloaded Q of 460 at 28.70 GHz. If the line lengths of the resonator are deembedded to the coupling gap, the resonator presents an impedance of $Z_{res} = 144$ Ω at resonance. By adding the correct line length, the phase of the reflection coefficient is rotated around the Smith chart so that it crosses the stability circle at resonance. The equivalent added line length is the original line length of the resonator structure including the transition plus 500 μm of 50 Ω CPW line on silicon. Figure 11.17 shows the measured response for the micromachined resonator with added line length (feedline) to achieve the correct phase. Note that the reflection coefficient crosses the input plane stability only around the resonance peak.

By loading the gate and source to create instability, a negative impedance is presented at the drain of the transistor. An output matching circuit was designed to satisfy the conditions of

$$R_L < -R_{in}, \quad (11.11)$$

$$X_L = -X_{in}, \quad (11.12)$$

where R_L and X_L are the effective resistance and reactance looking into the output matching network and R_{in} (negative) and X_{in} are the resistance and reactance looking into the drain of the transistor at 28.7 GHz. The simulated impedance at the drain is $-12 + j50$ Ω for small signal conditions at 28.7 GHz. The matching network was implemented by a 725 μm long 50 Ω line to a pair of 800 μm long open-circuited

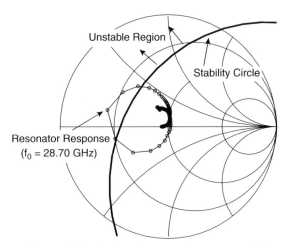

Figure 11.17. Input plane stability circle and measured resonator reflection coefficient from 20 to 24 GHz. The points are plotted at 50 MHz intervals.

stubs. The matching network presents a simulated impedance of $4 + j50\,\Omega$, and a measured impedance of $8.1 + j55.3\,\Omega$ at the drain of the transistor at 28.7 GHz.

11.4.2 Measured Results

The fabricated micromachined oscillator is shown in Figure 11.18. The total dimensions of the assembled circuit are 6.8 mm × 8 mm and 1.4 mm thick. The oscillator was measured using 150 μm pitch Picoprobes connected to HP11612B Bias Networks and the output signal was measured with an HP8564E Spectrum Analyzer. The transistor oscillated at a bias condition of $V_{GS} = -0.3$ V, $V_{DS} = 2$ V, and $I_{DS} = 10$ mA. The oscillation frequency was 28.6536 GHz in close agreement with the design value of 28.7 GHz. The output power was 0.6 dBm. This results in a 5.7% dc-RF efficiency (Fig. 11.19).

The measured phase noise is -92 dBc/Hz at 100 kHz and -122 dBc/Hz at 1 MHz offset. The slope of the phase noise below 1 MHz shows that the phase noise is rolling off as $1/f_m^3$, where f_m is the frequency offset from the carrier. This is characteristic of up-converted FM flicker noise [27]. The manufacturer of the HEMT does not specify any data for flicker noise of the FHR20X.

HEMT DROs have resulted in phase noise of -102 dBc/Hz at 100 kHz and -117 dBc/Hz at 1 MHz offset at 30 GHz [6], and -68 dBc/Hz at a 100 kHz offset for a DRO at 38 GHz in Wilson [28]. The micromachined oscillator showed a phase noise that was 10 dB worse than that of Funabashi et al. [6] at 100 kHz offset frequency and 5 dB better at 1 MHz. (Note that the reported phase noise of Funabashi et al. [6] does not follow either the $1/f$ phase noise profile of -30 dB/decade or the pure white FM phase noise profile of -20 dB/decade and no measurement details were provided in the published work.) The performance of the

IMPLEMENTATION OF MEMBRANE-SUSPENDED COMPONENTS 405

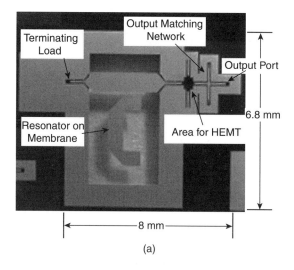

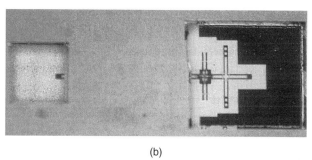

Figure 11.18. Fabricated micromachined oscillator (a) with the cover removed (photo taken before the HEMT was mounted) and (b) packaged with device mounted and ground plane cover assembled.

micromachined resonator is equivalent and even better than published results using dielectric resonators and HEMT devices.

11.5 IMPLEMENTATION OF MEMBRANE-SUSPENDED COMPONENTS AT OTHER FREQUENCY BANDS

The technology presented above at 28–30 GHz can easily be scaled to higher frequencies with virtually no change in the fabrication or design procedure. In fact, excellent filters have been demonstrated at 37 GHz for point-to-point telecommunications [18] (two-poles, two-zeros with 3% bandwidth filter and 2.5 dB insertion loss), and at 60 GHz for indoor telecommunications [29] (four-pole elliptic filter with 8% bandwidth and 1.5 dB insertion loss). The only requirement is to take into account the inductance due to the transition between the CPW line and the

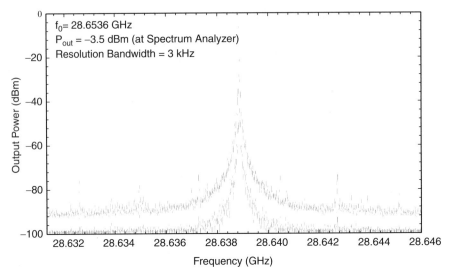

Figure 11.19. Measured micromachined oscillator spectrum. The power is defined at the spectrum analyzer port.

suspended microstrip resonator. The Q of the resonator actually increases at higher frequencies, and a Q of 550 was measured for a conductor width of 800 μm at 60 GHz (see Fig. 11.7). The maximum practical Q of the suspended resonators is 700–800 at 60–100 GHz, and, therefore, this technology is suitable for multipole filters having a bandwidth of 3% and above. Also, oscillators can be built at 60–100 GHz with a Q of 600–700 using millimeter-wave HEMT and HBT devices. Again, the advantage of this approach is the elimination of manual placement and tuning in oscillator design, which is quite hard at millimeter-wave frequencies. This technology is therefore ideally suited for low-loss filters and low phase noise oscillators in 20–60 GHz systems, 77 GHz automotive radars, and 80–100 GHz point-to-point telecommunications systems and short-range radars.

Membrane-suspended resonators can also be used at 10 GHz and above for high-performance filters and oscillators. In fact, at these frequencies, one can use the newly developed silicon transistors for excellent $1/f$ phase noise characteristics. The transition between the suspended resonator and the active circuit (transistors, etc.) can easily be modeled at 10–20 GHz and actually does not seriously affect the circuit design. However, at these frequencies, one can obtain similar resonator Q (400–500) from shielded microstrip resonators on quartz substrates. Also, the resonator dimensions become quite large at 10 GHz and below, which results in large membranes especially for multipole filters. Therefore, it is our opinion that this technology is of limited use in complicated filters at 10–15 GHz but is very useful for manual-free 10 GHz oscillators and above. Also, it offers excellent performance for both filters and oscillators at 20 GHz and above.

11.6 CONCLUSIONS

This chapter presents a micromachined membrane-suspended technology suitable for high-Q resonators at 10–100 GHz. The high-Q resonators are used to build state-of-the-art components (filters, diplexers, and oscillators) for a 28 GHz front-end. The components were fabricated monolithically and the measured results agree very well with simulations. The micromachined filter and diplexer show rejection to the -80 dB level with no leakage in the carrier substrates, which is due to the micropackaging technique used. Also, excellent oscillator performance was achieved at 28 GHz with no manual tuning whatsoever. The technology can easily be applied to 20–100 GHz systems with virtually no change in the design or fabrication methods.

All of the front-end components were fabricated on the same carrier substrate, and, therefore, no "off-chip" transitions are needed to build a medium power transmit–receive front-end. One can use the silicon substrate as the carrier wafer, build all the high-Q micromachined passive components and connecting t-lines on this silicon substrate, and place the active components (transistors, diodes, MMICs, etc.) using hybrid techniques. The entire front-end is less than 1 mm thick and can be assembled automatically using fast pick-and-place machines. This results in very compact and low-cost transceivers for millimeter-wave communications systems.

REFERENCES

1. D. A. Gray, "A broadband wireless access system at 28 GHz," in *1997 Wireless Communications Conferences Proceeding*, pp. 1–7, 1997.
2. H. H. Meinel, "Commercial applications of millimeter-waves. History, present status and future trands," *IEEE Transactions on Microwave Theory and Techniques*, Vol. 43, No. 7, 1639–1653 (July 1995).
3. J. Burns, "The application of millimetre-wave technology for personal communication networks in the United Kingdom and Europe: a technical and regulatory overview," *IEEE MTT-S International Microwave Symposium Digest*, 635–638 (1994).
4. G. L. Matthaei, L. Young, and E. M. T. Jones, *Microwave Filters, Impedance-Matching Networks, and Coupling Structures*, Artech House, Norwood, MA, 1980.
5. S. Nagano and S. Ohnaka, "A low-noise 80 GHz silicon IMPATT oscillator highly stabilized with a transmission cavity," *IEEE Transactions on Microwave Theory and Techniques*, Vol. 22, No. 12, 1152–1159 (Dec. 1974).
6. M. Funabashi, T. Inoue, K. Ohata, K. Maruhashi, K. Hosoya, M. Kuzuhara, K. Kanckawa, and Y. Kobayashi, "A 60 GHz MMIC stabilized frequency source composed of a 30 GHz DRO and a doubler," *IEEE MTT-S International Microwave Symposium Digest*, 71–74 (1995).
7. S. Chen, S. Tadayon, T. Ho, K. Pande, P. Rice, J. Adair, and M. Ghahremani, "U-band MMIC HBT DRO," *IEEE Microwave and Guided Wave Letters*, Vol. 4, No. 2, 50–52 (Feb. 1994).

8. H. Wang, K. Chang, L. Tran, J. Cowles, T. Block, E. Lin, G. Dow, A. Oki, D. C. Streit, and B. Allen, "Low phase noise millimeter wave frequency sources using InP-based HBT MMIC technology," *IEEE Journal of Solid-State Circuits*, Vol. 31, No. 10, 1419–1425 (Oct. 1996).
9. G. M. Rebeiz, L. P. B. Katehi, T. M. Weller, C. Y. Chi, and S. V. Robertson, "Micromachined membrane filters for microwave and millimeter-wave applications," *International Journal of Microwave and Millimeter-Wave CAE*, No. 7, 149–166 (1997).
10. C. Y. Chi and G. M. Rebeiz, "Conductor-loss limited stripline resonator and filters," *IEEE Transactions on Microwave Theory and Techniques*, Vol. 44, No. 4, 626–630 (Apr. 1996).
11. O. Tabata, R. Asahi, H. Funabashi, K. Shimaoko, and S. Sugiyama, "Anisotropic etching of silicon in TMAH solutions," *Sensors and Actuators A*, Vol. X, No. 34, 51–57 (1992).
12. L. M. Landsberger, S. Naseh, M. Kahrizi, and M. Paranjape, "On hillocks generated during anisotropic etching of Si in TMAH," *Journal of Microelectromechanical Systems*, Vol. 5, No. 2, 106–116 (June 1996).
13. *Zeland IE3D, Releas 4.12*, 1997.
14. E. Tuncer, B.-T. Lee, and D. P. Neikirk, "Interconnect series impedance determination using a surface ribbon method," in *3rd Topical Meeting on Electrical Performance of Electronic Packaging*, pp. 250–252, Nov. 1994.
15. S. Kim, E. Tuncer, B.-T. Lee, and D. P. Neikirk, *Surface Impedance Method for Interconnect Analysis*, http:// weewave.mer.utexas.edu/MedHome.html, 1997.
16. C. T.-C. Nguyen, L. P. B. Katehi, and G. M. Rebeiz, "Micromachined devices for wireless communications," *Proceedings of the IEEE*, Vol. 86, 1756–1768 (1998).
17. S. Wolf and R. N. Tauber, *Silicon Processing for the VLSI Era: Volume 1-Process Technology*, Lattice Press, Sunset Beach, CA, 1986.
18. P. Blondy, A. R. Brown, D. Cros, and G. M. Rebeiz, "Low loss micromachined filters for millimeter-wave telecommunication systems," *IEEE MTT-S International Microwave Symposium Digest*, 1181–1184 (1998).
19. A. R. Brown, P. Blondy, and G. M. Rebeiz, "Microwave and millimeter-wave high-Q micromachined resonators," *International Journal of Microwave and Millimeter-Wave CAE*, Vol. 9, 326–337 (July 1999).
20. R. B. Marks and D. F. Williams, *Multical v1.00*, NIST, Aug. 1995.
21. J. S. Hong and M. J. Lancaster, "Couplings of microstrip square open-loop resonators for cross-coupled planar microwave filters," *IEEE Transactions on Microwave Theory and Techniques*, Vol. 44, No. 12, 2099–2109 (Dec. 1996).
22. S. B. Cohn, "Direct coupled resonator filters," *Proceedings of the IRE*, 187–196 (Feb. 1957).
23. A. E. Atia and A. E. Williams, "Narrow bandpass waveguide filters," *IEEE Transactions on Microwave Theory and Techniques*, Vol. 20, No. 4, 258–265 (Apr. 1972).
24. *Sonnet EM, Release 5.1a*, 1997.
25. R. B. Marks, "A multiline method of network analyser calibration," *IEEE Transactions on Microwave Theory and Techniques*, Vol. 39, No. 7, 1205–1215 (July 1991).
26. P. G. Wilson and R. D. Carver, "An easy-to-use FET DRO design procedure suited to most CAD programs," *IEEE MTT-S International Microwave Symposium Digest*, 1033–1036 (1989).

27. U. L. Rohde, *Microwave and Wireless Synthesizers: Theory and Design*, Wiley, New York, 1997.
28. P. G. Wilson, "Monolithic 38 GHz dielectric resonator oscillator," *IEEE MTT-S International Microwave Symposium Digest*, 831–834 (1991).
29. P. Blondy, A. R. Brown, D. Cros, and G. M. Rebeiz, "Low loss micromachined filters for millimeter-wave telecommunication systems," *IEEE Transactions on Microwave Theory and Techniques*, Vol. 46, No. 12, 2283–2288 (Dec. 1998).

12

TRANSCEIVER FRONT-END ARCHITECTURES USING VIBRATING MICROMECHANICAL SIGNAL PROCESSORS

CLARK T.-C. NGUYEN
Department of Electrical Engineering and Computer Science
University of Michigan, Ann Arbor

12.1 INTRODUCTION

The need for passive off-chip components has long been a key barrier against communication transceiver miniaturization. In particular, the majority of the high-Q bandpass filters commonly used in the radio frequency (RF) and intermediate frequency (IF) stages of heterodyning transceivers are realized using off-chip, mechanically resonant components, such as crystal filters and SAW devices. Due to higher quality factor Q, such technologies greatly outperform comparable filters implemented using transistor technologies, in insertion loss, percent bandwidth, and achievable rejection [1–8]. High Q is further required to implement local oscillators or synchronizing clocks in transceivers, both of which must satisfy strict phase noise specifications. Again, off-chip elements (e.g., quartz crystals) are utilized for this purpose. Being off-chip components, the above mechanical devices must interface with integrated electronics at the board level, and this constitutes an important bottleneck against the miniaturization of superheterodyne transceivers. For this reason, recent attempts to achieve single-chip transceivers for paging and cellular communications have utilized alternative architectures [9,10], that attempt to

---

RF Technologies for Low Power Wireless Communications
Edited by Tatsuo Itoh, George Haddad, and James Harvey
ISBN 0-471-38267-1 Copyright © 2001 by John Wiley & Sons, Inc.

eliminate the need for off-chip high-Q components via higher levels of transistor integration. Unfortunately, without adequate front-end selectivity, such approaches have suffered somewhat in overall performance, to the point where they so far are usable only in less demanding applications. Given this, and recognizing that future communications needs will most likely require higher levels of performance, single-chip transceiver solutions that retain high-Q components and that preserve super-heterodyne-like architectures are desirable.

Recent demonstrations of vibrating beam micromechanical ("μmechanical") resonator devices with frequencies in the VHF range and Q's in the tens of thousands [11,12] have sparked a resurgence of research interest in communications architectures using high-Q passive devices. Much of the interest in these devices derives from their use of integrated circuit (IC)-compatible microelectromechanical systems (MEMSs) fabrication technologies [8,13] to greatly facilitate the on-chip integration of ultra-high-Q passive tanks together with active transistor electronics, allowing substantial size reduction. In essence, MEMS technology may eventually allow replacement of off-chip SAW and crystal technologies by on-chip vibrating micromechanical resonators with comparable Q and performance. Indeed, reductions in size and board-level packaging complexity, as well as the desire for the high performance attainable by superheterodyne architectures, are the principal drivers for this technology.

Although size reduction is certainly an advantage of this technology (commonly dubbed "RF MEMS"), it merely touches upon a much greater potential to influence general methods for signal processing. In particular, since they can now be integrated (perhaps on a massive scale) using MEMS technology, vibrating μmechanical resonators (or μmechanical links) can now be thought of as tiny circuit elements, much like resistors or transistors, in a new mechanical circuit technology. Like a single transistor, a single mechanical link does not possess adequate processing power for most applications. However, again like transistors, when combined into larger (potentially, VLSI) circuits, the true power of μmechanical links can be unleashed, and signal processing functions with attributes previously inaccessible to transistor circuits may become feasible. This in turn can lead to architectural changes for communication transceivers. MEMS technology may in fact make its most important impact not at the component level, but at the system level, by offering alternative transceiver architectures that emphasize selectivity to substantially reduce power consumption and enhance performance.

This chapter focuses on communications subsystem performance enhancements potentially attainable via use of vibrating micromechanical resonator circuits in MEMS-based transceiver architectures. Section 12.2 begins by reviewing the issues involved with transceiver miniaturization, specifically focusing on the need for high Q. Section 12.3 follows with an overview of various micromechanical signal processing circuits, describing micromechanical tanks, filters, mixer–filters, and switches and providing sufficient detail to facilitate the incorporation of such circuits into communications subsystems. Specific transceiver architectures using such devices are then presented in Sections 12.4 and 12.5, with a focus on the specific performance enhancements afforded by each approach. Section 12.6 then serves as a reminder that this is still a fledgling technology, emphasizing the present limitations of this technology and giving a sprinkling of the specific research problems that must be solved before the architectures of Sections 12.4 and 12.5 can become a reality. Sections 12.7 and 12.7.4.1 finally close the chapter with details on techniques for combining mechanical and transistor circuits onto single chips.

12.2 MINIATURIZATION OF TRANSCEIVERS

To illustrate more concretely the specific transceiver functions that can benefit from micromechanical implementations (to be discussed), Figure 12.1 presents the system-level schematic for a typical superheterodyne wireless transceiver. As implied in the figure, several of the constituent components can already be miniaturized using integrated circuit transistor technologies. These include the low noise amplifiers (LNAs) in the receive path, the solid-state power amplifier (SSPA) in the transmit path, synthesizer phase-locked loop (PLL) electronics, mixers, and lower frequency digital circuits for baseband signal demodulation. Due to noise, power, and frequency considerations, the SSPA (and sometimes the LNAs) are often implemented using compound semiconductor technologies (i.e., GaAs). Thus, they often occupy their own chips, separate from the other mentioned transistor-based components, which are normally realized using silicon-based bipolar and complementary metal oxide semiconductor (CMOS) technologies. However, given the rate of improvement of silicon technologies (silicon–germanium included [14]), it is not implausible that all of the above functions could be integrated onto a single chip in the foreseeable future.

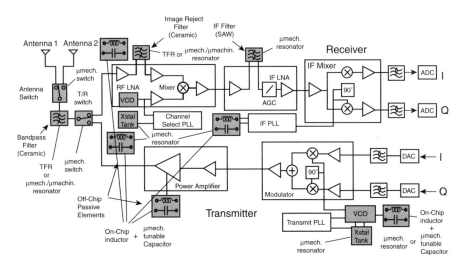

Figure 12.1. System-level schematic detailing the front-end design for a typical wireless transceiver. The off-chip, high-Q, passive components targeted for replacement via micromechanical versions are indicated by shading in the figure. (From reference [8])

Unfortunately, placing all of the above functions onto a single chip does very little toward decreasing the overall superheterodyne transceiver size, which is dominated not by transistor-based components, but by the numerous passive components indicated in Figure 12.1. The presence of so many frequency-selective passive components is easily justified when considering that communications systems designed to service large numbers of users require numerous communication channels, which in many implementations (e.g., time division multiple access (TDMA)) must have small bandwidths and must be separable by transceiver devices used by the system. The requirement for small channel bandwidths results in a requirement for extremely selective filtering devices for channel selection and extremely stable (noise-free) local oscillators for frequency translation. For the vast majority of cellular and cordless standards, the required selectivity and stability can only be achieved using high-Q components, such as discrete inductors, discrete tunable capacitors (i.e., varactors), and SAW and quartz crystal resonators, all of which interface with IC components at the board level. The needed performance cannot be achieved using conventional IC technologies, because such technologies lack the required Q. It is for this reason that virtually all commercially available cellular or cordless phones contain numerous passive SAW and crystal components.

12.2.1 The Need for High Q in Oscillators

For any communications application, the stability of the oscillator signals used for frequency translation, synchronization, or sampling is of the utmost importance. Oscillator frequencies must be stable against variations in temperature, against aging, and against any phenomena, such as noise or microphonics, that cause instantaneous fluctuations in phase and frequency. The single most important parameter that dictates oscillator stability is the Q of the frequency-setting tank (or of the effective tank for the case of ring oscillators). For a given application, and assuming a finite power budget, adequate long- and short-term stability of the oscillation frequency is ensured only when the tank Q exceeds a certain threshold value.

The correlation between tank Q and oscillator stability can be illustrated heuristically by considering the simple oscillator circuit depicted in Figure 12.2(a). Here, a series resonant oscillator is shown, comprised of a sustaining amplifier and an LC tank connected in a positive feedback loop. For proper start-up and steady-state operation, the total phase shift around the loop must sum to zero. Thus, if at the oscillation frequency the amplifier operates nominally with a $0°$ phase shift from its input to its output, then the tank must also have a $0°$ phase shift across its terminals. Given this, and referring to any one of the tank response spectra shown in Figure 12.2(b) or 12.2(c), this oscillator is seen to operate nominally at the tank resonance frequency. If, however, an external stimulus (e.g., a noise spike or a temperature fluctuation) generates a phase shift $-\Delta\theta$ across the terminals of the sustaining amplifier, the tank must respond with an equal and opposite phase shift $\Delta\theta$ for sustained oscillation. As dictated by the tank transfer functions of Figure 12.2, any tank phase shift must be accompanied by a corresponding operating frequency shift Δf. The magnitude of Δf for a given $\Delta\theta$ is largely dependent on the Q of the resonator

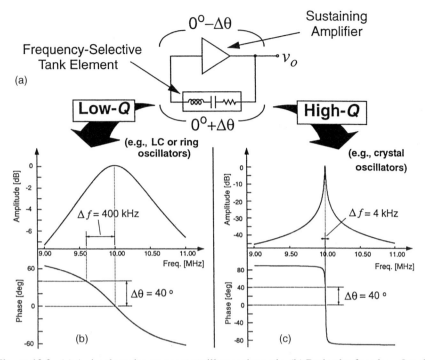

Figure 12.2. (a) A simple series resonant oscillator schematic. (b) Bode plot for a low-Q tank, indicating the Δf for a given $\Delta \theta$. (c) Similar to part (b), but for a high-Q tank. (From reference [8])

tank. Comparison of Figure 12.2(b) with 12.2(c) clearly shows that a given phase shift incurs a much smaller frequency deviation on the tank with the higher Q. Thus, the higher the tank Q, the more stable the oscillator against phase-shifting phenomena.

To help quantify the above heuristic concepts, one important figure of merit for oscillators is the phase noise power present at frequencies close to the carrier frequency. Typical phase noise requirements range from -128 dBc/Hz at 600 kHz deviation from a 915–980 MHz carrier in European Global System for Mobile Telecommunications (GSM) cellular phones, to -150 dBc/Hz at 67 kHz carrier deviations in X-Band, Doppler-based radar systems [15]. Through a more rigorous analysis of Figure 12.2 (assuming linear operation), the $1/f^2$ phase noise of a given oscillator can be described by the expression [16]

$$\left(\frac{N_{\text{op}}}{C}\right)_{f_m} = \frac{FkT}{C} \frac{1}{8Q^2} \left(\frac{f_o}{f_m}\right)^2 \quad [\text{dBc/HZ}], \tag{12.1}$$

where $(N_{op}/C)_{f_m}$ is the phase noise power density-to-carrier power ratio at a frequency f_m offset from the carrier frequency, F is the noise figure of the active device evaluated using the total oscillator power P, C is the carrier power delivered to the load, and f_o is the carrier frequency. From Eq. (12.1), phase noise is seen to be inversely proportional to the square of Q and directly proportional to the amplifier noise figure F. Given that F can often be reduced by increasing the operating power P of the sustaining amplifier, and that C increases or decreases with P, Eq. (12.1) can then be interpreted as implying that power and Q can be traded to achieve a given phase noise specification. Given the need for low power in portable units, and given that the synthesizer containing the reference oscillator and voltage control oscillator (VCO) is often a dominant contributor to total transceiver power consumption, modern transceivers could benefit greatly from technologies that yield high-Q tank components.

12.2.2 The Need for High Q in Filters

Tank Q also greatly influences the ability to implement extremely selective IF and RF filters with small percent bandwidth, small shape factor, and low insertion loss. To illustrate, Figure 12.3 presents simulated frequency characteristics under varying resonator tank Q's for a 0.3% bandwidth bandpass filter centered at 70 MHz, realized using the typical LC resonator ladder configuration shown in the inset. As shown, for a resonator tank Q of 10,000, very little insertion loss is observed. However, as tank Q decreases, insertion loss increases very quickly, to the point where a tank Q of 1000 leads to 20 dB of insertion loss—too much even for IF filters, and quite unacceptable for RF filters. As with oscillators, high-Q tanks are required for RF and IF filters alike, although more so for the latter, since channel selection is

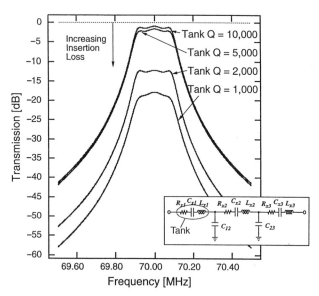

Figure 12.3. Simulated frequency characteristics for a three-resonator, 0.3% bandwidth, 70 MHz bandpass filter under varying tank Q's. (From reference [8])

done predominantly at the IF in superheterodyne receivers. In general, the more selective the filter, the higher the resonator Q required to achieve a given level of insertion loss. In particular, the above 0.3% bandwidth filter example applies for IF filters, which, because of their high selectivity, are best implemented with resonator Q's exceeding 5000; RF preselect or image-reject filters, on the other hand, typically require only 3% bandwidths and can thus be implemented using resonators with Q's on the order of 500–1000.

12.3 MICROMECHANICAL CIRCUITS

Although mechanical circuits, such as quartz crystal resonators and SAW filters, provide essential functions in the majority of transceiver designs, their numbers are generally suppressed due to their large size and finite cost. Unfortunately, when minimizing the use of high-Q components, designers often trade power for selectivity (i.e., Q) and, hence, sacrifice transceiver performance. As a simple illustration, if the high-Q IF filter in the receive path of a communications subsystem is removed, the dynamic range requirement on the subsequent IF amplifier, IQ mixer, and A/D converter circuits increases dramatically, forcing a corresponding increase in power consumption. Similar trade-offs exist at RF, where the larger the number or greater the complexity of high-Q components used, the smaller the power consumption in surrounding transistor circuits.

By shrinking dimensions and introducing batch fabrication techniques, MEMS technology provides a means for relaxing the present constraints on the number and complexity of mechanical circuits, perhaps with implications not unlike those that integrated circuit technology had on transistor circuit complexity. Before exploring the implications, specific μmechanical circuits are first reviewed, starting with the basic building block elements used for mechanical circuits, then expanding with descriptions of the some of most useful linear and nonlinear mechanical circuits.

12.3.1 The Micromechanical Beam Element

To date, the majority of μmechanical circuits most useful for communications applications in the very high frequency (VHF) range have been realized using μmechanical flexural-mode beam elements, such as shown in Figure 12.4 with clamped–clamped boundary conditions [11,12,18]. Although several micromachining technologies are available to realize such an element in a variety of different materials, surface micromachining has been the preferred method for μmechanical communications circuits, mainly due to its flexibility in providing a variety of beam end conditions and electrode locations, and its ability to realize very complex geometries with multiple levels of suspension.

Figure 12.5 summarizes the essential elements of a typical surface-micromachining process tailored to produce a clamped–clamped beam. In this process, a series of film depositions and lithographic patterning steps—identical to similar steps used in planar IC fabrication technologies—are utilized to first achieve the

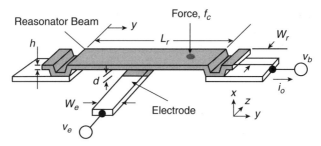

Figure 12.4. Perspective-view schematic of a clamped–clamped beam μmechanical resonator in a general bias and excitation configuration.

crosssection shown in Figure 12.5(a). Here, a sacrificial oxide layer supports the structural polysilicon material during deposition, patterning, and annealing. In the final step of the process, the wafer containing cross-sections similar to Figure 12.5(a) is dipped into a solution of hydrofluoric acid, which etches away the sacrificial oxide layer without significantly attacking the polysilicon structural material. This leaves the freestanding structure shown in Figure 12.5(b), capable of movement in multiple dimensions, if necessary, and, more importantly, capable of vibrating with high Q and good temperature stability, with temperature coefficients on the order of -10 ppm/°C [17]. Figure 12.6 presents the scanning electron micrograph (SEM) of a clamped–clamped beam polysilicon micromechanical resonator designed to operate at 17.4 MHz.

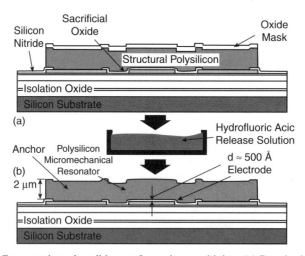

Figure 12.5. Cross sections describing surface micromachining. (a) Required film layers up to the release etch step. (b) Resulting free-standing beam following a release etch in hydrofluoric acid.

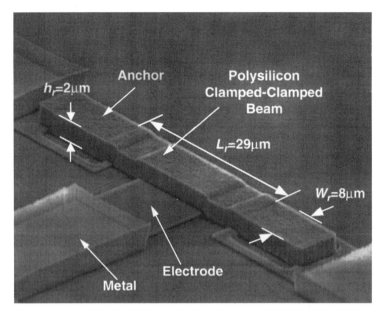

Figure 12.6. SEM of a 17.4 MHz polysilicon clamped–clamped beam micromechanical resonator with metallized electrodes.

For communications applications, clamped–clamped [18] and free–free [11] flexural-mode beams with Q's on the order of 10,000 (in vacuum) and temperature coefficients on the order of -12 ppm/°C have been popular for the VHF range, while thin-film bulk acoustic resonators [19–21] ($Q \approx 1000$) have so far addressed the ultrahigh frequency (UHF) range. To simplify the discussion, and because they have so far been the most amenable to the implementation of mechanical circuits, the remainder of this section focuses on clamped–clamped beam µmechanical resonators.

12.3.2 Clamped–Clamped Beam Micromechanical Resonators

As previously mentioned, Figure 12.4 presents the perspective–view schematic for a clamped-clamped beam µmechanical resonator, indicating key dimensions and showing a general bias and excitation configuration. As shown, this device consists of a beam anchored (i.e., clamped) at both ends, with an electrode underlying its central locations. Both the beam and electrode are constructed of conductive materials, with doped polycrystalline silicon being the most common to date. Note that both electrical and mechanical inputs are possible for this device.

For frequency reference, filtering, and mixing applications, the vibrational resonance frequency f_o of this flexural-mode mechanical beam is of great interest.

The fundamental resonance frequency of the clamped–clamped beam of Figure 12.4 is given by the expression [18]

$$f_o = \frac{1}{2\pi}\sqrt{\frac{k_r}{m_r}} = 1.03\kappa\sqrt{\frac{E\,h}{\rho L_r^2}}[1 - g(V_P)]^{1/2}, \quad (12.2)$$

where E and ρ are the Young's modulus and the density of the structural material, respectively; h and L_r are specified in Figure 12.4; the function g models the effect of an electrical spring stiffness k_e that appears when electrodes and voltages are introduced and that subtracts from the mechanical stiffness k_m; and κ is a scaling factor that models the effects of surface topography in actual implementations [18]. From Eq. (12.2), geometry clearly plays a major role in setting the resonance frequency, and, in practice, attaining a specified frequency amounts to a computer-aided design (CAD) layout of the proper dimensions. Table 12.1 presents expected resonance frequencies for various beam dimensions, modes, and structural materials for a free-free version of this beam (described in Section 12.6.1), showing a wide range of attainable frequencies, from VHF to UHF.

12.3.2.1 Electrical Excitation As shown in Figure 12.4, this device accepts two electrical inputs, v_e and v_b, applied to the electrode and beam, respectively. In this configuration, the difference voltage $(v_e - v_b)$ is effectively applied across the electrode-to-resonator capacitor gap, generating a force between the stationary electrode and movable beam given by

$$F_d = \frac{\partial E}{\partial x} = \frac{1}{2}(v_e - v_b)^2\frac{\partial C}{\partial x} = \frac{1}{2}(v_b^2 - 2v_bv_e + v_e^2)\frac{\partial C}{\partial x}, \quad (12.3)$$

where x is displacement (with direction indicated in Figure 12.4), and $(\partial C/\partial x)$ is the change in resonator-to-electrode capacitance per unit displacement, approximately

TABLE 12.1. μMechanical Resonator Frequency Design[a]

Frequency (MHz)	Material	Mode	h_r (μm)	W_r (μm)	L_r (μm)
70	Silicon	1	2	8	14.54
110	Silicon	1	2	8	11.26
250	Silicon	1	2	4	6.74
870	Silicon	2	2	4	4.38
870	Diamond	2	2	4	8.88
1800	Silicon	3	1	4	3.09
1800	Diamond	3	1	4	6.16

[a] Determined for free-free beams using Timoshenko methods that include the effects of finite h and W_r [11].

given by (neglecting fringing fields and static beam bending)

$$\frac{\partial C}{\partial x} = -\frac{\varepsilon_o W_r W_e}{d_o^2}, \qquad (12.4)$$

where d_o is the electrode-to-resonator gap spacing under static (nonresonance) conditions, and ε_o is the permittivity in vacuum. When using the resonator as a tank or filter circuit (as opposed to a mixer, to be discussed later), a dc-bias voltage V_P is applied to the conductive beam, while an ac excitation signal $v_i = V_i \cos \omega_i t$ is applied to the underlying electrode. In this configuration, Eq. (12.3) reduces to

$$F_d = \frac{\partial C}{\partial x}\left(\frac{V_P^2}{2} + \frac{V_i^2}{4}\right) - V_P \frac{\partial C}{\partial x} V_i \cos \omega_i t + \frac{\partial C}{\partial x}\frac{V_i}{4} \cos 2\omega_i t. \qquad (12.5)$$

The first term in Eq. (12.5) represents an off-resonance dc force that statically bends the beam, but that otherwise has little effect on its signal processing function, especially for VHF and above, for which the beam stiffness is very large. The second term constitutes a force at the frequency of the input signal, amplified by the dc-bias voltage V_P, and is the main input component used in high-Q tank and filter applications. When $\omega_i = \omega_o$ (the radian resonance frequency), this force drives the beam into resonance, with a zero-to-peak displacement amplitude at location y given by

$$x(y) = -\frac{QF_d}{jk_{\text{reff}}(y)} = -\frac{Q}{jk_{\text{reff}}(y)} V_P \frac{\partial C}{\partial x} v_i, \qquad (12.6)$$

where $k_{\text{reff}}(y)$ is an effective stiffness at location y to be determined later in this section via integration over the electrode width. Motion of the beam creates a dc-biased (via V_P) time-varying capacitance between the electrode and resonator that sources an output current given by

$$i_o = -V_P \frac{\partial C}{\partial x} \frac{\partial x}{\partial t}. \qquad (12.7)$$

When plotted versus the frequency of v_i, i_o traces out a bandpass biquad characteristic with a $Q \approx 10{,}000$ (cf. Figure 12.7 [18])—very suitable for reference oscillators and low-insertion-loss filters. Note, however, that this Q is only achievable under vacuum, where viscous gas damping is minimized [22]. Much lower Q's on the order of hundreds are seen under atmospheric pressure.

The third term in Eq. (12.5) represents a term capable of driving the beam into vibration when $\omega_i = \frac{1}{2}\omega_o$. If V_P is very large compared with V_i, this term is greatly suppressed, but it can be troublesome for bandpass filters in cases where very large interferers are present at half the passband frequency. In these cases, a μmechanical notch filter at $\frac{1}{2}\omega_o$ may be needed.

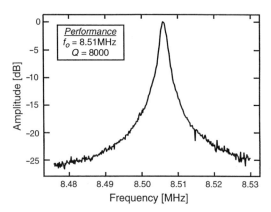

Figure 12.7. Frequency characteristic for an 8.5 MHz clamped–clamped beam polysilicon μmechanical resonator measured under 70 mtorr vacuum using a dc-bias voltage $V_P = 10$ V, a drive voltage of $v_i = 3$ mV, and a transresistance amplifier with a gain of 33 KΩ to yield an output voltage v_o. Amplitude $= v_o/v_i$. (From reference [18])

12.3.2.2 Equivalent Lumped-Parameter Mechanical Circuit

For the purposes of mechanical circuit design, it is often convenient to define an equivalent lumped-parameter mass–spring–damper mechanical circuit for this resonator (cf. Figure 12.8), with element values that vary with location on the resonator. With reference to Figure 12.9, the equivalent mass at a location y on the resonator is given by [23]

$$m_r(y) = \frac{KE_{\text{tot}}}{\frac{1}{2}[v(y)]^2} = \frac{\rho W_r h \int_0^{L_r}[X_{\text{mode}}(y')]^2 dy'}{[X_{\text{mode}}(y)]^2} \tag{12.8}$$

where

$$X_{\text{mode}}(y) = \zeta(\cos \beta y - \cosh \beta y) + (\sin \beta y - \sinh \beta y), \tag{12.9}$$

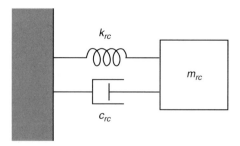

Figure 12.8. Lumped-parameter mechanical equivalent circuit for the micromechanical resonator of Figure 12.4.

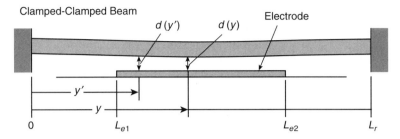

Figure 12.9. Resonator cross-sectional schematic for frequency-pulling and impedance analysis.

$\beta = 4.730/L_r$, and $\zeta = -1.01781$ for the fundamental mode, KE_{tot} is the peak kinetic energy in the system, $v(y)$ is the velocity at location y, and dimensional parameters are given in Figure 12.9. The equivalent spring stiffness follows readily from Eqs. (12.2) and (12.8) and is given by

$$k_r(y) = \omega_o^2 m_r(y), \tag{12.10}$$

where ω_o is the radian resonance frequency of the beam. Finally, the damping factor is given by

$$c_r(y) = \frac{\sqrt{k_m(y) m_r(y)}}{Q_{nom}} = \frac{\omega_{nom} m_r(y)}{Q_{nom}} = \frac{k_m(y)}{\omega_{nom} Q_{nom}}, \tag{12.11}$$

where

$$k_m(y) = \omega_{nom}^2 m_r(y) \tag{12.12}$$

is the mechanical stiffness of the resonator alone, without the influence of applied voltages and electrodes to be discussed next, and Q_{nom} is the quality factor of the resonator under the same conditions.

12.3.2.3 Voltage-Tunable Electrical Stiffness As indicated in Eq. (12.2), where g is seen to be a function of dc-bias voltage V_P, the resonance frequency of this device is tunable via adjustment of V_P [24,25] and this can be used advantageously to implement filters with tunable center frequencies or to correct for passband distortion caused by finite planar fabrication tolerances. The dc-bias dependence of resonance frequency arises from a V_P-dependent electrical spring constant k_e that subtracts from the mechanical spring constant of the system k_m, lowering the overall spring stiffness $k_r = k_m - k_e$, thus lowering the resonance frequency according to the expression

$$f_o = \frac{1}{2\pi}\sqrt{\frac{k_m - k_e}{m_r}} = \frac{1}{2\pi}\sqrt{\frac{k_m}{m_r}}\left(1 - \left\langle\frac{k_e}{k_m}\right\rangle\right)^{1/12}$$

$$= 1.03\kappa\sqrt{\frac{E h}{\rho L^2}}\left(1 - \left\langle\frac{k_e}{k_m}\right\rangle\right)^{1/2}, \tag{12.13}$$

where k_m and m_r denote values at a particular location (usually the beam center location), and the quantity $\langle k_e/k_m \rangle$ must be obtained via integration over the electrode width W_e due to the location dependence of k_m.

The electrical spring stiffness k_e is generated by the nonlinear dependence of electrode-to-resonator gap capacitance $C(x)$ on displacement x and is dependent very strongly on the electrode-to-resonator gap spacing d. At a specific location y' centered on an infinitesimally small width of the electrode dy', the differential in electrical stiffness is given by [25]

$$dk_e(y') = V_P^2 \frac{\varepsilon_o W_r dy'}{[d(y')]^3}, \quad (12.14)$$

where the electrode-to-resonator gap distance d is now seen to also be location dependent, since the beam bends somewhat due to the dc-bias V_P applied between the electrode and resonator. Recognizing that for the fundamental mode the static and dynamic stiffnesses are virtually the same, and assuming a static bending shape due to the distributed dc force defined by the function $X_{\text{static}}(y)$, the gap distance can be expressed as

$$d(y) = d_o - \frac{1}{2} V_P^2 \varepsilon_o W_r \int_{L_{e1}}^{L_{e2}} \frac{1}{k_m(y')[d,y']^2} \frac{X_{\text{static}}(y)}{X_{\text{static}}(y')} dy', \quad (12.15)$$

where d_o is the static electrode-to-resonator gap with $V_P = 0$ V. In Eq. (12.15), the second term represents the static displacement of the resonator toward the electrode at a particular location y, evaluated by integration over the width of the electrode, from $y = L_{e1}$ to L_{e2}. For the common case where the electrode is centered with the resonator beam center, $L_{e1} = 0.5(L_r - W_e)$ and $L_{e2} = 0.5(L_r + W_e)$. Since the desired variable $d(y)$ appears on both sides of Eq. (12.15), one of them within an integral, Eq. (12.15) is best solved by first assuming $d(y) = d_o$ on the right side, solving for $d(y)$ on the left, then using this function again on the right, iterating until $d(y)$ converges. In addition, for most cases Eq. (12.15) is not overly sensitive to the function $X_{\text{static}}(y)$, so $X_{\text{mode}}(y)$ given by Eq. (12.9) can be substituted for $X_{\text{static}}(y)$ with little difference. It should be noted that more rigorous versions of Eq. (12.15) are attainable through strict static analysis, but these often take the form of polynomial expansions and are less intuitive than Eq. (12.15).

The quantity $\langle k_e/k_m \rangle$ may now be found by integrating over the electrode width, and it is given by

$$\left\langle \frac{k_e}{k_m} \right\rangle = g(d, V_p) = \int_{L_{e1}}^{L_{e2}} \frac{dk_e(y')}{k_m(y')}. \quad (12.16)$$

12.3.2.4 Pull-In Voltage, V_{PI}
When the applied dc-bias voltage V_P is sufficiently large, catastrophic failure of the device ensues, in which the resonator beam is pulled

down onto the electrode. This leads to either destruction of the device due to excessive current passing through the now shorted electrode-to-resonator path, or at least a removal of functionality if a dielectric layer (e.g., an oxide or nitride) is present above the electrode to prevent electrical contact between it and the conductive resonator beam.

Unlike low-frequency micromechanical structures, such as used in accelerometers or gyroscopes [26,27], the attractive electrostatic force between the electrode and this high-frequency resonator that incites pull-down now acts against a very large distributed stiffness that must be integrated over the electrode area to accurately predict the pull-down voltage V_{PI}. Thus, previously used closed-form expressions for V_{PI} [25] based on lumped-parameter analysis are no longer applicable. Rather, for resonators with the design of Figure 12.4 and beam lengths less than 50 µm, the procedure for determining V_{PI} entails finding the V_P that sets the resonance frequency equal to zero. With reference to Eq. (12.13), this amounts to setting Eq. (12.16) equal to unity and solving for the V_P variable.

12.3.2.5 Small-Signal Electrical Equivalent Circuit To conveniently model and simulate the impedance behavior of this µmechanical resonator in an electromechanical circuit, an electrical equivalent circuit is needed. As shown in Figure 12.4, both electrical and mechanical inputs and outputs are possible for this device, so the equivalent circuit must be able to model both. In addition, for physical consistency from both transducer and noise perspectives, a circuit model that directly uses the lumped mechanical elements summarized by Eqs. (12.8)–(12.11) is preferred. Figure 12.10 presents one of the more useful equivalent circuits used for linear mechanical circuit design [18,28], in which transformers model both electrical and mechanical couplings to and from the resonator, which itself is modeled by a core *LCR* circuit—the electrical analogy to a mass–spring–damper system—with element values corresponding to actual values of mass, stiffness, and damping as given by Eqs. (12.8)–(12.11). In this circuit, the current electromechanical analogy is utilized, summarized in Table 12.2.

When looking into the electrode port of the equivalent resonator circuit of Figure 12.10, a transformed *LCR* circuit is seen, with element values given by

$$L_x = \frac{m_{re}}{\eta_e^2}, \quad C_x = \frac{\eta_e^2}{k_{re}}, \quad R_x = \frac{\sqrt{k_{re}m_{re}}}{Q\eta_e^2} = \frac{C_{re}}{\eta_e^2}, \quad (12.17)$$

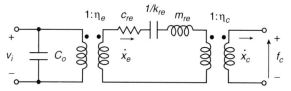

Figure 12.10. Equivalent circuit for a µmechanical resonator with both electrical (voltage v_i) and mechanical (force f_c) inputs and outputs.

TABLE 12.2. Mechanical-to-Electrical Correspondence in the Current Analogy

Mechanical Variable	Electrical Variable
Damping, c	Resistance, R
Stiffness^{-1}, k^{-1}	Capacitance, C
Mass, m	Inductance, L
Force, f	Voltage, V
Velocity, v	Current, I

where the subscript e denotes the electrode location at the very center of the resonator beam (i.e., at $y = L_r/2$). An expression for the electromechanical transformer turns ratio η_e can be obtained via an impedance analysis yielding the motional resistance R_x seen across the electrode-to-resonator gap at resonance. Pursuant to this, the voltage-to-displacement transfer function at a given location y (cf. Figure 12.9) at resonance is first found using phasor forms of Eqs. (12.4)–(12,6), (12.9), and (12.10), and integrating over the electrode width to yield

$$\frac{X}{V_i}(y) = \int_{L_{e1}}^{L_{e2}} \frac{QV_P\varepsilon_o W_r}{[d(y')]^2 k_r(y')} \frac{X_{\text{mode}}(y)}{X_{\text{mode}}(y')} dy'. \tag{12.18}$$

Using the phasor form of Eq. (12.7), the series motional resistance seen looking into the drive electrode is then found to be

$$R_x = \frac{V_i}{I_x} = \left(\int_{L_{e1}}^{L_{e2}} \frac{\omega_o V_P \varepsilon_o W_r}{[d(y)]^2} \cdot \frac{X}{V_i}(y) dy \right)^{-1}. \tag{12.19}$$

Inserting Eq. (12.18), factoring out $c_{re} = k_{re}/(\omega_0 Q)$, and extracting η_e yields

$$\eta_e = \left(\int_{L_{e1}}^{L_{e2}} \int_{L_{e1}}^{L_{e2}} \frac{V_P^2(\varepsilon_o W_r)^2}{[d(y')d(y)]^2} \frac{k_{re}}{k_r(y')} \frac{X_{\text{mode}}(y)}{X_{\text{mode}}(y')} dy' dy \right)^{1/2}. \tag{12.20}$$

Note that the effective integrated stiffness defined in Eq. (12.6) can also be extracted from Eq. (12.18), yielding

$$k_{\text{reff}}(y) = \left[\int_{L_{e1}}^{L_{e2}} \left(\frac{d_o}{d(y')} \right)^2 \frac{1}{k_r(y')} \frac{1}{W_e} \frac{X_{\text{mode}}(y)}{X_{\text{mode}}(y')} dy' \right]^{-1}. \tag{12.21}$$

The transformer turns ratio η_c in Figure 12.10 models the mechanical impedance transformation achieved by mechanically coupling to the resonator at a y location displaced from its center. As will be seen, such coupling is required when implementing filters with two or more resonators. Expressed in terms of a stiffness ratio, the equation for the mechanical transformer turns ratio when coupling at a distance l_c from an anchor takes the form

$$\eta_c = \sqrt{\frac{k_r(l_c)}{k_{re}}}. \tag{12.22}$$

Finally, for the equivalent circuit of Figure 12.10, it should be noted that the damping constant c_r is not inherently a function of the electrical stiffness k_e. Thus, when expressed in terms of the overall stiffness k_r of the system, the Q of the resonator must be adjusted so that c_r retains its original value given by Eq. (12.11). In terms of k_r and ω_o, then, expressions for c_r take on the form

$$c_r = \frac{\omega_o m_r}{Q} = \frac{k_r}{\omega_o Q} = \frac{\sqrt{k_r m_r}}{Q}, \tag{12.23}$$

where

$$Q = Q_{\text{nom}} \left(1 - \left\langle \frac{k_e}{k_m} \right\rangle \right)^{1/2}. \tag{12.24}$$

Note that the effective resonator quality factor Q is dependent on the electrical spring stiffness k_e and, thus, is also a function of the dc-bias voltage V_P. In this chapter, the variable Q denotes that defined by Eq. (12.24), while Q_{nom} is reserved for zero-bias conditions.

In the design of μmechanical circuits comprised of interlinked beams, the equivalent circuit in Figure 12.10 functions in a similar fashion to the hybrid-π small-signal equivalent circuit used for analog transistor circuit design. The main differences between mechanical links and transistors are the basic features that make them useful as circuit elements: while transistors exhibit high gain, mechanical links exhibit very large Q. By combining the strong points of both circuit elements, on-chip functions previously unachievable are now within the realm of possibilities. We now touch upon a few of these.

12.3.3 Micromechanical Filters

Among the more useful μmechanical circuits for communications are those implementing low-loss bandpass filters, capable of achieving frequency characteristics as shown in Figure 12.11, where a broader frequency passband than achievable by a single resonator beam is shown, with a sharper roll-off to the stopband (i.e., smaller shape factor).

To achieve the characteristic of Figure 12.11, a number of micromechanical resonators are coupled together by soft coupling springs [18,23,29,30], as illustrated schematically in Figure 12.12(a) using ideal mass–spring–damper elements. By

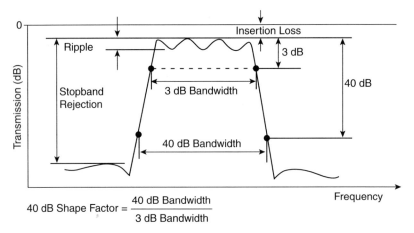

Figure 12.11. Parameters typically used for filter specification. (From reference [29])

linking resonators together using (ideally) massless springs, a coupled resonator system is achieved that now exhibits several modes of vibration. As illustrated in Figure 12.13 for the coupled three-resonator system of Figure 12.12, the frequency of each vibration mode corresponds to a distinct peak in the force-to-displacement frequency characteristic, and to a distinct physical mode shape of the coupled

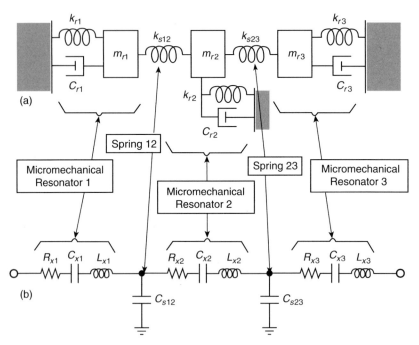

Figure 12.12. (a) Equivalent lumped-parameter mechanical circuit for a mechanical filter. (b) Corresponding equivalent *LCR* network.

MICROMECHANICAL CIRCUITS

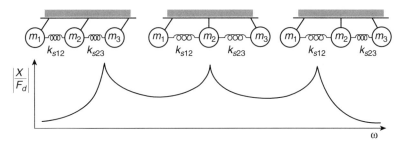

Figure 12.13. Mode shapes of a three-resonator micromechanical filter and their corresponding frequency peaks.

mechanical resonator system. In the lowest frequency mode, all resonators vibrate in phase; in the middle frequency mode, the center resonator ideally remains motionless, while the end resonators vibrate 180° out of phase; and finally, in the highest frequency mode, each resonator is phase-shifted 180° from its adjacent neighbor. Without additional electronics, the complete mechanical filter exhibits the jagged passband seen in Figure 12.13. As will be shown, termination resistors designed to lower the Q's of the input and output resonators by specific amounts are required to flatten the passband and achieve a more recognizable filter characteristic, such as in Figure 12.11.

In practical implementations, because planar IC processes typically exhibit substantially better *matching* tolerances than *absolute*, the constituent resonators in μmechanical filters are normally designed to be identical, with identical dimensions and resonance frequencies. For such designs, the center frequency of the overall filter is equal to the resonance frequency f_o of the resonators, while the filter passband (i.e., the bandwidth) is determined by the spacings between the mode peaks.

The relative placement of the vibration peaks in the frequency characteristic—and thus the passband of the eventual filter—is determined primarily by the stiffnesses of the coupling springs (k_{sij}) and of the constituent resonators at the coupling locations (k_r). Specifically, for a filter with center frequency f_o and bandwidth B, these stiffnesses must satisfy the expression [23]

$$B = \left(\frac{f_o}{k_{ij}}\right)\left(\frac{k_{sij}}{k_r}\right), \qquad (12.25)$$

where k_{ij} is a normalized coupling coefficient found in filter cookbooks [31]. Note from Eq. (12.25) that filter bandwidth is not dependent on the absolute values of resonator and coupling beam stiffness; rather, their ratio k_{sij}/k_r dictates bandwidth. Thus, the procedure for designing a mechanical filter involves two main steps (not necessarily in this order): first, design of a mechanical resonator with resonance frequency f_o and adjustable stiffness k_r; and second, design of coupling springs with appropriate values of stiffness k_{sij} to enable a desired bandwidth within the adjustment range of resonator k_r's.

To take advantage of the maturity of LC ladder filter synthesis techniques, the enormous database governing LC ladder filter implementations [31], and the wide availability of electrical circuit simulators, realization of the μmechanical filter of Figure 12.12(a) often also involves the design of an LC ladder version to fit the desired specification. The elements in the LC ladder design are then matched to lumped mechanical equivalents via electromechanical analogy, where inductance, capacitance, and resistance in the electrical domain equate to mass, compliance, and damping, respectively, in the mechanical domain. Figure 12.12(b) explicitly depicts the equivalence between the filter's lumped mass–spring–damper circuit and its electrical equivalent circuit. As shown, for this particular electromechanical analogy (the current analogy), each constituent resonator corresponds to a series LCR tank, while each (massless) coupling spring ideally corresponds to a shunt capacitor, with the whole coupled network corresponding to an LC ladder bandpass filter. It should be emphasized that the circuit in Figure 12.12(b) corresponds to the ideal mechanical circuit of Figure 12.12(a), in which the resonators are modeled by simple lumped elements, and coupling springs are considered massless. As will be seen, additional circuit complexity will be needed to model actual filters, where coupling springs generalize to transmission lines, and resonators must be modeled by circuits similar to that of Figure 12.10.

12.3.3.1 A Two-Resonator HF–VHF Micromechanical Filter Figure 12.14 shows the perspective-view schematic of a practical two-resonator micromechanical filter [18] capable of operation in the HF to VHF range. As shown, the filter consists of two μmechanical clamped–clamped beam resonators, coupled mechanically by a soft spring, all suspended 0.1 μm above the substrate. Conductive (polysilicon) strips underlie each resonator, the center ones serving as capacitive transducer electrodes positioned to induce resonator vibration in a direction perpendicular to the substrate, the flanking ones serving as tuning electrodes capable of voltage-controlled tuning of resonator frequencies, as governed by Eq. (12.13). The resonator-to-electrode gaps are determined by the thickness of a sacrificial oxide spacer during fabrication and can thus be made quite small (e.g., 0.1 μm or less) to maximize electromechanical coupling.

The filter is excited in a similar fashion to that described in Section 12.3.3, with a dc-bias voltage V_P applied to the conductive mechanical network, and an ac signal applied to the input electrode, but this time through an appropriately valued source resistance R_Q that loads the Q of the input resonator to flatten the passband [18]. The output resonator of the filter must also see a matched impedance to avoid passband distortion, and the output voltage v_o is generally taken across this impedance. As will be seen in Section 12.3.3.2, the required value of input/output (I/O) port termination resistance can be tailored for different applications, and this can be advantageous when designing low-noise transistor circuits succeeding the filter, since such circuits can then be driven by optimum values of source resistance to minimize the noise figure [32].

From a signal flow perspective, the operation of the above filter can briefly be summarized as follows:

MICROMECHANICAL CIRCUITS

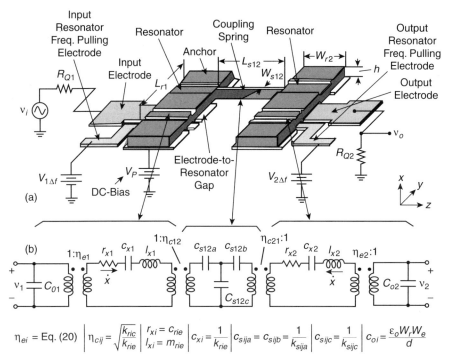

Figure 12.14. (a) Perspective-view schematic of a symmetrical two-resonator VHF μmechanical filter with typical bias, excitation, and signal conditioning electronics. (b) Electrical equivalent circuit for the filter in (a) along with equations for the elements [18]. Here, m_{rie}, k_{rie}, and c_{rie} denote the mass, stiffness, and damping of resonator i at the beam center location, and η_e and η_c are turns ratios modeling electromechanical coupling at the inputs and mechanical impedance transformations at low-velocity coupling locations. (From reference [18])

1. An electrical input signal is applied to the input port and converted to an input force by the electromechanical transducer (which for the case of Figure 12.14(a) is capacitive) that can then induce mechanical vibration in the x direction.
2. Mechanical vibration comprises a mechanical signal that is processed in the mechanical domain—specifically, the signal is rejected if outside the passband of the filter and passed if within the passband.
3. The mechanically processed signal appears as motion of the output resonator and is reconverted to electrical energy at the output transducer, ready for processing by subsequent transceiver stages.

From the above, the name "micromechanical signal processor" clearly suits this device. Details of the design procedure for micromechanical filters now follow.

12.3.3.2 HF–VHF Filter Design

As can be surmised from Figure 12.12(b), the network topologies for the mechanical filters of this work differ very little from those of their purely electronic counterparts and in principle, can be designed at the system level via a procedure derived from well-known, coupled resonator ladder filter synthesis techniques. In particular, given the equivalent LCR element values for a prototype μmechanical resonator, it is possible to synthesize a mechanical filter entirely in the electrical domain, converting to the mechanical domain only as the last step. However, although possible, such a procedure is not recommended, since knowledge and ease of design in both electrical and mechanical domains can greatly reduce the effort required.

The design procedure for the two-resonator micromechanical filter of Figure 12.14 can be itemized as follows:

1. Design and establish the μmechanical resonator prototype to be used, choosing necessary geometries for the needed frequency and ensuring that enough electrode-to-resonator transducer coupling is provided to allow for predetermined termination resistor values. For a given resonator, with predetermined values of W_r, h, W_e, V_P, and R_Q, this amounts to solving for the resonator length L_r and electrode-to-resonator gap spacing d that simultaneously satisfy Eqs. (12.13) and (12.25) and the equation for the needed termination resistor value [18,29]:

$$R_{Qi} = \left(\frac{Q_i}{q_i Q_{fltr}} - 1\right)\frac{c_{rie}}{Q_i \eta_{ei}} = \left(\frac{Q_i}{q_i Q_{fltr}} - 1\right) R_{xi}, \qquad (12.26)$$

where Q_i is the uncontrolled quality factor of resonator i; $Q_{fltr} = f_0/B$ is the filter quality factor; q_i is a normalized q parameter obtained from a filter cookbook [31]; η_{ei} is given by Eq. (12.20); c_{rie} is given by Eq. (12.23), where the subscript e denotes the center location of the resonator over the electrode; and R_{xi} is the series motional resistance of end resonator i, given in Eq. (12.17). (Note that as mentioned previously, the resonators in a micromechanical filter are often designed to be identical, so the i subscript notation can actually be dropped from virtually all variables in Eq. (12.26), except q_i and R_{Qi}. The i subscripts are included for all variables just for completeness.) If the filter is designed symmetrically, with $q_1 = q_2$, and with resonator Q's much greater than Q_{fltr}, the required value of the I/O port termination resistance for both end resonators becomes (dropping i subscripts)

$$R_Q \cong \frac{\sqrt{k_{re} m_{re}}}{q_1 Q_{fltr} \eta_e^2} = \frac{k_{re}}{\omega_o q_1 Q_{fltr} \eta_e^2}, \qquad (12.27)$$

where e denotes the center location of the resonator beam. Of the variables in Eq. (12.27), the electromechanical coupling factor η_e is often the most convenient parameter to adjust for a desired value of termination resistance. Given from Eq. (12.20) that $\eta_e \approx (V_P/d^2)$, termination impedance R_Q requirements and bias voltage V_P limitations often dictate the electrode-to-resonator gap spacing for a

MICROMECHANICAL CIRCUITS   433

TABLE 12.3. Two-Resonator μMechanical Filter Electrode-to-Resonator Gap Spacing Design[a]

Frequency	Gap Spacing, d, for R_Q of:				
	300 Ω	500 Ω	1000 Ω	2000 Ω	5000 Ω
70 MHz[b]	160 Å	178 Å	207 Å	243 Å	301 Å
870 MHz[c]	68 Å	77 Å	92 Å	109 Å	137 Å

[a] Determined with $Q = 10,000$, $W_e = 0.54$, $V_P = 10$ V, using Timoshenko methods and ignoring beam topography.
[b] CCBeam, polysilicon, $L_r = 14.92$ μm, $W_r = 8$ μm, $h = 2$ μm, BW = 200 kHz
[c] CCBeam, diamond, $L_r = 5.97$ μm, $W_r = 8$ μm, $h = 2$ μm, BW = 1.25 MHz.

particular resonator design. This can be seen in Table 12.3, which summarizes the needed gap spacings to achieve various values of R_Q for micromechanical filters centered at 70 and 870 MHz, and with $Q = 10,000$, $B = 1.25$ MHz, and $V_P = 10$ V.

2. Choose a manufacturable value of coupling beam width W_{s12} and design coupling beam(s) corresponding to a "quarter-wavelength" of the filter center frequency. Here, the coupling beam is recognized as an acoustic transmission line that can be made transparent to the filter when designed with quarter-wavelength dimensions [18,23,29]. For a flexural-mode coupling beam, neglecting rotational movements at the resonator attachment points, quarter-wavelength dimensions are achieved when the coupling beam width W_{s12} and length L_{s12} are chosen to satisfy the expression [29]

$$H_6 = \sinh\alpha\cos\alpha + \cosh\alpha\sin\alpha = 0, \quad (12.28)$$

where $\alpha = L_{s12}[\rho W_{s12} h\omega^2/(EI_{s12})]^{0.25}$, $I_{s12} = W_{s12} h^3/12$, and needed dimensions are given in Figure 12.14(a). Note that in choosing W_{s12} and L_{s12} to satisfy Eq. (12.28), the coupling beam stiffness k_{s12} is constrained to a particular value, given by [29]

$$k_{s12} = -\frac{EI_{s12}\alpha^3(\sin\alpha + \sinh\alpha)}{L_{s12}^3(\cos\alpha\cosh\alpha - 1)}. \quad (12.29)$$

Note that this also constrains the ability to set the bandwidth of the filter via the coupling beam dimensions and thus necessitates an alternative method for setting bandwidth.

3. Determine the coupling location(s) on the resonators corresponding to the filter bandwidth of interest. This procedure is based on two important properties of this filter and the resonators comprising it: first, the filter bandwidth B is determined not by absolute values of stiffness but rather by a ratio of stiffnesses (k_{s12}/k_{rc}), where the subscript c denotes the value at the coupling location; and second, the value of resonator stiffness k_{rc} varies with location (in particular, with location velocity) and so can be set to a desired value by simply choosing an appropriate coupling beam attachment point. Using Eqs. (12.8), (12.10), (12.25), and (12.29), an expression that can be solved for the location l_c on the resonator beam where the

coupling beam should be attached can be written as

$$X_{\text{mode}}(l_c) = \left(2\pi B \frac{k_{12}}{k_{s12}} \rho W_r h \int_o^{L_r} [X(y')]^2 dy'\right)^{1/2}. \qquad (12.30)$$

Figure 12.15 illustrates how the choice of coupling beam attachment point can greatly influence the bandwidth of a mechanical filter. In Figure 12.15(a), the coupling beam is attached at the highest velocity point, where the resonator presents its smallest stiffness, resulting in a very wide filter bandwidth. On the other hand, Figure 12.15(b) depicts coupling at a lower velocity point closer to the resonator anchors, where the resonator presents a much higher stiffness, leading to a much smaller percent bandwidth, as dictated by Eq. (12.25). In effect, the bandwidth of the filter is set not by choosing the coupling beam stiffness k_{s12}, but rather by choosing an appropriate value of resonator stiffness k_{rc} to satisfy Eq. (12.25), given a k_{s12} constrained by quarter-wavelength design.

4. Generate a complete equivalent circuit for the overall filter and verify the design using a circuit simulator. Figure 12.14(b) presents the equivalent circuit for the two-resonator micromechanical filter of Figure 12.14(a) along with equations for the elements. As shown, each of the outside resonators is modeled via circuits such as shown in Figure 12.10. The coupling beam actually operates as an acoustic transmission line and, thus, is modeled by a *T*-network of energy storage elements. Consistent with Figure 12.10 and the discussion in item 3 above, transformers are used between the resonator and coupling beam circuits of Figure 12.14(b) to model the velocity transformations that arise when attaching the coupling beams at locations offset from the center of the resonator beam. The whole circuit structure of Figure 12.14(b) can be recognized as that of the *LC* ladder network for a bandpass filter.

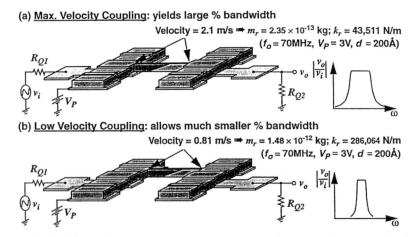

Figure 12.15. Filter schematics showing (a) maximum velocity coupling to yield a large percent bandwidth and (b) low-velocity coupling to yield a smaller percent bandwidth.

MICROMECHANICAL CIRCUITS

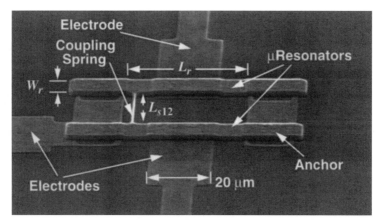

Figure 12.16. SEM of a fabricated 7.81 MHz two-resonator micromechanical filter. (From reference [18])

Further details on the design of micromechanical filters can be found in the literature [18,23,29].

12.3.3.3 HF Micromechanical Filter Performance   Figure 12.16 presents the SEM of a symmetrical (i.e., $R_{Q1} = R_{Q2}$) 7.81 MHz micromechanical filter using the design of Figure 12.14 and constructed of phosphorous-doped polycrystalline silicon [18]. The measured spectrum for a terminated version of this filter is shown in Figure 12.17 (solid curve), showing a bandwidth of 18 kHz, which is very close to

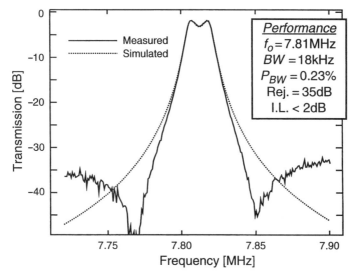

Figure 12.17. Measured spectrum for a terminated 7.81 MHz μmechanical filter with excessive input/output shunt capacitance. Here, $Q_{fltr} = 435$. (From reference [18])

the design value. The insertion loss is only 1.8 dB, which is impressive for a bandpass filter with a percent bandwidth of 0.23% ($Q_{\text{fltr}} = 435$) and which can be attributed to the high Q of the constituent μmechanical resonators. Designed and measured μmechanical filter characteristics are summarized in Table 12.4. It should be noted that although the analytical design calls for 19.6 kΩ termination resistors, only 12.2 kΩ resistors were used in the actual measurement to minimize phase lags caused by board-level parasitic capacitance.

In addition to the measured frequency response, Figure 12.17 also presents a simulated spectrum (dotted line) using the equivalent circuit described by Figure 12.14(b) with element values derived from the "Simulated" column of Table 12.4 and summarized in Table 12.5. This simulation attempts to mimic the measured frequency characteristic in the passband. As such, it includes shunt parasitic capacitors $C_{Pn} = 100$ fF at the input and output nodes to model board-level parasitics that interact with termination resistors R_{Qn} and generate increased passband ripple. It should be noted, however, that a few adjustments were necessary to attain the degree of matching shown, with the more important adjustments indicated in boldface. In particular, note that the target gap spacing of 1300 Å was not used to generate the "Simulated" column in Table 12.4, nor the values in Table 12.5. Rather, a larger gap spacing of 1985 Å was used that accounts for depletion in the resonator beam induced by the V_P-induced electric field between the nondegenerately doped n-type beam and the n-type electrode [18]. This value of gap spacing was semiempirically determined by matching measured plots of resonator f_o versus V_P with simulations based on Eq. (12.13), using d_o and κ as fitting parameters.

In addition, as indicated in boldface in Table 12.4, the coupling location l_c was adjusted to match bandwidths, and the resonator Q and the filter termination resistance R_Q were adjusted to match the measured insertion loss. In particular, the value of R_Q needed to match the simulated insertion loss and passband ripple was 14.5 kΩ, not the 12.2 kΩ actually used for the measurement.

The l_c adjustment is not unreasonable, since the coupling beam has a finite width of 0.75 μm, and the exact coupling location is not necessarily at the center of the coupling beam but could be anywhere along its finite width. Furthermore, torsional motions of the coupling beam can also influence the actual mechanical coupling, thus changing the effective l_c. The adjustment in Q seen in Table 12.4 is also plausible, since a small number of resonators in the filter fabrication run exhibited lower Q than the 8000 measured in Figure 12.7. The small deviation in R_Q also should not be alarming, given some uncertainty in the actual electrode-to-resonator gap distance for this process.

Note that although the simulation matches the measurement very well in the passband, it deviates substantially in its transition to the stopband. In particular, the measured curve features loss poles not modeled by the theory of Section 12.3.3.2 that substantially improve the shape factor of the filter. The loss poles in Figure 12.17 result largely from action of a feedthrough capacitor $C_{P(fd)}$ that connects the input and output electrodes, and that influences the filter frequency characteristic in a similar fashion to the introduction of loss poles via bridging capacitors in crystal filter design [33]. In the present experiment, $C_{P(fd)}$ is actually a parasitic element;

MICROMECHANICAL CIRCUITS

TABLE 12.4. HF Micromechanical Filter Summary [18]

Parameter	Des./Meas.[a,c]	Simulated[b,c,d]	Units
Coupling location, l_c	4.08	**4.48**	μm
Coupling velocity, v_c	$0.12v_{max}$	**$0.14v_{max}$**	m/s
Center frequency, f_o	7.81	7.81	MHz
Frequency modification factor, κ	(0.87915) [0.9]	0.87915	—
Bandwidth, B	18	18	kHz
Percent bandwidth, B/f_o	0.23	0.23	%
Passband ripple, PR	1.5	1.5 [0.5]	dB
Insertion loss, IL	1.8	1.8 [1.35]	dB
20 dB Shape factor	2.31	[2.54]	—
Stopband rejection, SR	35	—	dB
Sp.-free dynamic range, $SFDR$	~78	~78	dB
Resonator Q	8000	**6000**	—
Structural layer thickness, h	1.9	1.9	μm
μRes. beam length, L_r	40.8	40.8	μm
μRes. beam width, W_r	8	8	μm
Coupling beam length, L_{s12}	20.35	20.35	μm
Needed L_{s12} for λ/4	(22.47)	—	μm
Coupling beam width, W_{s12}	0.75	0.75	μm
Coupling beam stiffness, k_{s12a}	(−82.8)	−82.8	N/m
Coupling beam stiffness, k_{s12c}	(113.4)	113.4	N/m
Resonator mass @ I/O, m_{re}	(5.66×10^{-13})	5.66×10^{-13}	kg
Resonator stiffines @ I/O, k_{re}	(1362)	1362	N/m
Resonator mass @ l_c, m_{rc}	(3.99×10^{-11})	2.84×10^{-11}	kg
Resonator stiffness @ l_c, k_{rc}	(96,061)	68,319	N/m
Integrated μres. stiffness, k_{reff}	(1434)	1434	N/m
Young's modulus, E	150	150	GPa
Density of polysilicon, ρ	2300	2300	kg/m³
Electrode-to-μRes. gap, d_o	1300	1300	Å
Gap d_o adjusted for depletion	(1985)	1985	Å
Electrode width, W_e	20	20	μm
Filter dc-bias, V_P	35	35	V
Frequency pulling voltage, $V_{\Delta f}$	0.12	0	V
Q-control resistors, R_Q	12.2 (19.6)	**14.5** [19.6]	kΩ

[a] Numbers in parentheses indicate caluclated or semiempirical values.
[b] Boldface numbers indicate significant deviations needed to match simulated curves with measured curves.
[c] Numbers in square brackets indicate values expected from an ideal simulation with no parasitics and perfect termination. The value for κ in the "Des./Meas." column was obtained via finite-element simulation using ANSYS.
[d] Top 11 rows represent simulation outputs; the rest are used as inputs for simulation.

TABLE 12.5. HF μMechanical Filter Circuit Element Values for the Data of Table 12.4 [18]

Parameter	Value	Units
Coupling location, l_c	4.48	μm
$C_{o1} = C_{o2}$	7.14	fF
$l_{x1} = l_{x2}$	5.66×10^{-13}	H
$c_{x1} = c_{x2}$	0.000734	F
$r_{x1} = r_{x2}$	4.62×10^{-9}	Ω
$c_{s12a} = c_{s12b}$	-0.0121	F
c_{s12c}	0.00882	F
$\eta_{e1} = \eta_{e2}$	1.20×10^{-6}	C/m
$\eta_{c12} = \eta_{c21}$	7.08	C/m

that is, loss poles were introduced inadvertently. For fully integrated filters, in which μmechanics and circuits are fabricated side-by-side on a single chip, parasitic capacitors are expected to be much smaller. In this case, the feedthrough capacitor $C_{P(fd)}$ can then be purposefully designed into the filter if loss poles are desired.

12.3.4. Micromechanical Mixer–Filters

As indicated by Eq. (12.3), the voltage-to-force transducer used by the described resonators is nonlinear, relating input force F_d to input voltage $(v_e - v_b)$ by a square law. When $v_b = V_P$, this nonlinearity is suppressed, leading to a dominant force that is linear with v_e given by the second term of Eq. (12.5). If, however, signal inputs are applied to both v_e and v_b, a square law mixer results. In particular, if an RF signal $v_{RF} = V_{RF} \cos \omega_{RF} t$ is applied to electrode e, and a local oscillator signal $v_{LO} = V_{LO} \cos \omega_{LO} t$ to electrode b, then Eq. (12.3) contains the term

$$F_d = \cdots - \frac{1}{2} V_{RF} V_{LO} \frac{\partial C}{\partial x} \cos(\omega_{RF} - \omega_{LO})t + \cdots, \quad (12.31)$$

which clearly indicates a mixing of *voltage* signals v_{RF} and v_{LO} down to a *force* signal at frequency $\omega_{IF} = (\omega_{RF} - \omega_{LO})$. If the above transducer is used to couple into a μmechanical filter with a passband centered at ω_{IF}, an effective mixer–filter device results that provides both a mixer and filtering function in one passive, micromechanical device.

Figure 12.18(a) presents the schematic for a symmetrical μmechanical mixer–filter [34], showing the bias and input scheme required for downconversion and equating this device to a system-level functional block. As shown, since this device provides filtering as part of its function, the overall mechanical structure is exactly that of a μmechanical filter. The only differences are the applied inputs and the use of a nonconductive coupling beam to isolate the IF port from the LO. Note that if the source providing V_P to the second resonator is ideal (with zero source resistance) and

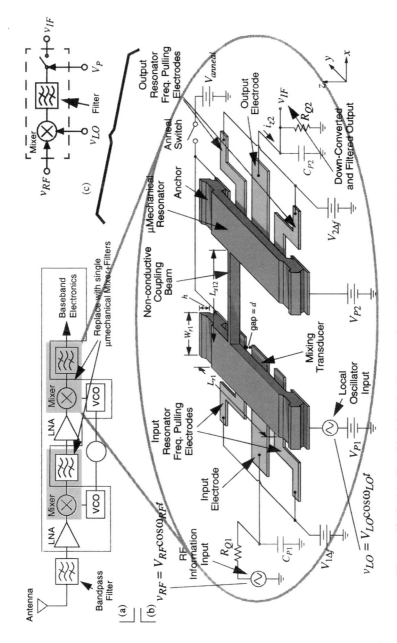

Figure 12.18. (a) Simplified block diagram of a wireless receiver, indicating (with shading) the components replaceable by mixer-filter devices. (b) Schematic diagram of the described μmechanical mixer–filter, depicting the bias and excitation scheme needed for downconversion. (c) Equivalent block diagram of the mixer–filter scheme.

the series resistance in the second resonator is small, LO signals feeding across the coupling beam capacitance are shunted to ac ground before reaching the IF port. In reality, finite resistivity in the resonator material allows some amount of LO-to-IF leakage.

The mixer conversion gain/loss in this device is determined primarily by the relative magnitudes of the dc-bias V_P applied to the resonator and the local oscillator amplitude V_{LO}. Using Eq. (12.31), assuming R_Q resistors given by Eq. (12.26), and assuming that the filter structure presents a large input impedance to both v_{RF} and v_{LO} (since their frequencies are off-resonance), the expression for conversion gain/loss takes the form

$$G_{\text{conv}} = \frac{1}{L_{\text{conv}}} = \frac{V_{LO}}{V_P} \rightarrow 20\log\left(\frac{V_{LO}}{V_P}\right) \quad [\text{dB}]. \tag{12.32}$$

Note that conversion gain is possible if $V_{LO} > V_P$.

The single-sideband (SSB) noise figure for this device derives from a combination of mixer conversion loss, filter insertion loss, and an additional 3 dB that accounts for noise conversion from two bands (RF and image) to one and can be expressed as

$$NF = L_{\text{conv}}|_{\text{dB}} + L_{\text{fltr}}|_{\text{dB}} + 3 \quad [\text{dB}], \tag{12.33}$$

where $L_{\text{fltr}}|_{\text{dB}}$ is the filter insertion loss in dB. Possible (and somewhat generous) values might be $L_{\text{conv}}|_{\text{dB}} = 0$ db (with $V_{LO} = V_P$) and $L_{\text{fltr}}|_{\text{dB}} = 0.5$ dB, leading to $NF = 3.5$ dB—very good calculated performance for a combined mixer and filter using passive components.

12.3.5 Micromechanical Switches

The mixer–filter device described above is one example of a μmechanical circuit that harnesses nonlinear device properties to provide a useful function. Another very useful mode of operation that further utilizes the nonlinear nature of the device is the μmechanical switch. Figure 12.19 presents an operational schematic for a single-

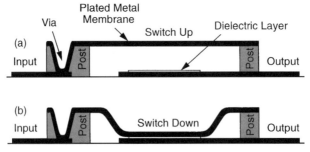

Figure 12.19. Cross-sectional schematics of a typical μmechanical switch: (a) switch up and (b) switch down [35].

pole, single-throw µmechanical switch [35], seen to have a structure very similar to that of the previous resonator devices: a conductive beam or membrane suspended above an actuating electrode. The operation of the switch of Figure 12.19 is fairly simple: to achieve the "on-state," apply a sufficiently large voltage across the beam and electrode to pull the beam down and short it (in either a dc or ac fashion) to the electrode.

In general, to minimize insertion loss, the majority of switches use metals as their structural materials. It is their metal construction that makes µmechanical switches so attractive, allowing them to achieve "on-state" insertion losses down to 0.1 dB— much lower than FET counterparts, which normally exhibit ~ 2 dB of insertion loss. In addition to exhibiting such low insertion loss, µmechanical switches are extremely linear, with IIP3's greater than 66 dBm [8], and can be designed to consume no dc power (as opposed to FET switches, which sink a finite current when activated).

Chapter 10 in this book covers micromechanical switches in much greater detail.

12.4 RF RECEIVER FRONT-END ARCHITECTURES USING MEMS

Having surveyed a subset of the mechanical circuits most useful for communications applications, we now consider methods by which these circuits are best incorporated into communications subsystems. Three approaches to using micromechanical vibrating resonators are described in order of increasing performance enhancement: (1) direct replacement of off-chip high-Q passives; (2) use of an RF channel-select architecture using a large number of high-Q micromechanical resonators in filter banks and switchable networks; and (3) use of an all-mechanical RF front-end.

In proposing these architectures, certain liberties are taken in an attempt to account for potential advances in micromechanical resonator technology. For example, in the RF channel-select architecture, µmechanical circuits are assumed to be able to operate at UHF with Q's on the order of 10,000. Given that TFRs already operate at UHF (but with Q's of 1000), and that 100 MHz free–free beam µmechanical resonators presently exhibit Q's around 8000, the above assumed performance values may, in fact, not be far away. At any rate, the rather liberal approach taken in this section is largely beneficial, since it better conveys the potential future impact of MEMS technology and provides incentive for further advancements in this area. Nevertheless, in order to keep in check the enthusiasm generated here, assumed performances in this section are briefly reevaluated in the next, with an eye toward practical implementation issues.

12.4.1 Direct Replacement of Off-Chip High-Q Passives

Perhaps the most direct way to harness µmechanical circuits is via direct replacement of the off-chip ceramic, SAW, and crystal resonators used in RF preselect and image-reject filters, IF channel-select filters, and crystal oscillator references. A schematic depicting this approach was shown previously in Figure 12.1, and now in a condensed form in Figure 12.20. In addition to high-Q components, Figures 12.1 and 12.20

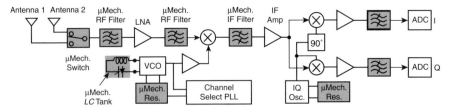

Figure 12.20. System block diagram of a superheterodyne receiver architecture showing potential replacements via MEMS-based components. (On-chip μmechanics are shaded.)

also show the use of other MEMS-based passive components, such as medium-Q micromachined inductors and tunable capacitors [36] used in VCOs and matching networks, as well as low-loss (~ 0.1 dB) μmechanical switches [35] that not only provide enhanced antenna diversity but that can also yield power savings by making TDD (rather than FDD) more practical in future transceivers.

Of course, the main benefits from the above approach to using MEMS are size reduction and, given the potential for integration of MEMS with transistor circuits, the ability to move more components onto the silicon die. A limited number of performance benefits also result from replacement of existing high-Q passives by μmechanical ones, such as the ability to tailor the termination impedances required by RF and IF filters (cf. Table 12.3). Such impedance flexibility can be beneficial when designing low-noise amplifiers (LNAs) and mixers in CMOS technology, which presently often consume additional power to impedance match their outputs to 50 Ω off-chip components. If higher impedances can be used, for example, at the output of an LNA, significant power savings are possible. As an additional benefit, since the source impedance presented to the LNA input is now equal to R_Q, it can now be tailored to minimize noise figure (NF).

Although beneficial, the performance gains afforded by mere direct replacement by MEMS are quite limited when compared to more aggressive uses of MEMS technology. More aggressive architectures will now be described.

12.4.2 An RF Channel-Select Architecture

To fully harness the advantages of μmechanical circuits, one must first recognize that due to their micro-scale size and zero dc power consumption, μmechanical circuits offer the same system complexity advantages over off-chip discrete components that planar ICs offer over discrete transistor circuits. Thus, to maximize performance gains, μmechanical circuits should be utilized on a massive scale.

Perhaps one of the simplest ways to harness the small size of micromechanical circuits is to add multiband reconfigurability to a transceiver by adding a preselect and image-reject filter for each communications standard included. Due to the small size of micromechanical filters, this can be done with little regard to the overall size of the transceiver.

Although the above already greatly enhances the capability of today's wireless transceivers, it in fact only touches upon a much greater potential for performance

RF RECEIVER FRONT-END ARCHITECTURES USING MEMS   443

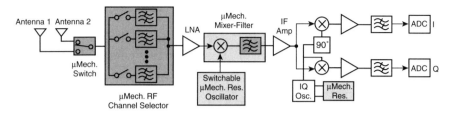

Figure 12.21. System block diagram for an RF channel-select receiver architecture utilizing large numbers of micromechanical resonators in banks to trade Q for power consumption. (On-chip μmechanics are shaded.)

enhancement. In particular, it does not utilize micromechanical circuits to their fullest complexity. Figure 12.21 presents the system-level block diagram for a possible receiver front-end architecture that takes full advantage of the complexity achievable via μmechanical circuits. The main driving force behind this architecture is power reduction, attained in several of the blocks by trading power for high selectivity (i.e., high-Q). The key power saving-blocks in Figure 12.21 are now described.

12.4.2.1 Switchable RF Channel-Select Filter Bank If channel selection (rather than preselection) were possible at radio frequencies (rather than just at intermediate frequencies), then succeeding electronic blocks in the receive path (e.g., LNA, mixer) would no longer need to handle the power of alternate channel interferers. Thus, their dynamic range can be greatly relaxed, allowing substantial power reductions. In addition, the rejection of adjacent channel interferers also allows reductions in the phase noise requirements of local oscillator (LO) synthesizers, providing further power savings.

To date, RF channel selection has been difficult to realize via present-day technologies. In particular, low-loss channel selection at RF would require tunable resonators with Q's in the thousands. Unfortunately, however, high Q often precludes tunability, making RF channel selection via a single RF filter a very difficult prospect.

On the other hand, it is still possible to select individual RF channels via many nontunable high-Q filters, one for each channel, and each switchable by command. Depending on the standard, this could entail hundreds or thousands of filters—numbers that would be absurd if off-chip macroscopic filters are used, but that may be perfectly reasonable for microscale, passive, μmechanical filters, such as described in Section 12.3.3.

Figure 12.22 presents one fairly simple rendition of the key system block that realizes the desired RF channel selection. As shown, this block consists of a bank of μmechanical filters with all filter inputs connected to a common block input and all outputs to a common block output, and where each filter passband corresponds to a single channel in the standard of interest. In the scheme of Figure 12.22, a given filter is switched on (with all others off) by decoder-controlled application of an appropriate dc-bias voltage to the desired filter. (Recall from Eqs. (12.3) and (12.7)

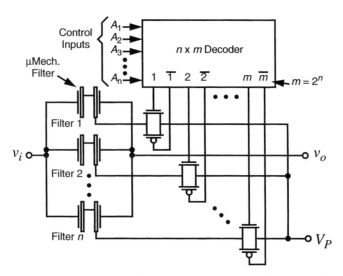

Figure 12.22. System/circuit diagram for an RF channel-select micromechanical filter bank.

that the desired force input and output current are generated in a μmechanical resonator only when a dc-bias V_P is applied; that is, without V_P, the input and output electrodes are effectively open-circuited.)

The potential benefits afforded by this RF channel selector can be quantified by assessing its impact on the LNA linearity specification imposed by the IS-98-A interim standard for CDMA cellular mobile stations [37]. In this standard, the required IIP3 of the LNA is set mainly to avoid desensitization in the presence of a single tone (generated by AMPS [37]) spaced 900 kHz away from the CDMA signal center frequency. Here, reciprocal mixing of the local oscillator phase noise with the 900 kHz offset single tone and cross-modulation of the single tone with leaked transmitter power outputs dictate that the LNA IIP3 exceeds $+7.6$ dBm [37]. However, if an RF channel-select filter bank such as shown in Figure 12.22 precedes the LNA and is able to reject the single tone by 40 dB, the requirement on the LNA then relaxes to IIP3 ≤ -29.3 dBm (assuming the phase noise specification of the local oscillator is *not* also relaxed). Given the well-known noise and linearity versus power trade-offs available in LNA design [38,39], such a relaxation in IIP3 can result in nearly an order of magnitude reduction in power. In addition, since RF channel selection relaxes the overall receiver linearity requirements, it may become possible to put more gain in the LNA to suppress noise figure (*NF*) contributions from later stages, while relaxing the required *NF* of the LNA itself, leading to further power savings.

Turning to oscillator power, if the single-tone interferer is attenuated to 40 dB, then reciprocal mixing with the local oscillator is also greatly attenuated, allowing substantial reduction in the phase noise requirement of the local oscillator. Requirement reductions can easily be such that on-chip solutions to realization of the receive path VCO (e.g., using spiral inductors and pn-diode tunable capacitors) become plausible.

12.4.2.2 Switchable Micromechanical Resonator Synthesizer Although the µmechanical RF channel selector described above may make possible the use of existing on-chip technologies to realize the receive path VCO, this approach is not recommended, since it denies the system from achieving much greater power reduction factors that may soon be available through MEMS technology. In particular, given that power and Q can often be interchanged when designing for a given oscillator phase noise specification, a better approach to implementing the VCO would be to use µmechanical resonators (with orders of magnitude higher Q than any other on-chip tank) to set the VCO frequency. In fact, with Q's as high as achievable via µmechanics, the basic design methodologies for oscillators must be reevaluated. For example, in the case where the oscillator and its output buffer contribute phase noise according to Leeson's equation [40], where the $1/f^2$-to-white phase noise corner occurs at $f_o/2Q$, a tank $Q > 1500$ is all that would be required to move the $1/f^2$-to-white phase noise corner close enough to the carrier that only white phase noise need be considered for CDMA cellular applications, where the phase noise power at frequency offsets from 285 to 1515 kHz is most important. If only white noise is important, then only the output buffer noise need be minimized, and sustaining amplifier noise may not even be an issue. If so, the power requirement in the sustaining amplifier might be dictated solely by loop gain needs (rather than by phase noise needs), which for a µmechanical resonator-based VCO with $R_x \approx 40\,\Omega$, $L_x \approx 84\,\mu H$, and $C_x \approx 0.5$ fF, might be less than 1 mW.

To implement a tunable local oscillator synthesizer, a switchable bank is needed, similar to that of Figure 12.22 but using µmechanical resonators, not filters, each corresponding to one of the needed LO frequencies, and each switchable into or out of the oscillator sustaining circuit. Note that because µmechanical resonators are now used in this implementation, the Q and thermal stability (with compensation electronics) of the oscillator may now be sufficient to operate without the need for locking to a lower frequency crystal reference. The power savings attained upon removing the PLL and prescaler electronics needed in past synthesizers can obviously be quite substantial. In effect, by implementing the synthesizer using µmechanical resonators, synthesizer power consumption can be reduced from the ~ 90 mW dissipated by present-day implementations using medium-Q, L, and C components [41], to something in the range of only 1–4 mW. Again, all this is attained using a circuit topology that would seem absurd if only macroscopic high-Q resonators were available, but that becomes plausible in the micromechanical arena.

12.4.2.3 Micromechanical Mixer–Filter The use of a µmechanical mixer–filter in the receive path of Figure 12.21 eliminates the dc power consumption associated with the active mixer normally used in present-day receiver architectures. This corresponds to a power savings on the order of 10–20 mW. In addition, if multiple input electrodes (one for RF, one for matching) are used for the mixer–filter, the RF input can be made to appear purely capacitive to the LNA (i.e., at the RF), and the

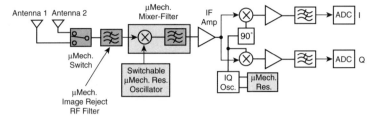

Figure 12.23. System block diagram for an all-MEMS RF front-end receiver architecture. (On-chip μmechanics are shaded.)

LNA would no longer require a driver stage to match a certain impedance. Rather, an inductive load can be used to resonate the capacitance, as in Shaeffer and Lee [38], allowing power savings similar to that discussed in Section 12.4.1.

12.4.3 An All-MEMS RF Front-End Receiver Architecture

In discussing the above MEMS-based architecture, one very valid question may have arisen: If μmechanical filters and mixer–filters can truly post insertion losses consistent with their high-Q characteristics, then is an LNA really required at radio frequencies? It is this question that inspires the receiver architecture shown in Figure 12.23, which depicts a receive path comprised of a relatively wideband image-reject μmechanical RF filter followed immediately by a narrowband IF mixer-filter that then feeds subsequent IF electronics. The only active electronics operating at RF in this system are those associated with the local oscillator, which, if it uses a bank of μmechanical resonators, may be able to operate at less than 1 mW. If plausible, the architecture of Figure 12.23 clearly presents enormous power advantages, completely eliminating the power consumption of the LNA and active mixer of Figure 12.20—a total power savings on the order of 40 mW—and, together with the additional 85 mW of power savings from the micromechanical LO, substantially increasing mobile phone standby times.

To assess the plausibility of this all-MEMS front-end, one can determine whether or not this scheme yields a reasonable noise figure requirement at the input node of the IF amplifier in Figure 12.23. An expected value for RF image-reject filter insertion loss is $IL \approx 0.2$ dB, assuming that three resonators are used, each with $Q = 5000$. Using the value for mixer–filter noise figure $NF_{mf} = 3.5$ dB projected in Section 12.3.4, the total combined noise figure $NF_{f+mf} = 3.7$ dB Given IS-98-A's requirement that the receiver noise figure $NF_{RX} \leq 7.8$ dB (with a 2 dB conservative design buffer), the needed value at the IF amplifier input is $NF_{IF} \leq 4.1$ dB, which can be reasonable if the IF amplifier gain can be increased to suppress the noise of succeeding stages. IF-baseband strips for GSM with $NF \approx 3.8$ dB are, in fact, already available [42].

Although the all-MEMS front-end architecture of Figure 12.23 may at first seem the most preposterous of the bunch, early versions of the primary filtering and mixing devices required for its implementation have already been demonstrated. In particular, TFR image-reject filters have been demonstrated at ultra high frequencies with insertion losses of less than 3 dB [19]. It should be noted, however, that the first demonstrated mixer–filter based on polysilicon clamped–clamped beam μmechanical resonators achieved $NF_{mf} = 15$ dB [34]—quite worse than the 3.5 dB used in the above calculation, and in fact, a value that precludes the use of the architecture shown in Figure 12.23. It is not unreasonable, however, to expect that future renditions of mixer–filters, perhaps using more appropriate resonators (e.g., higher Q free–free beams, rather than clamped–clamped beams), might be able to achieve the projected 3.5 dB.

12.5 AN RF TRANSMITTER ARCHITECTURE USING MEMS

Due to a lack of sufficient in-band power handling capability, very little consideration has been given to date to the possibility of using μmechanical resonators in the transmit path. However, research efforts are presently underway to remedy this, and, if successful, equally compelling MEMS-based transmit architectures can also be proposed.

Figure 12.24 depicts one rendition, in which an RF channel selector is placed after the power amplifier (PA) in the transmit path. This channel selector might utilize a similar circuit as that of Figure 12.22, but using μmechanical resonators with sufficient power handling capability. Assuming for now that such devices are possible, this transmit topology could provide enormous power savings. In particular, if a high-Q, high-power filter with less than 1 dB of insertion loss could follow the PA, cleaning all spurious outputs, including those arising from spectral regrowth, then more efficient PA designs can be utilized, despite their nonlinearity. For example, a PA previously restricted by linearity considerations to 30% efficiency in present-day transmitter architectures may now be operable closer to its maximum efficiency, perhaps 50%. For a typical transmit power of 600 mW, this efficiency increase corresponds to 800 mW of power savings. If a more efficient PA topology could be used, such as Class E, with theoretical efficiencies approaching 100%, the power savings could be much larger.

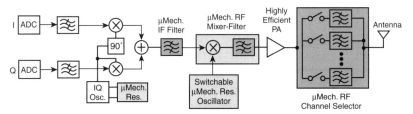

Figure 12.24. RF channel-select transmitter architecture, possible only if high-power μmechanical resonators can be achieved. Here, on-chip μmechanical blocks are shaded, and the PA is not necessarily implemented on-chip.

In addition to the MEMS-based channel-select RF filter bank, the architecture of Figure 12.24 also features a micromechanical upconverter that uses a mixer–filter device, such as described in Section 12.3.4, to upconvert and filter the information signal before directing it to the power amplifier.

12.6 RESEARCH ISSUES

As stated at the beginning of Section 12.4, the transceiver architectures described above rely to some extent on performance characteristics not yet attained by μmechanical resonators, but targeted by ongoing research efforts. Specifically, μmechanical devices with the following attributes have been assumed: (1) adequate Q at ultra high frequencies; (2) sufficient linearity and power handling capability; (3) usable port impedances; and (4) massive scale integration methods.

12.6.1 Frequency and Q

As previously mentioned, since TFRs can already operate in the 3–7 GHz range with Q's of ~ 1000, vibrating micromechanical high-Q tanks operating at UHF already exist. However, although their ~ 200 μm diameters are much smaller than corresponding dimensions on macroscopic counterparts, TFRs are still much larger than the micromechanical beams described in Section 12.3, which can have lengths less than 10 μm at UHF. Since tiny size is paramount in many of the proposed micromechanical architectures of Sections 12.4 and 12.5, especially those calling for banks of numerous high-Q filters or resonators, the ultimate frequency range of micromechanical beam resonators is of great interest. Table 12.1 (in Section 12.3.2) showed that, from a purely geometric perspective, the frequencies required by the architectures of Section 12.4, from 10 MHz to 2.5 GHz, are reasonable for beam elements. Geometry, however, is only one of many important considerations. Indeed, the applicable frequency range of micromechanical resonators will also be a function of several other factors, including:

1. Quality factor, which may change with frequency for a given material, depending on frequency-dependent energy loss mechanisms [43].
2. Linearity and power handling ability, which may decrease as the size of a given resonator decreases [18].
3. Series motional resistance R_x (cf. Figure 12.14), which must be minimized to allow impedance matching with other transceiver components and to alleviate filter passband distortion due to parasitics [17,18,44,45].
4. Absolute and matching tolerances of resonance frequencies, which will both be functions of the fabrication technology and of frequency trimming or tuning strategies [46].
5. Stability of the resonance frequency against temperature variations, mass loading, aging, and other environmental phenomena.

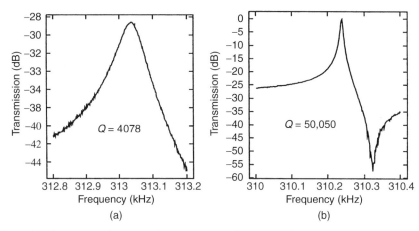

Figure 12.25. Measured transconductance spectra for (a) a $POCl_3$-doped resonator and (b) an implant-doped version, both after furnace annealing. (From reference [29])

Each of the above phenomena are currently under study. In particular, assuming adequate vacuum can be achieved, the ultimate quality factor will be strongly dependent on the material type, and even the manufacturing process. For example, surface roughness or surface damage during fabrication may play a role in limiting quality factor. In fact, preliminary results comparing the quality factor achievable in diffusion-doped polysilicon structures (which exhibit substantial pitting of the poly surface) versus implant-doped ones indicate that the latter exhibit almost an order of magnitude higher Q at frequencies near 10 MHz. Figure 12.25 presents measured transconductance spectra for two comb-driven folded-beam micromechanical resonators fabricated in the same polycrystalline material, but doped differently—one $POCl_3$-doped, the other phosphorous implant-doped—using the process sequences summarized in Table 12.6 [46]. The difference in Q is very intriguing and is consistent with a surface roughness-dependent dissipation mechanism.

From a design perspective, one Q-limiting loss mechanism that becomes more important with increasing frequency is loss to the substrate through anchors. The

TABLE 12.6. Doping Recipes

$POCl_3$	Implant
(i) Deposit 2 μm LPCVD fine-grained polysilicon @ 588°C	(i) Deposit 1 μm LPCVD fine-grained polysilicon @ 588°C
(ii) Dope 2.5 h @ 950°C in $POCl_3$ gas	(ii) Implant phosphorus: Dose = 10^{16} cm^{-2}, Energy = 90 keV
(iii) Anneal for 1 h @ 1100°C in N_2 ambient	(iii) Deposit 1 μm LPCVD fine-grained polysilicon @ 588°C
	(iv) Anneal for 1 h @ 1100°C in N_2 ambient

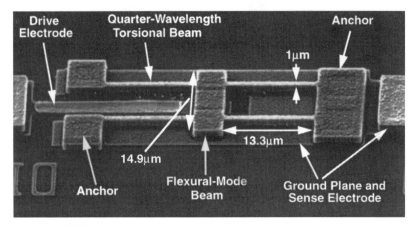

Figure 12.26. SEM of a free–free beam virtually levitated micromechanical resonator with relevant dimensions for $f_o = 71$ MHz. (From reference [11])

frequency dependence of this mechanism arises because the stiffness of a given resonator beam generally increases with resonance frequency, giving rise to larger forces exerted by the beam on its anchors during vibration. As a consequence, more energy per cycle is radiated into the substrate via the anchors. Antisymmetric resonance designs, such as balanced tuning forks, could prove effective in alleviating this source of energy loss.

Alternatively, anchor loss mechanisms can be greatly alleviated by using "anchorless" resonator designs, such as shown in Figure 12.26. This recently demonstrated device utilizes a free–free beam (i.e., xylophone) resonator suspended by four torsional supports attached at flexural node points. By choosing support dimensions corresponding to a quarter-wavelength of the free–free beam's resonance frequency, the impedance presented to the beam by the supports can be effectively nulled out, leaving the beam virtually levitated and free to vibrate as if it had no supports [11]. Figure 12.27 presents the frequency characteristic for a 92.25 MHz version of this μmechanical resonator, showing a Q of nearly 8000—still plenty for channel-select RF applications. (Note that the excessive loss in the spectrum of Figure 12.27 is an artifact of improper impedance matching between the resonator output and the measurement apparatus. In addition, this resonator used a conservative electrode-to-resonator gap spacing of $d \approx 2000$ Å, so a rather large V_P was needed to provide a sufficient output level.)

What loss mechanisms await at gigahertz frequencies for flexural-mode resonators is, as yet, unknown. In particular, there is concern that frequency-dependent material loss mechanisms may cause Q to degrade with increasing frequency. Again, however, Q's of over 1000 at UHF (and beyond) have already been achieved via thin-film bulk acoustic resonators based on longitudinal resonance modes and piezoelectric structural materials. It is hoped that μmechanical resonators based on chemical vapor deposited (CVD) materials can retain Q's of at least 8000 at similar frequencies.

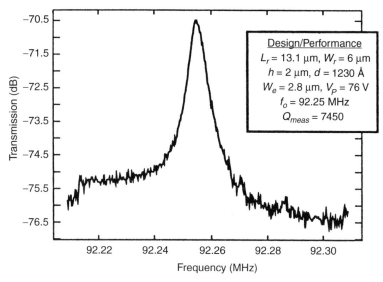

Figure 12.27. Frequency characteristic for a fabricated 92.25 MHz free–free beam micromechanical resonator. (From reference [11])

12.6.2 Linearity and Power Handling

Macroscopic high-Q filters based on ceramic resonator or SAW technologies are very linear in comparison with the transistor blocks they interface with in present-day transceivers. As a result, their contributions to the total IIP3 budget can generally be ignored in the majority of designs. In scaling the sizes of high-Q filtering devices to the microscale, however, linearity considerations must now be reconsidered, since past experience often says that the smaller the device, the less power it can handle.

For the capacitively driven μmechanical resonator of Figure 12.4, an approximate expression for the magnitude of the in-band force component at ω_o arising from third-order intermodulation of two out-of-band interferers at $\omega_1 = \omega_o + \Delta\omega$ and $\omega_2 = \omega_o + 2\Delta\omega$ can be derived by considering nonlinearities in the input capacitive transducer. Assuming that resonator displacements are small enough that stiffening nonlinearity can be neglected, such a derivation yields [18]

$$F_{IM_3} = V_i^3 \left[\frac{1}{4} \frac{(\varepsilon_o A_o)^2}{d_o^5} \frac{V_P}{k_{\text{reff}}} (2\Theta_1 + \Theta_2^*) \right. \\
\left. + \frac{3}{4} \frac{(\varepsilon_o A_o)^3}{d_o^8} \frac{V_P^3}{k_{\text{reff}}^2} \Theta_1 (\Theta_1 + 2\Theta_2^*) + \frac{3}{2} \frac{(\varepsilon_o A_o)^4}{d_o^{11}} \frac{V_P^5}{k_{\text{reff}}^3} \Theta_1^2 \Theta_2^* \right], \tag{12.34}$$

where $\Theta_1 = \Theta(\omega_1)$, $\Theta_2 = \Theta(\omega_2)$, and

$$\Theta(\omega) \; \frac{1}{1 - (\omega/\omega_{u3dB})^2 + j\omega/(Q\omega_{u3dB})}, \tag{12.35}$$

where $\omega_{u3\text{dB}} = \omega_o + B/2$ is the 3 dB frequency at the upper edge of the filter passband. Equating Eq. (12.34) with the in-band force component (i.e., the second term of Eq. (12.5)), then solving for V_i, the IIP3 for a 70 MHz µmechanical resonator is found to be around 12 dBm [18]. This is adequate for virtually all receive path functions, except for those in standards that allow simultaneous transmit and receive (such as CDMA), where the RF preselect filter is required to reject out-of-band transmitter outputs to alleviate cross-modulation phenomena [37]. For such situations, at least at present, a more linear filter must precede the filter bank of Figure 12.22 if cross-modulation is to be sufficiently suppressed. This additional filter, however, can now have a very wide bandwidth, as it has no other purpose than to reject transmitter outputs. Thus, it may be realizable with very little insertion loss using on-chip (perhaps micromachined) inductor and capacitor technologies [8].

It should be noted that the above hindrances exist mainly for systems using simultaneous transmit and receive. Burst mode, quasi-time-duplexed systems, such as GSM, should be able to use the micromechanical RF channel selector by itself, without the need for a transmit-reject filter.

It should also be mentioned that higher power handling micromechanical resonators are also presently being investigated. Among approaches being taken are the use of alternative geometries (e.g., no longer flexural mode) and the use of alternative transduction (e.g., piezoelectric, magnetostrictive). Such research efforts are aimed at not only out-of-band transmit power rejection, but on in-band handling of transmit power as well, with a goal of realizing the RF channel-select transmit architecture described in Section 12.5.

12.6.3 Resonator Impedance

Thin-film bulk acoustic resonators can already impedance match to conventional antennas, so if their frequency, Q, yield, size, and integration capacity are adequate for a given architecture (e.g., the all-MEMS architecture of Section 12.4), then they present a very good solution. If higher Q is needed, however, then µmechanical resonators may be better suited for the given application. From Table 12.3, RF µmechanical filters should be able to match to 300 Ω impedances, provided their electrode-to-resonator gaps can be reduced to $d \approx 70$ Å. Since electrode-to-resonator gaps are achieved via a process very similar to that used to achieve MOS gate oxides [18], such gaps are not unreasonable. However, device linearity generally degrades with decreasing d, so practical designs must balance linearity with impedance requirements [18].

In cases where linearity issues constrain the minimum d to a value larger than that needed for impedance matching (assuming a fixed V_P), several µmechanical filters with identical frequency characteristics may be used to divide down the needed value of termination impedance. For example, ten of the filters in the fourth column of Table 12.3 can be hooked up in parallel to realize an $R_Q = 2000/10 = 200$ Ω. Note that the use of numerous filters in parallel also increases the power handling threshold. For example, if a given micromechanical filter were designed to handle 10 mW of power while retaining adequate linearity, then ten of them will handle 100 mW.

Once again, the ability to use numerous high-Q elements in complex micromechanical circuits without regard to size greatly extends the applicable range of micromechanical signal processors. Given a suitable massive-scale trimming technique, the above parallel-filter solution may work well even in the transmit path, perhaps making plausible some of the more aggressive power saving transmit architectures, such as that of Figure 12.24.

12.6.4 Massive-Scale Integration

Massive-scale manufacturing technology capable of combining MEMS and transistor circuits onto single chips constitutes the fourth major research issue mentioned at the beginning of this section. The importance and breadth of this topic, however, demands a section of its own, which now follows.

12.7 CIRCUITS/MEMS INTEGRATION TECHNOLOGIES

Although a two-chip solution that combines a MEMS chip with a transistor chip can certainly be used to interface μmechanical circuits with transistor circuits, such an approach becomes less practical as the number of μmechanical components increases. For instance, practical implementations of the switchable filter bank in Figure 12.21 require multiplexing support electronics that must interconnect with each μmechanical device. If implemented using a two-chip approach, the number of chip-to-chip bonds required could become quite cumbersome, making a single-chip solution desirable.

In the pursuit of single-chip systems, several technologies that merge micromachining processes with those for integrated circuits have been developed and implemented over the past several years. These technologies can be categorized into three major approaches: mixed circuit and micromechanics, precircuits, and postcircuits. Each is now described.

12.7.1 Mixed Circuit and Micromechanics

In the mixed circuit/micromechanics approach, steps from both the circuit and the micromachining processes are intermingled into a single process flow. Of the three approaches, this one has so far seen the most use. However, it suffers from two major drawbacks: (1) many passivation layers are required (one needed virtually every time the process switches between circuits and μmechanics); and (2) extensive redesign of the process is often necessary if one of the combined technologies changes (e.g., a more advanced circuit process is introduced). Despite these drawbacks, mixed circuit/micromechanics processes have unquestionably made a sizable commercial impact. In particular, Analog Devices' BiMOSII process

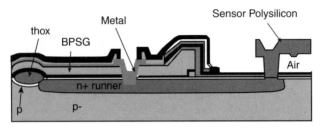

Figure 12.28. Cross section of the sensor area in Analog Devices' BiMOSII process [47].

(Figure 12.28 [47]), which has successfully produced a variety of accelerometers in large volume, is among the most successful examples of mixed circuit/micromechanics processes.

12.7.2 Precircuits

In the precircuits approach, micromechanics are fabricated in a first module, then circuits are fabricated in a subsequent module, and no process steps from either module are intermingled. This process has a distinct advantage over the mixed process above in that advances in each module can be accommodated by merely replacing the appropriate module. Thus, if a more advanced circuit process becomes available, the whole merging process need not be redesigned; rather, only the circuits module need be replaced. An additional advantage is that only one passivation step is required after the micromechanics module.

One of the main technological hurdles in implementing this process is the large topography leftover by micromechanical processes, with features that can be as high as 9 μm, depending on the number and geometry of structural layers. Such topographies can make photoresist spinning and patterning quite difficult, especially if submicron circuit features are desired. These problems, however, have been overcome by researchers at Sandia National Laboratories, whose *iMEMS* process (Figure 12.29) performs the micromechanics module in a trench, then planarizes features using chemical mechanical polishing (CMP) before doing the circuits module [48].

12.7.3 Postcircuits

The postcircuit approach is the dual of precircuits, in which the circuits module comes first, followed by the micromechanics module, where again, no process steps from either module are interspersed. This process has all the advantages of precircuits, but with relaxed topography issues, since circuit topographies are generally much smaller than micromechanical ones. As a result, planarization is often not necessary before micromechanics processing. Postcircuit processes have an additional advantage in that they are more amenable to multifacility processing, in which a very expensive fabrication facility (perhaps a foundry) is utilized for the

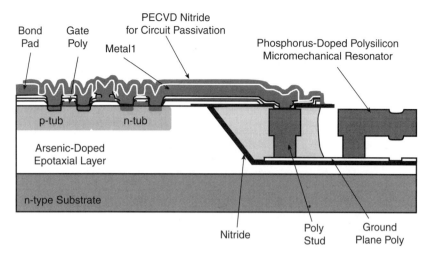

Figure 12.29. Cross section of Sandia's *iMEMS* process [48].

circuits module, and relatively lower capital micromechanics processing is done in-house at the company site (perhaps a small start-up). Such an arrangement may be difficult to achieve with a precircuits process because IC foundries may not permit "dirty" micromachined wafers into their ultraclean fabrication facilities.

Postcircuit processes have taken some time to develop. The main difficulty has been that aluminum-based circuit metallization technologies cannot withstand subsequent high-temperature processing required by many micromechanics processes—especially those that must achieve high Q. Thus, compromises in either the circuits process or the micromechanics process have been necessary, undermining the overall modularity of the process. The MICS process (Figure 12.30 [17, 49]), which used tungsten metallization instead of aluminum to withstand the high temperatures used in a following polysilicon surface micromachining module, is a good example of a postcircuit process that compromises its metallization technology. More recent renditions of this process have now been introduced that retain aluminum metallization, while substituting lower temperature poly-SiGe as the structural material, with very little (if any) reduction in micromechanical performance [51].

12.7.4 Other Integration Approaches

There are a number of other processes that can to some extent be placed in more than one of the above categories. These include front bulk-micromachining processes using deep reactive-ion etching (DRIE) [50] or anisotropic wet etchants [52] and other processes that slightly alter conventional CMOS processes [53]. In addition, bonding processes, in which circuits and μmechanics are merged by bonding one onto the wafer of the other, are presently undergoing a resurgence [54]. In particular, the advent of more sophisticated aligner-bonder instruments are now making

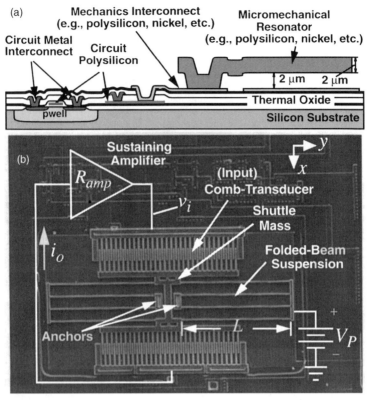

Figure 12.30. (a) Cross section of the MICS process [17]. (b) Overhead view of a fully integrated micromechanical resonator oscillator fabricated using MICS [17].

possible much smaller bond pad sizes, which soon may enable wafer-level bonding with bond pad sizes small enough to compete with fully planar processed merging strategies in interface capacitance values. If the bond capacitance can indeed be lowered to this level with acceptable bonding yields, this technology may well be the ultimate in modularity.

From a cost perspective, which technology is best depends to a large extent on how much of the chip area is consumed by MEMS devices in the application in question. For cases where the MEMS utilizes only a small percentage of the chip area, bonding approaches may be more economical, since a larger number of MEMS chips can be achieved on a dedicated wafer. For cases where MEMS devices take up a large amount of chip area, or where node capacitance must be minimized for highest performance, planar integration may make more sense.

12.7.4.1 Vacuum Encapsulation From a broader perspective, the integration techniques discussed above are really methods for achieving low-capacitance packaging of microelectromechanical systems. From the discussion in Section 12.3.2, another level of packaging is required to attain high-Q vibrating μmechanical resonators: vacuum encapsulation. Although the requirement for vacuum is unique to vibrating μmechanical resonators, the requirement for encapsulation is nearly universal for all of the micromechanical devices discussed in this chapter. In particular, some protection from the environment is necessary, if only to prevent contamination by particles (or even by molecules), or to isolate the device from electric fields or feedthrough currents.

The need for encapsulation is, of course, not confined to communications devices but also extends to the vast majority of micromechanical applications, for example, inertial navigation sensors. For many micromechanical applications, the cost of the encapsulation package can be a significant (often dominating) percentage of the total cost of the product. Thus, to reduce cost, packaging technologies with the highest yield and largest throughput are most desirable. Pursuant to this philosophy, wafer-level packaging approaches—some based on planar processing, some based on bonding—have been the focus of much research in recent years. Figure 12.31 presents cross sections that summarize one approach to *wafer-level* vacuum encapsulation [55], in which planar processing is used to realize an encapsulating cap. Although this and other encapsulation strategies have shown promise [57–60], there is still much room for improvement, especially given the large percentage of total product cost attributed to the package alone. Research to reduce the cost (i.e., enhance the yield and throughput) of encapsulation technologies continues.

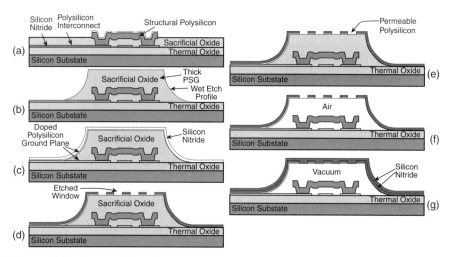

Figure 12.31. Process flow for vacuum-encapsulating a micromechanical resonator via planar processing. (a) Cross section immediately after the structural poly etch. (b) Deposit and pattern a thick, reflown PSG. (c) Deposit upper ground plane polysilicon and first nitride cap film. (d) Pattern etch windows in the cap. (e) Deposit permeable polysilicon [55]. (d) Etch sacrificial oxide (i.e., release structures) using HF, which accesses the sacrificial oxide through the permeable polysilicon, then dry via supercritical CO_2 [56], yielding the cross section in (f). (g) Seal shell under vacuum via a second cap nitride deposition done via LPCVD. Details for this process can be found in Lebouitz et al. [55].

12.8 CONCLUSIONS

Vibrating μmechanical resonators constitute the building blocks for a new integrated mechanical circuit technology in which high Q serves as a principal design parameter that enables more complex circuits. By combining the strengths of integrated μmechanical and transistor circuits, using both in massive quantities, previously unachievable functions become possible that enable transceiver architectures with projections for orders of magnitude performance gains. In particular, with the addition of high-Q μmechanical circuits, paradigm-shifting transceiver architectures that trade power for selectivity (i.e., Q) become possible, with the potential for substantial power savings and multiband reconfigurability. To reap the benefits of these new architectures, however, further advancements in device frequency, linearity, and manufacturability are required. Research efforts are ongoing, and it is hoped that this chapter has provided enough background information and research foresight to instigate new efforts toward making mechanical circuits commonplace in the near future.

REFERENCES

1. H. Khorramabadi and P. R. Gray, "High-frequency CMOS continuous-time filters," *IEEE Journal of Solid-State Circuits*, Vol. SC-19, No. 6, 939–948 (Dec. 1984).
2. K. B. Ashby, I. A. Koullias, W. C. Finley, J. J. Bastek, and S. Moinian, "High Q inductors for wireless applications in a complementary silicon bipolar process," *IEEE Journal of Solid-State Circuits*, Vol. 31, No. 1, 4–9 (Jan. 1996).
3. N. M. Nguyen and R. G. Meyer, "Si IC-compatible inductors and LC passive filters," *IEEE Journal of Solid-State Circuits*, Vol. SC-25, No. 4, 1028–1031 (Aug. 1990).
4. P. R. Gray and R. G. Meyer, "Future directions in silicon IC's for RF personal communications," in *Proceedings of the 1995 IEEE Custom Integrated Circuits Conference*, Santa Clara, CA, May 1–4, 1995, pp. 83–90.
5. R. A. Sykes, W. L. Smith, and W. J. Spencer, "Monolithic crystal filters," in *1967 IEEE International Convention Record*, Pt. II, Mar. 20–23, 1967, pp. 78–93.
6. R. C. Rennick, "An equivalent circuit approach to the design and analysis of monolithic crystal filters," *IEEE Transactions on Sonics and Ultrasonics*, Vol. SU-20, 347–354 (Oct. 1973).
7. C. K. Campbell, *Surface Acoustic Wave Devices for Mobile Wireless Communications*, Academic Press, New York, 1998.
8. C. T.-C. Nguyen, L. P. B. Katehi, and G. M. Rebeiz, "Micromachined devices for wireless communications (invited)," *Proceedings of the IEEE*, Vol. 86, No. 8, 1756–1768 (Aug. 1998).
9. A. A. Abidi, "Direct-conversion radio transceivers for digital communications," *IEEE Journal of Solid-State Circuits*, Vol. 30, No. 12, 1399–1410 (Dec. 1995).
10. J. C. Rudell, J.-J. Ou, T. B. Cho, G. Chien, F. Brianti, J. A. Weldon, and P. R. Gray, "A 1.9-GHz wide-band IF double conversion CMOS receiver for cordless telephone applications," *IEEE Journal of Solid-State Circuits*, Vol. 32, No. 12, 2071–2088 (Dec. 1997).

11. K. Wang, A.-C. Wong, and C. T.-C. Nguyen, "VHF free-free beam high-Q micromechanical resonators," *IEEE/ASME Journal of Microelectromechanical Systems*, Vol. 9, No. 3, 347–360 (Sept. 2000).
12. A. N. Cleland and M. L. Roukes, "Fabrication of high frequency nanometer scale mechanical resonators from bulk Si crystals," *Applied Physics Letters*, Vol. 69, No. 18, 2653–2655 (Oct. 28, 1996).
13. J. M. Bustillo, R. T. Howe, and R. S. Muller, "Surface micromachining for microelectromechanical systems," *Proceedings of the IEEE*, Vol. 86, No. 8, 1552–1574 (Aug. 1998).
14. J. D. Cressler et al., "Silicon–germanium heterojunction bipolar technology: the next leap for silicon?" in *Digest of Technical Papers*, 1994 ISSCC, San Francisco, CA, Feb. 1994.
15. N. Slawsby, "Frequency control requirements of radar," in *Proceedings of the 1994 IEEE International Frequency Control Symposium*, June 1–3, 1994, pp. 633–640.
16. W. P. Robins, *Phase Noise in Signal Sources*. Peter Peregrinus, Ltd., London, 1982.
17. C. T.-C. Nguyen and R. T. Howe, "An integrated CMOS micromechanical resonator high-Q oscillator," *IEEE Journal of Solid-State Circuits*, Vol. 34, No. 4, 440–445 (Apr. 1999).
18. F. D. Bannon III, J. R. Clark, and C. T.-C. Nguyen, "High-Q HF microelectromechanical filters," *IEEE Journal of Solid-State Circuits*, Vol. 35, No. 4, 512–526 (Apr. 2000).
19. K. M. Lakin, G. R. Kline, and K. T. McCarron, "Development of miniature filters for wireless applications," *IEEE Transactions on Microwave Theory and Techniques*, Vol. 43, No. 12, 2933–2939 (Dec. 1995).
20. S. V. Krishnaswamy, J. Rosenbaum, S. Horwitz, C. Yale, and R. A. Moore, "Compact FBAR filters offer low-loss performance," *Microwaves & RF*, 127–136 (Sept. 1991).
21. R. Ruby and P. Merchant, "Micromachined thin film bulk acoustic resonators," in *Proceedings of the 1994 IEEE International Frequency Control Symposium*, Boston, MA, June 1–3, 1994, pp. 135–138.
22. W. E. Newell, "Miniaturization of tuning forks," *Science*, Vol. 161, 1320–1326 (Sept. 1968).
23. R. A. Johnson, *Mechanical Filters in Electronics*, Wiley, New York, 1983.
24. R. T. Howe and R. S. Muller, "Resonant microbridge vapor sensor," *IEEE Transactions on Electron Devices*, Vol. ED–33, 499–506 (1986).
25. H. Nathanson, W. E. Newell, R. A. Wickstrom, and J. R. Davis, Jr., "The resonant gate transistor," *IEEE Transactions on Electron Devices*, Vol. ED–14, No. 3, 117–133, (Mar. 1967).
26. B. E. Boser and R. T. Howe, "Surface micromachined accelerometers," *IEEE Journal of Solid-State Circuits*, Vol. 31, No. 3, 366–375 (Mar. 1996).
27. N. Yazdi, F. Ayazi, and K. Najafi, "Micromachined inertial sensors," *Proceedings of the IEEE*, Vol. 86, No. 8, 1640–1659 (Aug. 1998).
28. H. A. C. Tilmans, "Equivalent circuit representation of electromechanical transducers: I. Lumped-parameter systems," *Journal of Micromechanical Microengineering*, Vol. 6, 157–176 (1996).
29. K. Wang and C. T.-C. Nguyen, "High-order medium frequency micromechanical electronic filters," *IEEE Journal of Microelectromechanical Systems*, Vol. 8, No. 4, 534–557 (Dec. 1999).
30. L. Lin, C. T.-C. Nguyen, R. T. Howe, and A. P. Pisano, "Micro electromechanical filters for signal processing," in *Technical Digest, IEEE Micro Electromechanical Systems Workshop*, Travemunde, Germany, pp. 226–231, Feb. 4–7, 1992.

31. A. I. Zverev, *Handbook of Filter Synthesis*, Wiley, New York, 1967.
32. P. R. Gray and R. G. Meyer, *Analysis and Design of Analog Integrated Circuits*, 2nd ed., Wiley, New York, 1984.
33. M. S. Lee, "Polylithic crystal filters with loss poles at finite frequencies," *Proceedings of the 1975 IEEE International Symposium on Circuits and Systems*, Apr. 21–23, 1975, pp. 297–300.
34. A.-C. Wong, H. Ding, and C. T.-C. Nguyen, "Micromechanical mixer + filters," *Technical Digest, IEEE International Electron Devices Meeting*, San Francisco, CA, Dec. 6–9, 1998, pp. 471–474.
35. C. Goldsmith, J. Randall, S. Eshelman, T. H. Lin, D. Denniston, S. Chen, and B. Norvell, "Characteristics of micromachined switches at microwave frequencies," *IEEE MTT-S Digest*, 1141–1144 (June 1996).
36. D. J. Young and B. E. Boser, "A micromachined variable capacitor for monolithic low-noise VCOs," *Technical Digest, 1996 Solid-State Sensor and Actuator Workshop*, Hilton Head Island, SC, June 3–6, 1996, pp. 86–89.
37. W. Y. Ali-Ahmad, "RF system issues related to CDMA receiver specifications," *RF Design*, 22–32 (Sept. 1999).
38. D. K. Shaeffer and T. H. Lee, "A 1.5-V, 1.5-GHz CMOS low noise amplifier," *IEEE Journal of Solid-State Circuits*, Vol. 32, No. 5, 745–759 (May 1997).
39. R. G. Meyer and A. K. Wong, "Blocking and desensitization in RF amplifiers," *IEEE Journal of Solid-State Circuits*, Vol. 30, No. 8, 994–946 (Aug. 1995).
40. D. B. Leeson, "A simple model of feedback oscillator noise spectrum," *Proceedings of the IEEE*, Vol. 54, 329–330 (Feb. 1966).
41. J. F. Parker and D. Ray, "A 1.6-GHz CMOS PLL with on-chip loop filter," *IEEE Journal of Solid-State Circuits*, Vol. 33, No. 3, 337–343 (Mar. 1998).
42. P. Orsatti, F. Piazza, and Q. Huang, "A 70-MHz CMOS IF-baseband strip for GSM," *IEEE Journal of Solid-State Circuits*, Vol. 35, No. 1, 104–108 (Jan. 2000).
43. V. B. Braginskky, V. P. Mitrofanov, and V. I. Panov, *Systems with Small Dissipation*, University of Chicago Press, Chicago, 1985.
44. F. D. Bannon III and C. T.-C. Nguyen, "High frequency microelectromechanical IF filters," *Technical Digest, 1996 IEEE Electron Devices Meeting*, San Francisco, CA, Dec. 8–11, 1996, pp. 773–776.
45. K. Wang and C. T.-C. Nguyen, "High-order micromechanical electronic filters," *Proceedings, 1997 IEEE International Micro Electro Mechanical Systems Workshop*, Nagoya, Japan, Jan. 26–30, 1997, pp. 25–30.
46. K. Wang, A.-C. Wong, W.-T. Hsu, and C. T.-C. Nguyen, "Frequency-trimming and Q-factor enhancement of micromechanical resonators via localized filament annealing," *Digest of Technical Papers, 1997 International Conference on Solid-State Sensors and Actuators*, Chicago, IL, June 16–19, 1997, pp. 109–112.
47. T. A. Core, W. K. Tsang, and S. J. Sherman, "Fabrication technology for an integrated surface-micromachined sensor," *Solid State Technology*, 39–47 (Oct. 1993).
48. J. H. Smith, S. Montague, J. J. Sniegowski, J. R. Murray, et al., "Embedded micromechanical devices for the monolithic integration of MEMS with CMOS," *Proceedings, IEEE International Electron Devices Meeting*, Washington, DC, Dec. 10–13, 1995, pp. 609–612.

49. J. M. Bustillo, G. K. Fedder, C. T.-C. Nguyen, and R. T. Howe, "Process technology for the modular integration of CMOS and polysilicon microstructures," *Microsystem Technologies*, Vol. 1, 30–41 (1994).

50. T. J. Brosnihan, J. M. Bustillo, A. P. Pisano, and R. T. Howe, "Embedded interconnect and electrical isolation for high-aspect-ratio SOI inertial instruments," *Digest of Technical Papers, 1997 International Conference on Solid-State Sensors and Actuators*, Chicago, IL, June 16–19, 1997, pp. 637–640.

51. A. E. Franke, D. Bilic, D. T. Chang, P. T. Jones, T.-J. King, R. T. Howe, and G. C. Johnson, "Post-CMOS integration of germanium microstructures," *Technical Digest, 12th International IEEE MEMS Conference*, Orlando, FL, Jan. 17–21, 1999, pp. 630–637.

52. H. Baltes, O. Paul, and O. Brand, "Micromachined thermally based CMOS microsensors," *Proceeding of the IEEE*, Vol. 86, No. 8, 1660–1678 (Aug. 1998).

53. G. K. Fedder, S. Santhanam, M. L. Reed, S. C. Eagle, D. F. Guillou, M. S.-C. Lu, and L. R. Carley, "Laminated high-aspect-ratio microstructures in a conventional CMOS process," *Sensors and Actuators*, Vol. A57, No. 2, 103–110 (Mar. 1997).

54. A. Singh, D. A. Horsley, M. B. Cohn, A. P. Pisano, and R. T. Howe, "Batch transfer of microstructures using flip-chip solder bonding," *Journal of Microelectromechanical Systems*, Vol. 8, No. 1, 27–33 (Mar. 1999).

55. K. S. Lebouitz, A. Mazaheri, R. T. Howe, and A. P. Pisano, "Vacuum encapsulation of resonant devices using permeable polysilicon," *Technical Digest, 12th International IEEE MEMS Conference*, Orlando, FL, Jan. 17–21, 1999, 470–475.

56. G. T. Mulhern, D. S. Soane, and R. T. Howe, "Supercritical carbon dioxide drying of microstructures," in *Digest of Technical Papers, 7th International Conference on Solid-State Sensors and Actuators (Transducers'93)*, Yokohama, Japan, June 1993, pp. 296–299.

57. R. Legtenberg and H. A. C. Tilmans, "Electrostatically driven vacuum-encapsulated polysilicon resonators," *Sensors and Actuators*, Vol. A45, 57–66 (1994).

58. L. Lin, K. M. McNair, R. T. Howe, and A. P. Pisano, "Vacuum-encapsulated lateral microresonators," in *Digest of Technical Papers, 7th International Conference on Solid-State Sensors and Actuators (Transducers'93)*, Yokohama, Japan, June 7–10, 1993, pp. 270–273.

59. M. B. Cohn, Y. Liang, R. T. Howe, and A. P. Pisano, "Wafer-to-wafer transfer of microstructures for vacuum packaging," in *Technical Digest, Solid-State Sensor and Actuator Workshop*, Hilton Head Island, SC, June 3–6, 1996, pp. 32–35.

60. S. Mack, H. Baumann, and U. Gosele, "Gas tightness of cavities sealed by silicon wafer bonding," in *Proceedings, 10th International Workshop on Micro Electro Mechanical Systems*, Nagoya, Japan, Jan. 26–30, 1997, pp. 488–493.

INDEX

Amplifier, nonlinearity, 3, 5
Abrupt emitter-base junction, 90
Absolute tolerance, 448
Acoustic transmission line, 433
ACPR, *see* Adjacent channel power ratio
Active integrated antenna (AIA), 214, 306, 326
   design methodology, 328
   measurement issue, 328–332
   SOM receiver, 336–339
Active load pull, 255
Actuation voltage, 357, 358
Additive white Gausian noise (AWGN), 231
Adjacent channel leakage power, 116
Adjacent channel power ratio (ACPR), 192, 231
AlGaAs/GaAs HEMTs, 104
All-MEMS front-end, 446
AM-AM, 236
AM-PM, 236
Analog-to-digital conversion, 37
Anchor loss, 450
Antenna pattern, 5
AWGN, *see* Additive white Gaussian noise

Bandgap engineering, 130
Bandpass, 245
Bandpass filter, 293, 427
Bandwidth, 11, 429
Base and collector thickness, 83

Base ballast resistors, 150
Base mobility, 91
Base push-out, 93
Base resistance, 81
Base station, 192
Base thickness, 91
Base transit properties of holes, 92
Base-collector barrier, 62
Base-collector capacitance, 81
Batch processing, 353
Battery life, 96, 230
Battery requirements, 1
Behavioral methodology, 245
BER, *see* Bit error rate
Beryllium out-diffusion, 91
Bias dependence of f_T and f_{max}, 88
Binary phase shift keying (BPSK), 13
Bipolar junction transistor (BJTs), 4, 270
Bit error rate (BER), 231
Breakdown voltage, 4, 81, 85, 91, 93, 96, 99, 101, 137

Carrier density fluctuation, 160
Cellular system, 9, 32
Channel decoder, 11
Channel encoder, 10
Channel thickness, 104
Chemical mechanical planarization (CMP), 271
Chireix power combiner, 212

Circuits/MEMS integration, 453
  mixed circuit and micromechanics, 453
  postcircuits, 454
  precircuits, 454
Circuit miniaturization, 350
Circular segment patch antenna, 219
Clamped-clamped beam, 417
Class A, 99, 194
Class AB, 107
Class B, 107, 219
Class E, 5, 203
Class F, 5, 203
Class S, 5, 212
CMOS, 5
Code division multiple access (CDMA), 109, 192
Coding, 3
Cognitive radio, 6, 351
Collector current density, 93
Collector delay, 89
Collector design, 4
Collector doping, 91
Collector ideality factor, 90
Collector ohmic contact self-aligned, 111
Collision avoidance systems (CASs), 116
Common aperture, 6
Common-base HBTs, 101
Complementary HBT, 4
  push-pull amplifier, 106
Compositional grading, 97
Constant envelope, 231
Conversion gain, 440
Conversion loss, 440
Convolutional codes, 25
Convolutional encoder, 27
Coplanar circuit, 113
Coplanar stripline, 266
Coplanar waveguide (CPWs), 265, 266, 268, 298
Coupler, 279
Coupling beam stiffness, 433
Coupling location, 433, 436
Coupling spring, 429
Crossover distortion, 109
Crosstalk, 272, 276, 279
Current electromechanical analogy, 425
Current gain modulation, 46
Cutoff frequency, f_T, 80, 142

Damping factor, 423
Data rate, 11
dc and microwave HBT characteristic, 90
dc-bias voltage, 421
dc current gain, 80, 84

dc-dc converter, 5, 195, 210
Decoded waveform, 27
Deep-level transient spectroscopy (DLTS), 180
Deinterleaved waveform, 27
Delta-sigma amplifiers, 212
Demodulated waveform, 27
Demodulator, 11
Device modeling, 101
Diffusion constant, 94
Digital modulation, 245
Digital signal processor (DSP), 201
Diplexer, 384, 385, 394, 397
Direct conversion, 37
Direct sampling, 37
Direct-sequence (DS) spread spectrum, 16
Distortion, 230
Distributed transmission line (DTML), 380
Doping gradient, 93, 101
Double heterojunction bipolar transistors (DHBT), 4, 55, 94, 97, 120
Drain mixer, 296
Drift field, 93
Drift-diffusion simulation, 102
DSP, *see* Digital signal processor
Dual bias control, 245
Duplex communication, 6
Dynamic range, 6
Dynamic supply voltage amplifier, 195

Early voltage, 93, 136
Effective air gap, 375-376
  derivation, 375
  effect on capacitance, 376
Efficiency, 3
Electrical equivalent circuit, 425
Electrical spring constant, 423
Electrical spring stiffness, 424
Electrode-to-resonator gap, 421
Electromagnetic bandgap structure, 5
Electromechanical transformer, 426
Emitter area, 96
Emitter-base heterojunction, 92
Emitter ledge passivation, 4, 40
Emitter-to-collector delay, 80
Envelope simulation, 249
Envelope tracking amplifier, 195
Epilayer, 90
Even-order harmonic, 108

F_{min}, 103
Fiber optic transmission, 117
Field-effect transistors (FETs), 270
Filter, 384, 385, 394

INDEX
465

Finite-difference time-domain (FDTD), 275
Finite ground coplanar waveguide (FGCPW) line, 357, 359, 368
Flicker (1/f) noise, 50, 160
Folded-slot antenna, 313
Folded suspension, 359, 361. *See also* Compliant beam-mechanical design
Fractional temperature coefficient, 393
Free-free beam, 450
Frequency-hopped spread spectrum, 17
Friis transmission formula, 329

GaAs, 4
GaAs-based complementary push-pull amplifier, 117
GaAs-based HBTs, 118, 198
GaAs-based PNP HBTs, 100
Gain saturation, 231
GaInP design, 117
GaN, 4
GaN HFET, 159-161
GaN transistor, 185
Generation-recombination (G-R) noise, 177
Graded-base junction, 112
Guided wave, 282
Gummel plot, 93
Gunn diode, 386

Harmonic balance, 247
Harmonic filter, 5
Harmonic tuning, 256, 283, 286
HEMT, 4, 270, 386, 400
Heterojunction bipolar transistors (HBTs), 4, 39, 270, 386
  optimization, 82
  reliability, 51
Hexagonal emitter, 92
High-frequency operation, 105
High isolation switching, 377–380
  connecting beams, 377–378
  parasitic inductance, 377
  resonant frequency, 378
  resonant switch, 377–380
High-linearity requirement, 107
High-Q, 411–412
High-Q filter, 6
Hole lifetime, 94
Hole mobility, 94
Hole transport, 94
Hooge parameter, 161, 166–167, 177
Hybrid push-pull amplifier, 107

Ideality factor, 80, 88, 92
IIP3, 146, 451–452

Impact ionization breakdown voltage, 87
IMPATT diode, 386
InAlAs , 90
InAlAs emitter, 91
InAlAs/InGaAs emitter-base heterojunction, 90
Inductively coupled plasma (ICP), 271
Inductor Q, 142
InGaAs channel, 105
InGaAs collector, 85
InGaAs/InP composite channel, 105
InP, 4, 79
InP-based HBTs, 118
InP-based PNP HBTs, 100
InP-based PNPs, 120
InP-based single and double HBTs, 119
InP collector, 85, 99
InP emitter, 90, 91
InP/InGaAs NPN HBTs, 87
Input back off (IBO), 237
Input impedance, 101
Insertion loss, 271, 274, 276, 277, 291, 385, 416, 436
Integrated capacitor, 266
Integration of NPN and PNP HBTs, 107
Integration with PIN diode, 95
Interference, 5
Interference suppression, 32
Interleaver, 27
Intermodulation, 239
Intermodulation (IM) product, 29, 35

Johnson figure of merit, 81

Kinetic energy, 423
Kirk effect, 64, 82, 93, 119
Knee voltage, 4, 56, 111–112

Large signal, 254
Large-signal modeling, 105
Large-signal operation, 98
Lateral undercut, 91
Leakage, 298, 299
Leaky wave, 283
Leaky wave antenna, 315-316
LINC, 5
LINC amplifier, 210
Linear amplification with nonlinear components, 5
Linearity, 96, 115, 230
Loaded Q, 391
Load-pull, 254
Load-pull characteristics, 101
Local oscillator, 384, 385

Local oscillator synthesizer, 445
Loss, 5
Lower velocity coupling, 434
Low-frequency noise, 103
Low-noise amplifier (LNA), 102, 144
Low-noise amplifier circuit, 103
Lowpass filter, 295
Low phase noise oscillator, 400
Low power electronics, 2
LPCVD, 387

Magnetic impedance surface, 6
Mass, 422
Mass-spring-damper, 422
Matching network, 5
Matching tolerance, 448
Maximum current density in HBTs, 81
Maximum frequency of oscillation, f_{max}, 80–81, 142
Maximum power density for HBTs, 82
Mechanical transformer, 427
Mechnical spring constant, 423
Membrane, 386, 387
Membrane-suspended resonator, 406
Memoryless, 245
MEMS-based transmit architecture, 447
MESFET, 102, 116
Microelectromechanical system (MEMS), 6, 411–457
Micromachined oscillator, 404
Micromachined resonator, 400
Micromachining, 6, 386
Micromechanical circuit, 417
Micromechanical filter, 427, 430, 435
Micromechanical resonator, 417
Micromechanical strain gauge, 363. *See also* Compliant beam-residual stress
Micromechanical switch, 440
Micropackaging, 395
Microstrip leaky wave antenna, 316–317
Microstrip line, 265, 266, 268
Microstrip patch antenna, 289
MICs, 454
Minimum shift keying (MSK), 231
Mixed signal ICs, 271
Mixer-filter, 438
Mobile communications, 143
Mobility fluctuation, 160
Mode, 428
Modified fix point method, 249
Modulation, 3, 231
Modulator, 10
Module, 455

Monolithic microwave integrated circuit (MMIC), 95, 139, 266, 383
Motional resistance, 426
MSK, *see* Minimum shift keying
Multicarrier modulation, 24
Multicarrier transmitter, 3
Multidisplinary university research initiative (MURI), 2
Multipath fading, 10

Narrowband technique, 13
Noise, 5
Noise figure, 440
Noise in FET, 102
Noise performance, 102
Noise reduction, 182
Noise spectral density, 166, 183
Nonconstant envelope, 231
Noncontact ID, 339–341
Nonlinear amplification, 26
Nonlinear simulation, 245
Nonselective MOVPE, 107
Nonuniform heating, 97
NPN, 79
NPN and PNP amplifier, 108
NPN HBTs, 84–86
NPN single HBTs (SHBTs), 86

Odd-order harmonic, 108
Offset voltage, 4, 56, 97, 117
Ohmic loss, 5
On-edge Schottky diode potentiometer, 40
Optoelectronic integrated circuits (OEICs), 95
Orthogonal frequency division multiplexing (OFDM), 236
Oscillator, 414
Out-of-band rejection, 395
Output buffer, 111
Output power, 99

Packaging, 7
Parasitic charging time, 101
Passive component, 5
Patch antenna, 217, 307–313
Peak electric field, 81
Peak gain, 99
Peak linear output power, 97
Peak-to-average-ratio, 231
Percent bandwidth, 416
Phase array, 7
Phase conjugation, 6
Phase noise, 6, 385, 404, 415
Phase shifter, 283

INDEX 467

Photonic bandgap structures (PBG), 5, 224, 282,
π/4 QPSK, 236
Piezoeletric effect, 172-173
Planar filter, 6
Planar traveling wave antenna, 314
PNP, 91
PNP HBTs, 83, 111, 120
Poisson's ratio, 360, 364
Power, 101
Power added efficiency (PAE), 34, 97, 99, 101,
  103, 116, 287, 296, 297, 330
Power amplifier, 3, 27, 143, 296
Power application, 103
Power backoff, 105
Power characteristic of InP HBTs, 99
Power consumption, 230
Power density, 99
Power efficiency, 230
Power handling, 395
Power performance, 99
Power saturation, 5
Printed quasi-Yagi antenna, 318–326
Propagation loss, 10
Pseudomorphic high electron mobility transistor
  (PHEMT), 102
Pull-in voltage, 424. *See also* Actuation voltage
Push-pull AIA power amplifier, 332–336
Push-pull amplifier, 83, 107, 110–111, 113, 332
Push-pull operation, 4, 108

Quasi-band limited (QBL), 236
Quadrature phase shift keying (QPSK), 96, 116,
  117, 190, 231
Quality factor Q, 265, 267, 385, 386, 411–415
Quarter-wavelength, 433
Quasi-Yagi antenna, 6

Radiation, 5
Radiation efficiency, 289
Radiation wave, 282
Radio frequency integrated circuit (RFIC),
  267
Rayleigh fading, 15, 32
RC parasitics, 94
Reactive ion etching (RIE), 271, 387
Receive mode, 3
Reconfigurable aperture, 6, 350, 351
Resonance frequency, 420
Resonator, 7, 386, 389, 391, 394
Resonant switch, *see* High isolation switching
Retrodirective array, 6, 341–345
RF channel selection, 442
  filter bank, 443

switchable micromechanical resonator
  synthesizer, 445
RF channel selector, 447
RF MEMS, 350, 351, 353, 357
  phase shifter, 380
  switch, 354-380
    advantages, 353
    cantilever beam, 354, 355–356
      approximation, 361, 363
      residual stress, 355
      stress gradient, 355
    capacitance
      off, 365, 370
      on, 365–366, 370, 376
      ratio, 365–366
    compliant beam, 354, 356–380
      cross-axis sensitivity, 361
      fabrication, 368–369
      mechanical design, 358-368
      model, 369–371
      power-handling, 371–374
      self-biasing, 371, 374
      residual stress, 362–366
        axial stress, 362–365
        stress gradient, 362–365
      spring constant, 358, 360–361, 362. *See
        also* Cantilever beam-approximation
      switching speed, 366–368
        damping, 366–367
      fixed-fixed beam, 354–355
      types of contact, 354
RF micromachining, 350

Satellite communication, 116
Saturation velocity, 97
Second harmonic, 107
Second harmonic content, 109, 112, 115
Self-biasing, 108
Self-heating, 82
Shadowing, 10
Shape factor, 416
Shielding cavity, 394
SiGe, 4
Signal phasing, 97
Silicon, 387
Silicon nitride, 374–377
  dielectric strength, 374
  low pressure chemical vapor deposition
    (LPCVD), 374
  plasma enhanced chemical vapor deposition
    (PECVD), 369, 374
Silicon/silicon germanium heterojunction, 126
SIMPOL, 5

Simultaneous transmit/receive, 6
Single bias control, 240
Single heterojunction bipolar transistors (SHBT), 94, 117
Single-tone excitation, 113
Slot antenna, 299, 313–312
Slotline, 266
Slow-wave factor, 292
Slow wave structure, 6
Small signal, 254
Source decoder, 11
Source encoder, 10
Spectral efficiency, 236
Spectral regrowth, 5, 231
Spiral inductor, 266
Spread-spectrum, 10
Spurious passband, 293, 295
Standby mode, 3
Stiffness, 423
Stopband, 286
Stress, 387
Stripline, 268
Substrate mode, 5
Supercritical CO_2 drying, 369
Superheterodyne receiver, 36
Suppression filter, 32
Surface-micromachining, 417
Surface wave, 282, 283, 287
Surface wave antenna, 315
Suspended microstrip, 390
Switch, 6
Switching-mode amplifiers, 203
System, 116
System-on-a-chip, 5

Tapered slot antenna, 317–318
τ_b, 86
τ_{ec}, 86

Temperature coefficient, 391
Termination impedance, 452
Termination resistor, 432, 436
Tetherless communications, 349
Thermal diffusion, 102
Thermal expansion, 391
Thermal oxidation, 387
Thermal shunt, 82
Thermally stable cascode (TSC) HBT, 105
Thermionic emission, 88
Thin film resonator (TFR), 441, 448
Third-order intermodulation, 451
Third order intermodulation distortion (IMD3), 103
Three-dimensional layered integration, 6
Time domain, 256
Time-varying envelope, 107
Transit time, 101
Transmit mode, 3
Transmitter, 3
Trimming, 453
TRL calibration, 398
Tunneling B/C structure, 70
Turn-on voltage, 80

UHV/CVD, 128
Unloaded Q, 391

Vacuum encapsulation, 456
Van Atta array, 341
Very large scale integration (VLSI), 353
Vivaldi antenna, 317

Wide-bandgap collector, 111
Wide-bandgap compound semiconductor, 159
Wireless communication system, 10

Young's modulus, 360